Physical Metallurgy

Second Edition

Physical Metallurgy

Second Edition

William F. Hosford

CRC Press
Taylor & Francis Group
Boca Raton   London   New York

CRC Press is an imprint of the
Taylor & Francis Group, an **informa** business

CRC Press
Taylor & Francis Group
6000 Broken Sound Parkway NW, Suite 300
Boca Raton, FL 33487-2742

Printed in the United States of America on acid-free paper
10 9 8 7 6 5 4 3 2 1

International Standard Book Number: 978-1-4398-1360-7 (Hardback)

Library of Congress Cataloging-in-Publication Data

Hosford, William F.
    Physical metallurgy / William F. Hosford. -- 2nd ed.
        p. cm.
    Includes bibliographical references and index.
    ISBN 978-1-4398-1360-7 (hardcover : alk. paper)
    1. Physical metallurgy. I. Title.

TN690.H845 2010
669'.9--dc22   2009031931

Visit the Taylor & Francis Web site at
http://www.taylorandfrancis.com

and the CRC Press Web site at
http://www.crcpress.com

Contents

Preface

This text attempts to combine practical and theoretical aspects of physical metallurgy. It is assumed that the students have already had an introductory course in materials science and therefore understand the basic concepts of crystal structures, Miller indices, dislocations, binary phase diagrams, and Fick's first and second laws of diffusion. Portions of Chapters 3, 5, 7, and 8 may be skipped depending on the students' backgrounds. Likewise, if the students have already learned about corrosion, portions of Chapter 25 may be omitted.

However, this text goes into further depth in these areas than most introductory materials courses do. For example, ternary diagrams are introduced in Chapter 7 on phase diagrams, the Kirkendall effect and diffusion in multiphase systems are covered in Chapter 3 on diffusion, the thermodynamic basis for solid solubility is covered in Chapter 7 on phase diagrams, stacking faults are treated in Chapter 8 on dislocations, and hydrogen embrittlement is covered in Chapter 23 on corrosion.

After an introductory chapter on metals, the following chapter topics are common to all metals: solidification, diffusion, surfaces, solid solutions, intermediate phases, dislocations, annealing, and phase transformations. The middle of the text is focused on specific nonferrous metal systems including aluminum, copper, nickel, magnesium, titanium, and other metals. There are several chapters devoted to steels and one to cast irons.

The final four chapters on powder metallurgy, corrosion, welding, and magnetic alloys are optional, depending on the length of the term and the instructor's interests. There are two appendices: one on microstructural analysis and one on the Miller–Bravais system of indices for hexagonal crystals.

At the end of each chapter there is a miscellany section covering something interesting or historical that relates to the chapter. These are included solely for interest and are not part of the textual material. There are references, example problems, and homework problems throughout.

The second edition differs from the first edition by having a new chapter (Chapter 11) devoted to crystallographic textures and their effects. The chapters on aluminum and copper and low-carbon steels have been expanded, and there are some changes in the chapter on cast irons.

Author

William F. Hosford is a professor emeritus of Material Science and Engineering at the University of Michigan. He holds degrees in metallurgy and metallurgical engineering from Lehigh University, Yale University, and the Massachusetts Institute of Technology. He is the author of several books, including *Metal Forming: Mechanics and Metallurgy,* 3rd edition (with R. M. Caddell), *Mechanics of Crystals and Textured Polycrystals*, *Physical Metallurgy*, *Materials Science*, *Materials for Engineers*, and *Reporting Results: A Practical Guide for Engineers and Scientist*s.

William H.

1 Introduction

1.1 METALLIC ELEMENTS

The importance of metals to society cannot be over-emphasized. Historians have divided human existence into the stone age, the bronze age, the iron age, and are now using the term silicon age. Only a few metallic elements occur naturally in their elemental form. Most are mined as oxides or sulfides. This book is not concerned with the reduction of ores to metallic form. The shaping of most metals and alloys starts with casting, either into final shape or into ingots that are mechanically formed into the final shape. A few products are made by powder processing and in a very few cases solid metals are shaped by plating or vapor deposition.

Processing controls the microstructure and properties. However, certain fundamental properties depend on the elements. It is important for engineers to become familiar with the periodic table of elements. The crystal structures, the approximate melting points, and the approximate densities correlate well with the position in the periodic table.

Figure 1.1 shows the crystal structures of various elements. Alkali metals and many transition metals tend to be body-centered cubic. Most of the face-centered cubic (fcc) elements occur in the middle of the periodic table, (Ni, Cu, and the platinum metals) and in the heavier alkaline earths. Aluminum and lead are exceptions to these groupings. The hcp structure is found in the alkaline earth elements, a few of the heavier transition metals, and zinc and cadmium.

Figure 1.2 shows how the melting points correlate with position in the periodic table. Elements in the lower middle of the periodic table have the highest melting points, with W and Re having melting points over 3000°C. The only other elements with extremely high melting points are C and B. The alkali metals together with Ga and In melt at temperatures below 200°C.

Because the melting point is an indicator of the bonding strength in a metal, it correlates well with the modulus of elasticity and the coefficient of thermal expansion. (Figures 1.3 and 1.4).

Knowing whether a metal has a high or low melting point allows an engineer to make a very good guess at the elastic modulus and the coefficient of thermal expansion.

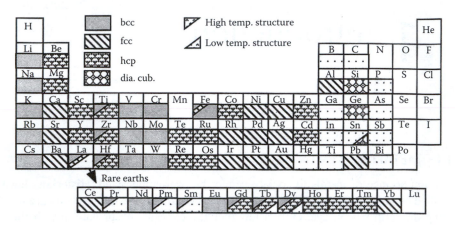

FIGURE 1.1   Periodic table showing the crystal structures of the metallic elements.

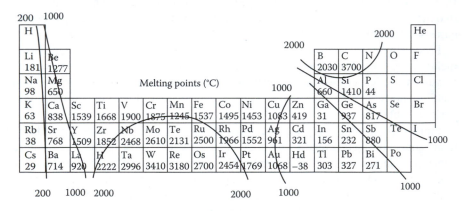

FIGURE 1.2   Periodic table showing the melting points of the metallic elements.

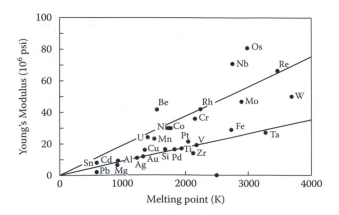

FIGURE 1.3   Correlation of Young's moduli with melting point.

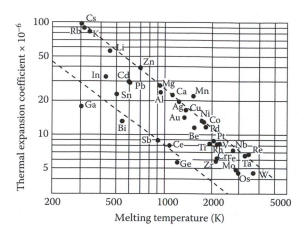

FIGURE 1.4 Inverse correlation of the coefficient of thermal expansion with melting point. Note that the close-packed metals all fall near the upper trend line, while the less close-packed elements fall below it.

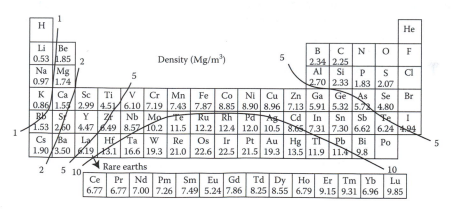

FIGURE 1.5 Periodic table showing the densities of the metallic elements.

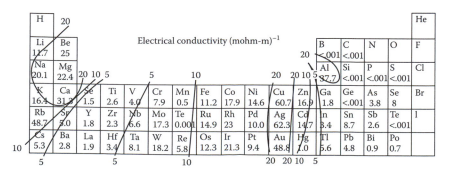

FIGURE 1.6 Periodic table showing the electrical conductivities of the metallic elements.

The densities of metals also correlate with position in the periodic table (Figure 1.5). The elements low in the periodic table (with high atomic numbers) and toward the center have highest densities.

Other properties, such as carbide-forming tendency, sulfide-forming tendency, oxidation tendency, and corrosion inhibition also correlate with position on the periodic table. Figure 1.6 shows that the elements with high electrical conductivities are clustered together in the periodic table.

1.2   PRICE AND ABUNDANCE

The abundance of elements in the earth's crust is shown in Table 1.1. Of the metals of engineering importance, aluminum, iron, magnesium, and titanium are the most abundant. Table 1.2 lists the prices (2009) of common metals. These prices depend on the condition of the metal (i.e., whether it is in a finished form or simply in a form that can be melted). It is interesting to note that there is little correlation between price and abundance. Lead is inexpensive because it occurs in concentrated deposits and is easily reduced. Titanium, on the other hand, is expensive because of the difficulty in reducing and processing it. Aluminum, although much more abundant than copper is more expensive because of the cost of reducing its ore. The prices of metals fluctuate, and tend to increase with inflation. However, the price of one metal relative to another tends to remain fairly constant as shown in Figure 1.7. It should be emphasized that the costs per pound are often not the

TABLE 1.1
Abundance of Elements in the Earth's Crust

Element	%	Element	%
Si	47	Zn	0.004
Al	8.1	Pb	0.002
Fe	5.0	Co	0.001
Ca	3.6	Be	0.001
Mg	2.1	Mo	0.0001
Ti	0.63	Sn	0.0001
Mn	0.10	Sb	0.00001
Cr	0.037	Cd	0.00001
Zr	0.026	Hg	0.00001
Ni	0.020	Bi	0.000001
V	0.017	Ag	0.000001
Cu	0.010	Pt	0.0000001
U	0.008	Au	0.0000001

TABLE 1.2
Metal Costs

2009 Metal Prices in Dollars per Pound[a]

Less than 1$/lb

Pig iron	0.14
Steel—hot rolled	0.23
Zinc	0.82
Aluminum	0.85

Between 1$/lb and 5$/lb

Lead	1.00
Magnesium	1.30
Silicon	1.5
Copper	2.78

Between 5$/lb and 25$/lb

Tin	6.652
Nickel	8.45
Bismuth	12.00
Titanium ingot	20.00

Over 25$/lb

Beryllium	17.83
Cobalt	42
Silver	300
Gold	18,000
Platinum	23,000

2009 Ferroalloy Prices per Pound of Contained Alloy

Ferrochromium (60–65% Cr)	1.00
Ferromanganese	1.55
Ferrosilicon	0.72
Ferrotitanium	3.75
Ferroniobium (columbium)	21.00

[a] Data from http//metalprices.com

most important issue in choosing between equally viable metals. The cost per volume is usually more relevant if the intention is to make the same part (same shape). There is a correlation between the price of an element and its production, as shown in Figure 1.8.

1.3 MISCELLANY

The seven metals known to ancient civilizations were gold, silver, mercury, copper iron, tin, and lead. Table 1.3 gives their symbols. Of these, only copper, gold, and

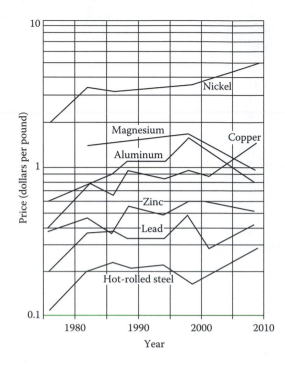

FIGURE 1.7 Price of several metals. Note that although the prices change, the price of one metal relative to another does not change much.

silver are found in their "native" state. In addition, metallic meteorites are iron–nickel alloys.

The building of the Egyptian pyramids did not occur until the discovery of copper mines in the Sinai (later called King Solomon's mines). Then copper-base alloys needed for the cutting tools to shape rock became available.

The Hittites were the first to use weapons made from iron. These made possible for them to conquer Egypt.

In Greek and Roman times, gold was used for ornaments, silver for ornaments and coinage, copper for vessels, and combined with tin for cast statues, tools, and weapons, iron for tools and weapons, lead for plumbing, and mercury in the extraction of gold from its ores.

PROBLEMS

1. a. Find the cheapest conductor based on the cost to meet a specified maximum resistance per length of conductor. The cost for a fixed resistance per length is proportional to $\rho_{res}\rho_{dens}C$, where ρ_{res} is the resistivity, ρ_{dens} is the density, and C is the cost per weight.
   b. At what price per pound would sodium make an even cheaper conductor? Sodium has a density of 0.967 mg/m^3 and a resistivity of 49.6 nohm-m.

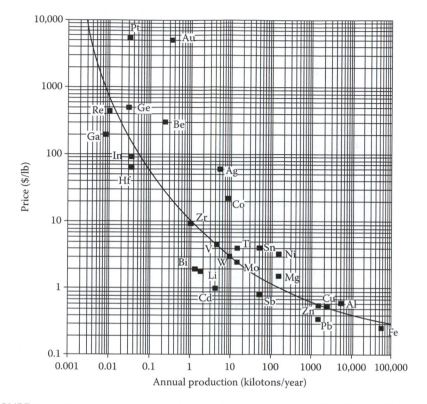

FIGURE 1.8 1992 production and price of various metals in the U.S. (From J. H. Brophy, *Advanced Materials and Processes*, 148, 78–80, 1995. With permission.)

TABLE 1.3
Seven Metals of Antiquity and Their Symbols

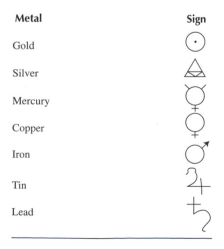

Metal	Sign
Gold	
Silver	
Mercury	
Copper	
Iron	
Tin	
Lead	

Metal	Density (mg/m³)	Resistivity (nohm-m)	Price ($/lb)
Iron	7.88	98	0.39
Magnesium	1.738	45	1.98
Copper	8.90	16.73	1.44
Aluminum	2.699	26.55	0.69

2. Vanadium is often added to steels in minor amounts to form carbo-nitrides.
   a. If vanadium were not available, what element would you suggest as a substitute to form similar carbo-nitrides?
   b. Assuming that each atom of the substitute is equally effective as a vanadium atom, what weight percent of the substitute would you recommend to replace 0.05 wt% V?
3. Prior to World War II, most high-speed steels contained about 18% W. A typical composition was 1% C, 4% Cr, 2% V, 18% W, and balance Fe. Because of the shortage of tungsten, a substitute had to be found for some or all of the tungsten.
   a. What element would you suggest as a substitute for tungsten?
   b. Assuming an equal effectiveness of the substitute and tungsten on an atom-for-atom basis, what percent of the substitute would be needed to substitute for each percent W?
4. Using the prices in Table 1.2, by what factor would the cost of material in a part increase if magnesium were substituted for aluminum without making any changes in dimensions?

REFERENCE

J. H. Brophy, *Advanced Materials and Processes*, 148, 78–80, 1995.

2 Freezing

2.1 INTRODUCTION

Almost all solid metals and alloys are produced from liquids by freezing (vapor and electrodeposition are the main exceptions). Often the solidified metal is used in the form of a casting with little or no additional shaping. On the other hand, the largest tonnage is produced in the form of wrought products. These are solidified as ingots or continuously cast billets before being shaped into plate, sheet, rod, and so on. For both castings and wrought products, what happens during freezing has a major influence on microstructure and properties of the final product. Of particular importance are porosity and segregation of alloying elements.

2.2 LIQUID METALS

The liquid state is intermediate between a gas and a crystal. Liquids do not have the high degree of order of crystals or the complete disorder of gasses. X-ray studies of liquids indicate some short-range order. The distance between near-neighbor atoms is very close to that in their crystalline state as shown in Figure 2.1. Table 2.1 shows that for close-packed metals, (hcp and fcc), melting causes a volume increase of 3.5–6%. For most bcc metals, the volume increase is 1–3%. Elements for which packing in the solid is not dense (Si, Ge, Bi, Ga, etc.) actually expand when they solidify.

When a metal freezes, its energy (enthalpy) decreases. The difference between the energy in the liquid and solid states is the latent heat of fusion, which is released to the surroundings. Similarly, when a metal vapor condenses, the latent heat of vaporization is released. Table 2.2 shows that for most metals, the heats of vaporization are 20–30 times as great as the heats of fusion. The difference is that all of the near-neighbor bonds are broken on vaporization, whereas melting statistically breaks only a fraction of a bond per atom. Assuming that the heats of fusion, ΔH_f, and vaporization, ΔH_v, are proportional to the number of near-neighbor bonds broken, and that vaporization breaks 8–12 bonds/atom, melting must break only about 1/2 of a bond/atom. The entropy change on melting, $\Delta S_f = \Delta H_f/T_m$, about 10 mJ/mol K and the entropy change on vaporization, $\Delta S_v = \Delta H_v/T_b$, is about 10 times as great. Latent heats of phase changes in the solid state are much smaller.

2.3 NUCLEATION

The first step in solidification is the nucleation of the first solid. Consider the formation of a tiny sphere of solid of radius, r, in the midst of a liquid at a temperature, T, lower than

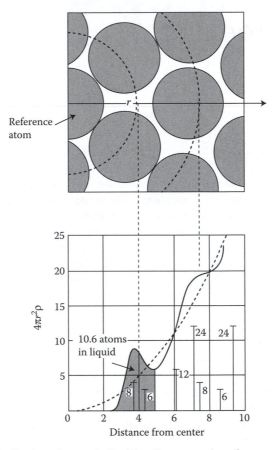

FIGURE 2.1 Distribution of atoms in liquid sodium around a reference atom as calculated from x-ray data. The dashed curve represents a uniform distribution. The vertical lines represent the distribution in solid bcc sodium. (From A. G. Guy, *Elements of Physical Metallurgy*, Addison-Wesley, 1959. With permission.)

the equilibrium melting–freezing temperature, T_m, (Figure 2.2) The surface area, A, of the sphere is $4\pi r^2$ and its volume, V, is $(4/3)\pi r^3$. There are two contributions to the change of free energy, ΔG, caused by the freezing of this sphere. The first is an increase of energy caused by the introduction of a new surface between the liquid and solid, $A\gamma_{LS}$, where γ_{LS}, is the energy per area of surface (surface tension). The second term is a reduction of energy because below T_m the solid has a lower energy. This term is proportional to the change of ΔG per volume, ΔG_v, and to the volume. The net change of free energy is

$$\Delta G = A\gamma_{LS} - V\Delta G_v$$
$$= 4\pi r^2 \gamma_{LS} - \left(\frac{4}{3}\right)\pi r^3 \Delta G_v. \qquad (2.1)$$

At very small radii, growth of solid increases the energy of the system (Figure 2.3). However, there is a critical radius, r^*, beyond which the energy decreases with

TABLE 2.1

Volume Change on Melting

Metal	Xtal Structure	% Volume Change on Melting
Li	bcc	1.65
Na	bcc	2.2
K	bcc	2.55
Rb	bcc	2.5
Cs	bcc	2.6
Fe	bcc	3.4
La	bcc	0.6
Nb	bcc	0.9
Al	fcc	6.0
Cu	fcc	4.15
Ag	fcc	3.8
Au	fcc	5.1
Pb	fcc	3.5
Mg	hcp	4.1
Zn	hcp	4.2
Cd	hcp	4.7
Sn	bct	2.8
Hg	rhomb	−1.6
Sb	rhomb	−0.95
Bi	rhomb	−3.35
Si	dia. Cub.	−12.0
Ge	dia. Cub.	−12.0

Source: Adapted from B. R. T. Frost in *Progress in Metal Physics*, Vol. 5, Pergamon Press, 1954.

further growth, so particles larger than $r*$ spontaneously grow. Thermal activation is required for particles to reach this critical radius. The critical radius and the critical activation energy, $\Delta G*$, can be found setting $d\Delta G/dr$ equal to zero.

$$\frac{d\Delta G}{dr} = 8\pi r\gamma_{LS} - 4\pi r^2 \Delta G_v = 0, \tag{2.2}$$

$$r* = \frac{2\gamma_{LS}}{\Delta G_v}, \tag{2.3}$$

and

$$\Delta G* = \left(\frac{16}{3}\right)\frac{\pi\gamma_{LS}^3}{\Delta G_v^2} \tag{2.4}$$

As with other thermally activated processes, the rate of nucleation, $\dot{N}$, can be expressed by an Arrhenius equation with $\Delta G*$ as the activation energy:

$$\dot{N} = \dot{N}_O \exp\left(\frac{-\Delta G*}{kT}\right) \tag{2.5}$$

TABLE 2.2

Entropy of Melting and Vaporization

Element	ΔS_f (J/mol K)	ΔS_v (J/mol K)	$\Delta S_v/\Delta S_f$
Al	11.5	105.0	9.15
Bi	20.7	97	4.7
Cd	11.3	95.5	8.4
Ca	7.7	101.5	13.2
Co	9.73	134	13.7
Cu	9.6	104.7	10.9
Ga	18.4	101.2	5.5
Au	9.25	106.9	11.6
Fe	7.71	124.6	16.4
Pb	7.93	96.7	12.2
Li	6.61	97.5	14.7
Mg	9.69	93.3	9.6
Hg	10.1	86.9	8.6
Mo	9.0	84.3	9.4
Re	9.59	103	10.7
Rb	6.96	79	11.3
Ag	11.7	116.5	10.0
Na	7.01	77.2	11.0
Ta	8.85	133.8	15.1
Sn	14.0	93.6	6.7
Ti	10.9	133.3	12.3
W	11.0	144	13.3
Zn	9.52	98.8	10.4

where $\dot{N}_0$ is a constant and k is Boltzmann's constant. Unlike most other thermally activated processes, the rate of nucleation does not increase with increasing temperature as a casual glance at Equation 2.5 might suggest. This is because ΔG_v and therefore ΔG^* become more negative as the temperature is lowered. For freezing at any temperature, $\Delta G_v = \Delta H_v - T\Delta S_v$, where the changes of enthalpy, ΔH_v and ΔS_v,

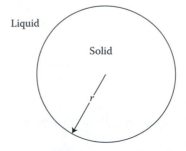

FIGURE 2.2   Spherical embryo of solid forming in a liquid.

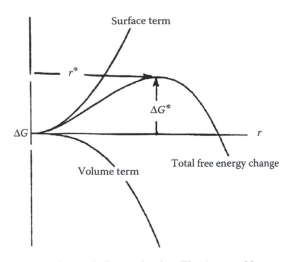

FIGURE 2.3 Free energy change during nucleation. The change of free energy, ΔG, increases with embryo size up to a critical radius, r^*. The critical free energy for nucleation is ΔG^*.

are almost independent of temperature. For freezing at the equilibrium melting temperature, T_m, $\Delta G_v = \Delta H_v - T_m\Delta S_v = 0$ so $\Delta H_v = T_m\Delta S_v$ and at any temperature

$$\Delta G_v = (T_m - T)\Delta S_v = \left(\frac{\Delta T}{T_m}\right)\Delta H_v \qquad (2.6)$$

Substituting into Equation 2.4, $\Delta G^* = (16/3)\pi r\gamma_{LS}^3/[(\Delta T/T_m)\Delta H_v]^2$; so ΔG^* is proportional to $1/(\Delta T)^2$. At T_m, $\Delta G_v = 0$, so the nucleation rate is zero. As the temperature drops below T_m, the term $\Delta G^*/kT$ decreases rapidly.

EXAMPLE PROBLEM 2.1

For copper, estimate the critical radius, r^*, and the activation energy, ΔG^*, at 0.01°C, 1.0°C, and 100°C below T_m. For copper, $T_m = 1083$°C, $\Delta H_f = 205$ kJ/kg, $\rho = 8.57$ mg/m³, and $\gamma_{LS} = 0.177$ J/m².

SOLUTION

$\Delta H_{vol} = (205 \text{ kJ/kg})(8.57 \times 10^3 \text{ kg/m}^3) = 1.75 \text{ GJ/m}^3$, $T_m = 1083 + 273 = 1356$ K,
$\Delta G_{vol} = (\Delta T/T_m)\Delta H_v = (1.29 \times 10^6)\Delta T$ J/m³ K.
  $r^* = 2\gamma_{LS}/\Delta G_v = 2(0.177)/(1.29 \times 10^6 \Delta T)$ (=0.27 μm for $\Delta T = 0.01$, 0.027 μm for $\Delta T = 1$, and 2.7 nm for $\Delta T = 100$).
  $\Delta G^* = (16/3)\pi r\gamma_{LS}^3/\Delta G_v^2 = 5.58 \times 10^{-14}/\Delta T^2$ (= 5.58×10^{-10} J at $\Delta T = 0.01$, 5.58×10^{-14} J at $\Delta T = 1$, and 5.58×10^{-18} J at $\Delta T = 100$).

The value of $\dot{N}_O$ in Equation 2.5 has been estimated to be about 10^{39} nuclei/m³s. The basis of this estimate will not be discussed here, but it should be noted that estimates of nucleation rates based on this number are such that very large subcooling would be required to produce any nuclei in any reasonable time. For the example of copper (Example Problem 2.1) at subcoolings of $\Delta T = 100$°C, Equation 2.5 predicts

TABLE 2.3
Data on Subcooling of Small Droplets

Material	T_m (K)	$\Delta S_f = H_f/T_m$ (cal/mol K)	ΔT_{max} (K)	$\Delta T_{max}/T_m$
Hg	234	2.38	46	0.197
Sn	506	3.41	110	0.218
Pb	601	2.04	80	0.133
Al	933	2.71	130	0.140
Ag	1234	2.19	227	0.184
Cu	1356	2.29	236	0.174
Ni	1725	2.43	319	0.185
Fe	1803	1.97	295	0.164
Pt	1828	2.25	332	0.182
Water	273	5.28	39	0.143

Source: From J. H. Hollomon and D. Turnbull, in *Solidification of Metals and Alloys*, AIME, 1951. With permission.

that $\dot{N} = 10^{39} \exp[-5.58 \times 10^{-18}]/(13 \times 10^{-24} \times 1266)] = 1 \times 10^{-147}$ nuclei/m³s. At this nucleation rate, it would take 3×10^{138} centuries to form one nucleus in a cubic meter of liquid. Such calculations do not agree with usual laboratory experience, but very large undercoolings have been observed in careful experiments on freezing of isolated drops. In these experiments, undercoolings of $\Delta T \approx 0.18 T_m$ have been reported for many liquids (Table 2.3).

Such large undercoolings are not observed in normal casting. Often the degree of undercooling is so small that it is not noticed. The reason for the difference between normal experience and theory is that the theory assumes that nucleation occurs homogeneously (i.e., randomly throughout the liquid), whereas nucleation normally occurs on preexisting solid surfaces. The importance of special nucleation sites is apparent when one watches the formation of gas bubbles in a carbonated beverage. The bubbles first form on a preexisting surface (liquid–glass), and at special places on the surface (probably small cracks in the glass or dirt particles) so that streams of bubbles arise from the same spot.

Although the formation of a new solid, S, requires new surfaces between it and the preexisting solid, Q, and between it and the liquid, L, the surface between L and Q is destroyed (Figure 2.4). This reduces the activation energy for nucleation.

$$\Delta G^*_{hetero} = \Delta G^*_{homo} \frac{(2 + \cos\theta)(1 - \cos\theta)^2}{4}, \tag{2.7}$$

where ΔG^*_{hetero} and ΔG^*_{homo} are the activation energies for such heterogeneous and homogeneous nucleation, respectively. ($\Delta G_{homo} = \Delta G^*$ in Equation 2.4) The wetting angle, θ, is given by

$$\cos\theta = \frac{(\gamma_{LQ} - \gamma_{QS})}{\gamma_{LS}}. \tag{2.8}$$

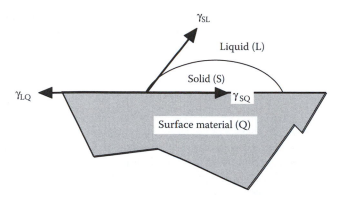

FIGURE 2.4 Heterogeneous nucleation on a preexisting surface, Q. As the new surfaces, LS and SQ, are formed, the surface LQ is lost.

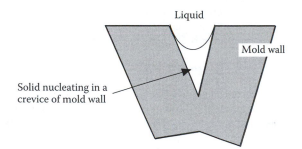

FIGURE 2.5 Heterogeneous nucleation in a crevice.

If the surface energy, γ_{LQ}, is high and γ_{SQ} is low, nucleation on the surface is energetically favorable. The preexisting solid surface of most importance in casting is usually the mold wall, where the temperature is the lowest. Crevices or cracks in the mold wall offer special nucleation sites, since they further lower the activation energy (Figure 2.5).

Sometimes special nucleating agents are added to the melt to produce a fine as-cast grain by promoting nucleation. To be an effective agent, the addition must be already solid at the freezing temperature and must have a low interfacial energy with the solid being nucleated. An example is the use of AgCl to promote rain by serving as a nucleus for ice crystals. The crystal structure of AgCl is such that its lattice spacing is very near that of ice. Practical grain refiners in metal castings include Zr and Fe additions to copper-base alloys, TiB in aluminum alloys, V in steels, and Zr in magnesium alloys. Most grain refiners are added to increase nucleation but some are added to restrict grain growth after solidification. Ultrasonics also promotes fine-grain castings, probably by breaking dendrite arms, which can serve as new nuclei.

2.4 GROWTH

Once a solid has been nucleated, the freezing rate (velocity of the liquid–solid interface) is controlled almost entirely by the rate of heat removal. The temperature of the solid–liquid interface remains very near the equilibrium freezing temperature. When there is substantial undercooling before nucleation, the temperature rises rapidly back to T_m as freezing occurs. This is because the latent heat, H_f, is large and its release will heat any undercooled liquid.

EXAMPLE PROBLEM 2.2

Suppose a melt is somehow undercooled by $0.18T_m$ before nucleation occurs and that as freezing occurs, release of the latent heat raises the temperature of the melt without heat transfer to the surroundings. Calculate the fraction, f, of the melt that freezes before the temperature of the liquid raises back to T_m. Assume a typical value for heat capacity, $C = 25$ J/mol and that $\Delta H_f = T_m \Delta S_f$, where $\Delta S_f = 10$ J/mol K.

SOLUTION

Let the total volume be 1 so that the volume frozen is f. The heat released is then $f\Delta H_f$. The amount of heat released in raising the temperature is $C\Delta T$. Equating these, $f\Delta H_f = C\Delta T$ or $f = C\Delta T/\Delta H_f$. Substituting $\Delta T = 0.18T_m$, $\Delta H_f = T_m \Delta S_f = 10T_m$, and $C = 25$, $f = 0.45$. Thus with the maximum possible subcooling, the temperature will raise to the melting point before the melt is half frozen.

EXAMPLE PROBLEM 2.3

Consider the freezing of pure aluminum. Estimate the temperature gradient in the solid near the liquid–solid interface if the interface moves at 5 mm/min. For aluminum, the thermal conductivity $k = 4.8$ Jm/(Ks), the latent heat of freezing is 397 kJ/kg, and the density is 2.7 mg/m³.

SOLUTION

The heat flux, $q = k \, dT/dx$ must equal the rate of latent heat release.
   $q = (397$ kJ/kg $\times 2.7 \times 10^3$ kg/m³$)(5 \times 10^{-3}$ m/m$)/(60$ s/m$) = 89.3$ J/s $= k \, dT/dx$,
$dT/dx = (89.3$ J/s$)/(4.8$ Jm/(Ks$)) = 18.6$ K/m.

2.5   INGOT STRUCTURE

Examination of a section of an ingot (Figure 2.6) reveals two or three distinct regions. The first of these is the outside, which has relatively small grains that are more-or-less randomly oriented. Inside of this is a zone of columnar (literally column-shaped) grains. With alloys, there may be a third zone of equiaxed and randomly oriented grains in the center. This zone is absent in pure metals.

The axes of the columnar grains are parallel to the direction of heat flow. For fcc and bcc metals, they are oriented parallel to <100>. With hcp metals, any direction in the (0001) plane may be parallel to the columnar axis. The degree of alignment becomes more perfect toward the center of the ingot. More perfectly oriented grains must grow slightly ahead of less perfectly oriented grains and cut them off.

One can argue that if some grains extend further into the liquid than others, there must be a difference in interface temperature and therefore some small degree of undercooling. The amount of undercooling is extremely small; hence this is not in conflict with the earlier statement that the interface is at a temperature very near T_m.

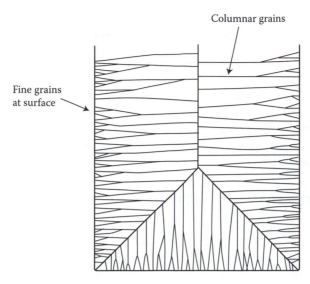

FIGURE 2.6 Cast structure of an ingot of a pure metal. At the center of an ingot of an alloy there would be a third zone of nearly equiaxed grains.

During the freezing of pure metals, the liquid–solid interface is planar and crystals grow by the advance of this interface. In contrast, freezing of alloys normally occurs by dendritic growth. (The word *dendrite* comes from the Greek *dendrites* meaning "tree-like.") The basic features of dendritic growth are the long thin crystals that grow into the liquid and thicken. Usually there are side arms (secondary arms) and sometimes tertiary arms (Figure 2.7). The primary arms and the secondary and tertiary arms are crystallo-graphically oriented, <100> being the direction in cubic metals. The reason for dendritic growth will be taken up later. The final columnar structure results from parallel growth of different colonies of dendrites and the gradual lateral growth between them. Whether columnar grains form from plane-front growth as in pure metals or by dendritic growth as in alloys, the final shape and orientations are the same. During the freezing of "nearly pure" metals, the liquid–solid interface may develop an intermediate morphology with cells, differing only very slightly in

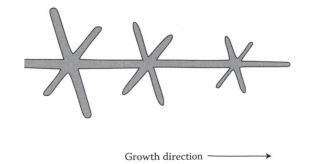

Growth direction

FIGURE 2.7 Schematic drawing of a dendrite with secondary and tertiary arms.

orientation. The centers of these cells extend slightly further into the liquid than the region near the boundaries where the solute concentrates.

The third region at the center of alloy castings consists of smaller grains that are randomly oriented and nearly equiaxed. As freezing progresses, the thermal gradient decreases and this causes the dendrites to become very long. The breakdown of columnar growth may be a result of fracturing of very long thin dendrite arms by convection currents in the liquid. The broken arms can act as nuclei for new grains. It is also possible that with the low thermal gradient and the segregation that occurs during freezing, new grains nucleate ahead of the liquid–solid interface. This topic will be discussed later.

2.6   SEGREGATION DURING FREEZING

To understand the segregation of solute during freezing, we will consider several models based on simplifying assumptions. In all cases, we will assume local equilibrium at the liquid–solid interface. If there were perfect mixing in both the liquid and the solid, the composition of each phase would be that given by the phase diagram. Convection in the liquid might provide such mixing in the liquid but the amount of diffusion required for homogenization of the solid would be unrealistic. Figure 2.8 treats the freezing of a binary alloy containing 10% B. The first solid to form contains 2% B. As the system is cooled, newly formed solid will contain ever-increasing amounts of B. An unrealistically slow cooling rate (years or centuries per degree) would be required to allow enough time for diffusion to eliminate these concentration gradients.

It is more reasonable to assume that there is no mixing in the solid but there is perfect mixing in the liquid. To aid the visualization, consider freezing in a horizontal mold from one end to the other as shown in Figure 2.9a. Let x denote the position of the solid–liquid interface and let length of the mold be L so that the fraction solid $f_S = x/L$ and the fraction liquid is $f_L = 1 - f_S = (L - x)/L$. Figure 2.9b shows the

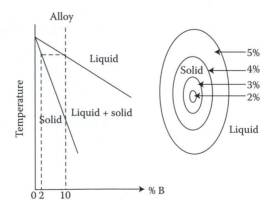

FIGURE 2.8 Schematic drawing showing a binary phase diagram (left) and contours of concentration in a region that has frozen.

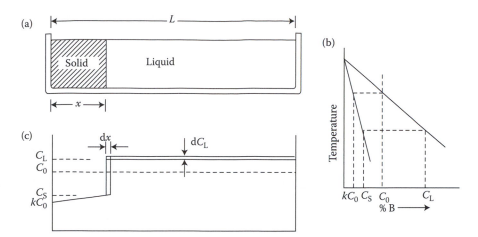

FIGURE 2.9 Directional solidification in a horizontal mold.

relevant portion of the phase diagram and Figure 2.9c shows that the composition profile at some time during freezing C_L is the composition of the liquid and C_S is the composition of the solid. In the following development, C_L and C_S should be in weight of B per volume. However, if the density differences are neglected, they can be expressed as wt% or atom%.

As the solid–liquid interface advances a distance, dx, the amount of solute rejected by the solid is $(C_L - C_S)dx$. This solute enriches the liquid composition by dC_L, making a mass balance,

$$(C_L - C_S)dx = (L - x)dC_L. \tag{2.9}$$

Usually the liquidus and solidus can be approximated by straight lines. In this case,

$$C_S = kC_L, \tag{2.10}$$

where k is the partition coefficient. Substituting $C_S = kC_L$ and $f_S = x/L$ and integrating between limits of $C_L = C_0$ at $f_S = 0$ and C_L at f_S,

$$kC_L \int \frac{df_S}{(1 - f_S)} = \frac{1}{(1 - k)} \int \frac{dC_L}{C_L},$$

$$\left[\frac{1}{(1 - k)} \right] \ln\left(\frac{C_L}{C_0} \right) = -\ln(1 - f_S) \quad \text{or} \tag{2.11}$$

$$C_L = C_0(1 - f_S)^{-(1 - k)} \quad \text{and} \quad C_S = kC_0(1 - f_S)^{-(1 - k)}.$$

Equation 2.11 is called the Scheil equation (Scheil, 1942).

Figure 2.10 shows how the composition of the ingot would change with f_S. It should be noted that the assumption of plane-front freezing from one end of the mold

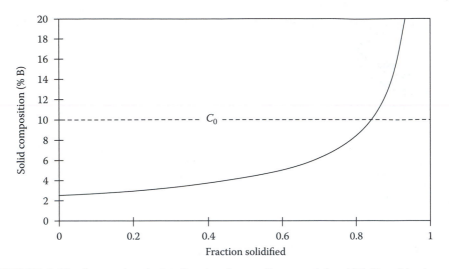

FIGURE 2.10 Segregation during freezing for an alloy containing 10% B and having a distribution coefficient, $k = 0.25$.

to the other is unnecessary. Equation 2.11 is valid regardless of the geometry of freezing. It even applies for dendritic growth.

EXAMPLE PROBLEM 2.4

Plot the nonequilibrium solidus (average composition of the solid) as a function of temperature for an alloy of A and B containing 2% B. Assume complete mixing in the liquid and no mixing in the solid. A melts at 1083°C and the liquidus temperature can be approximated by $T_L = 1083 - 6C_L$ or $C_L = (1083 - T_L)/6$. The distribution coefficient is about 0.20.

SOLUTION

The average composition of the solid, $\bar{C}_S$, can be found from a simple mass balance, $C_0 = \bar{C}_S f_S + C_L(1 - f_S)$ or $\bar{C}_S = [C_0 - C_L(1 - f_S)]/f_S$. Now substituting $f_S = 1 - (C_L/C_0)^{-1/(1-k)}$ from the Scheil equation, or $\bar{C}_S = [C_0 - C_L(C_L/C_0)^{-1/(1-k)}]/[1 - (C_L/C_0)^{-1/(1-k)}]$.
   Substituting $C_0 = 0.02$, $k = 0.2$, and $C_L = (1083 - T)/6$, $\bar{C}_S$ can be found as a function of T (Figure 2.11).

2.7 ZONE REFINING

It is possible to purify a metal by directional solidification. The first end to freeze can be cropped off of the ingot. Further purification can be achieved by adding this to other first ends, remelting, and continuing the process. However, William Pfann developed a continuous process, called *zone melting*, to achieve similar results. In this process, a molten zone is passed through the ingot as shown in Figure 2.12. Consider the first pass of a molten length B.
   As freezing progresses from x to $x + dx$, the amount of impurity rejected into the liquid, $(C_L - C_S)dx$, must be equal to the increased amount of impurity in the liquid,

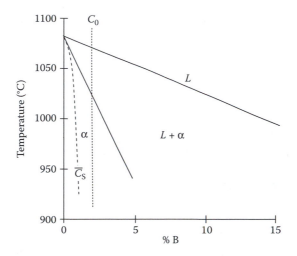

FIGURE 2.11 Plot of the nonequilibrium solidus for an alloy of 2%, superimposed on the phase diagram.

$B \, dC_L$. Substituting $C_S = kC_L$, $C_L(1-k)dx = B \, dC_L$, or $B \, dC_L/C_L = (1-k)dx$. Integrating from $C_L = C_0$ at $x = 0$, to the current conditions, $\ln(C_L/C_0) = (1-k)x/B$ or

$$C_S = C_0 \exp\left[\frac{(1-k)x}{B}\right]. \qquad (2.12)$$

The first pass in zone refining produces less purification than directional solidification. The advantage of the process is that further purification can be achieved by passing additional zones through the material. Figure 2.13 shows calculations of the purification by successive passes.

The actual purification is somewhat less than that predicted by Equations 2.11 and 2.12 (and by Figures 2.11 and 2.13) because mixing is not perfect in the liquid. Instead a thin boundary layer of impurity forms in the liquid just ahead of the interface (Figure 2.14).

The presence of this boundary layer means that the concentration of the impurity in the liquid is higher than calculated for perfect mixing. Therefore the concentration

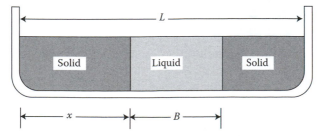

FIGURE 2.12 Zone refining. As a liquid zone is passed slowly from left to right, it collects impurities.

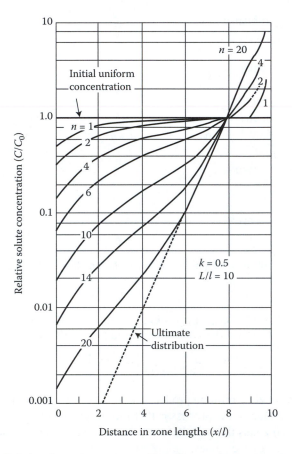

FIGURE 2.13 Relative impurity concentration after an increasing number of passes. (From W. G. Pfann, *Zone Melting*, Wiley, NY, 1958. With permission.)

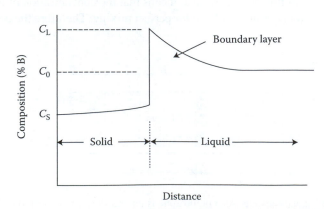

FIGURE 2.14 Formation of an impurity boundary layer ahead of the solid–liquid interface.

of the freezing solid is also less pure. The amount of purification is thus decreased by this boundary layer. Zone refining is most efficient in relatively pure metals, where the boundary layer builds up slowly.

Horizontal zone melting is possible if there is a suitable mold material that does not react with the melt. Graphite molds can be used for most nontransition metals. On the other hand, there are many metals for which there is no suitable mold material (e.g., Si, W, Mo, etc.). These materials can be zone melted using a floating zone technique. The crystal is held vertically and a single narrow zone is melted by an electron beam or an induction coil and slowly passed from one end to the other. This process is repeated until the desired purification is attained. Surface tension keeps the liquid from running out.

2.8 STEADY-STATE FREEZING

A steady-state condition may be reached with a boundary layer great enough so that the solid forming has the same composition, C_0, as the liquid beyond the boundary layer. This is illustrated in Figure 2.15. The interface composition in the boundary layer is C_0/k.

The thickness, t, of the boundary layer can be taken as

$$t = \frac{D}{v} \tag{2.13}$$

where D is the diffusivity of the solute in the liquid and v is the velocity of the solid–liquid interface.

2.9 DENDRITIC GROWTH

The boundary layer formed by rejection of the solute into the liquid may lead to a breakdown of plane-front growth. Where there is a boundary layer in the liquid

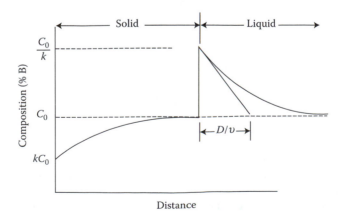

FIGURE 2.15 Steady-state freezing. The interface composition in the liquid is C_0/k, so the solid that forms has the composition, C_0.

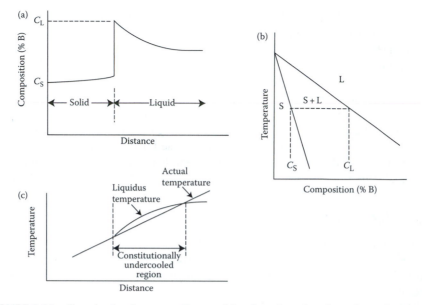

FIGURE 2.16 Constitutional supercooling resulting from boundary layer formation.

(Figure 2.16a) there must be a variation of the freezing temperature (liquidus temperature) just ahead of the interface as shown in Figure 2.16B. These temperatures are determined by comparing the local composition with the phase diagram (Figure 2.16c). The liquidus temperature and the actual temperature are plotted in Figure 2.16b. At the interface, the actual temperature must equal the liquidus temperature according to the concept of local equilibrium. Just ahead of the interface, the actual temperature is lower than the liquidus temperature for the local composition. Such a condition is called *constitutional supercooling* because it results from compositional (constitutional) variations. This situation is not stable. If any area of the liquid interface happens to extend slightly ahead of the other areas, it will freeze faster and grow rapidly into the undercooled liquid. This is the principal cause of dendritic growth.

Figure 2.17 shows that there is a critical thermal gradient for prevention of dendritic growth. There will be supercooling if the gradient is less than

$$\left(\frac{dT}{dx}\right)_{crit} = \frac{T_1 - T_3}{D/v}.$$ (2.14)

Some authors make a distinction between *cellular* growth and dendritic growth. In cellular growth, the primary arms extend into the liquid but there are no secondary dendrite arms. Figure 2.18 illustrates this. In either case, however, primary arms extend in the direction of heat flow. The tendency to cellular and dendritic growth increases as the thermal gradient, G, decreases and as the growth velocity, v, increases as illustrated in Figure 2.18. The tendency to cellular and dendritic growth increases with lower thermal gradients, G, and higher growth velocities, v.

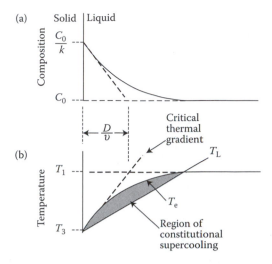

FIGURE 2.17 (a) The composition profile near the liquid–solid interface for steady-state freezing. (b) The temperature in the liquid near the interface, T_L, and the equilibrium liquidus temperature, T_e, corresponding to the local composition. The region where $T_L < T_e$ is supercooled so dendrites can form. No dendrites can form if the actual thermal gradient is greater than the critical gradient.

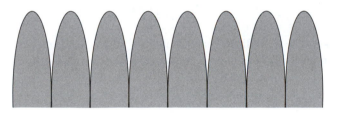

FIGURE 2.18 Cellular growth.

2.10 LENGTH AND SPACING OF DENDRITE ARMS

The length of dendrites, L, can be estimated from knowledge of the thermal gradient and the phase diagram. For a given alloy, the temperature at the tips of the dendrites is the liquidus temperature of the alloy and the temperature at the base of the dendrites is the solidus temperature of the last liquid to freeze. While this may be lower than the equilibrium solidus of the alloy, the separation of the liquidus and solidus temperatures give a rough indication of the relative tendency to form long or short dendrites,

$$L = \frac{T_L - T_S}{G} \tag{2.15}$$

Comparison of the Cu–Zn and Cu–Sn phase diagrams (Figure 2.19a and b) shows the reason why very long dendrites are formed in Cu–Sn castings but not in Cu–Zn castings.

(a)

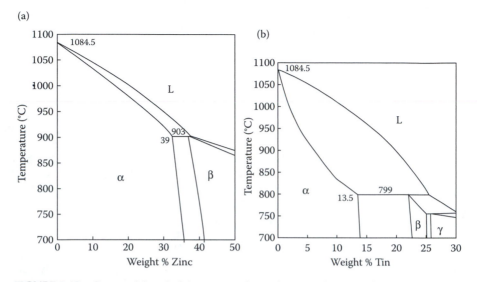

(b)

FIGURE 2.19   Copper-rich end of the copper–zinc and copper–tin phase diagrams. The much greater solid–liquid temperature range in copper–tin alloys causes much longer dendrites.

The spacing, λ, between secondary dendrite arms has been shown to increase with solidification time, t_f,

$$\lambda = k\left(\frac{dT}{dt}\right)^p \tag{2.16}$$

Data for Al-4.5% Cu indicate that p is about 1/3. Data for permanent mold casting of aluminum alloy A356 is shown in Figure 2.20. The exponent p is about 0.47.

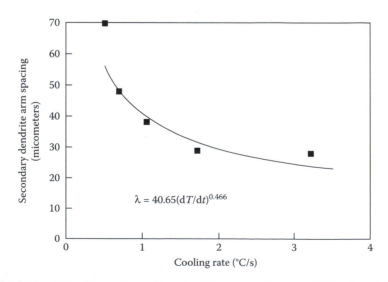

$$\lambda = 40.65(dT/dt)^{0.466}$$

FIGURE 2.20   Dependence of secondary dendrite arm spacing on solidification rate in aluminum alloy A356 data. (From T. F. Bower et al., in *Light Metals 2001*, Canadian Institute of Mining, Metallurgy and Petroleum, 2001. With permission.)

Obviously, rapid solidification (short freezing times or high velocities, v, of the liquid–solid interface) promotes finer dendrites.

EXAMPLE PROBLEM 2.5

Estimate the length of dendrites in the freezing of copper alloys containing, (a) 5% Zn, (b) 5% Sn, and (c) 0.05% Sn. In each case, assume a thermal gradient of $G = 0.5°C/mm$ and assume the temperatures at the tips and bases of the dendrites are the equilibrium for liquidus and solidus temperatures the alloy.

SOLUTION

From the phase diagrams, the freezing ranges, $\Delta T_{freezing}$, for the alloys are (a) 6°, (b) 130°, and (c) 1.3°. Taking the lengths as $\Delta T_{freezing}/G$, the lengths for the three alloys are (a) 12 mm, (b) 260 mm, and (c) 2.6 mm.

2.11 CONSEQUENCES OF DENDRITIC GROWTH

When dendrites form during solidification, the segregation is almost entirely interdendritic rather than macroscopic. The distances between the concentration minima and maxima are half of the distance between dendrite arms. Even this distance is usually too great for homogenization by diffusion to be practical. In the case of ingots, later hot or cold mechanical working will reduce the distance between the concentration minima and maxima, which may make homogenization possible. In contrast, homogenization would be virtually impossible for the center-to-surface segregation that would occur in the absence of dendritic growth.

Dendritic growth also affects the nature of porosity in castings. Most metals shrink about 4% as they solidify. Without dendritic growth, this shrinkage results in large cavities in the last regions to freeze. In ingots, a large *pipe* may form in the center as shown in Figure 2.21. Large internal cavities may form in castings of more complex shape. If the dendrites are long, liquid may not be able to feed through the interdendritic channels; hence the shrinkage may occur interdendritically on a microscopic scale. With such interdendritic shrinkage, the macroscopic shrinkage will be absent or greatly reduced. Interdendritic shrinkage causes no problem in casting of statues whereas macroscopic shrinkage would. On the other hand, castings with interdendritic porosity are not desirable where pressure tightness is required, because the interdendritic channels may cause gas or water leaks.

2.12 GAS SOLUBILITY AND GAS POROSITY

When gases dissolve in liquid metals, they usually dissolve monatomically. For example, the solution reactions may be written, $H_2 \rightarrow 2\underline{H}$, $N_2 \rightarrow 2\underline{N}$, $O_2 \rightarrow 2\underline{O}$, where the underlining signifies the element is in solution. For diatomic gases, Sievert's law, which is an application of the mass action principle, states that the solubility is proportional to the square root of the partial pressure of the gas. For example,

$$\underline{H} = k\sqrt{(P_{H_2})}, \tag{2.17}$$

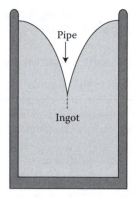

FIGURE 2.21   Shrinkage that occurs during freezing causes pipes to be formed in castings.

where $\underline{H}$ is the concentration of the dissolved hydrogen that is in equilibrium with the partial pressure of hydrogen gas and k is a temperature-dependent constant. In addition to monatomic gases, carbon monoxide and water vapor are soluble in metals ($CO \rightarrow \underline{C} + \underline{O}$) and $H_2O \rightarrow 2\underline{H} + \underline{O}$). Hydrogen is soluble in almost all liquid metals, but the solubility of nitrogen and carbon is limited in nontransition metals.

The solubilities of gases in solid metals are much lower than liquid metals. Figure 2.22 shows the solubility of hydrogen in copper and copper-aluminum alloys.

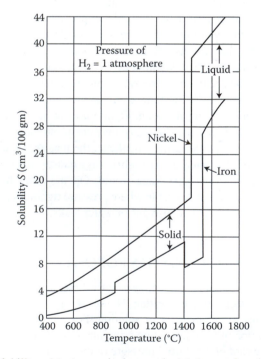

FIGURE 2.22   Solubility of hydrogen in iron and nickel as a function of temperature. (Adapted from A. G. Guy, *Elements of Physical Metallurgy*, Addison-Wesley, 1959.)

Because of the lower solubility in the solid, gas bubbles are released at the liquid–solid interface as the metal freezes. With long dendrites, the gas bubbles are trapped and the result is gas porosity. With short dendrites, elongated tubes of gas (*worms*) may form. Gas worms are very common in ice cubes made in household refrigerators. In some cases, gas bubbles rising in the melt can cause violent stirring.

Gas porosity can result in defective castings. There are several ways of preventing or lessening the problem. Most depend on removing the dissolved gas in the melt. With low-carbon steels, aluminum is added to react with dissolved oxygen. In copper alloys, porosity resulting from water vapor can be prevented by additions of phosphorus, which reacts with dissolved oxygen but hydrogen must also be removed. Dissolved gases may be removed by bubbling an insoluble gas such as Ar or N_2 through the melt. Since the partial pressure of hydrogen is very low in the bubble, hydrogen will transfer from the liquid to the bubble. Solid degassing agents may be used to generate gas bubbles by vaporization or by chemical decomposition. Mixed carbonates decompose to form CO_2 bubbles. Melting under vacuum or an inert atmosphere will also remove gases particularly if the melt is stirred. The porosity resulting from dissolved gas can be reduced by freezing under a high external pressure.

2.13   GROWTH OF SINGLE CRYSTALS

Directional solidification is one of the ways to make metal single crystals. With the Bridgman technique, a vertical mold is slowly lowered out of a furnace so that freezing starts at the bottom and slowly progresses upward. Because freezing starts at a point, only one crystal is nucleated. The furnace can be raised instead of lowering the mold or crystals can be grown horizontally by moving a furnace or a mold. The crystal orientation can be controlled with a seed crystal. Initially, melting is controlled so that the seed crystal partially melts.

Single crystals can also be grown by the Czochralski method. A seed crystal is lowered into melt so that it partially melts and then it is slowly withdrawn upward. The growing crystal is rotated about a vertical axis to help stir the liquid. The lack of a mold eliminates contamination from mold walls but makes impossible to grow crystals of fixed exterior shapes. This is the method is used to grow Si crystals for the semiconductor industry.

It is difficult to grow single crystals of iron or titanium because of the phase transformations in the solid state. Directional recrystallization is however an alternative method of obtaining single crystals.

2.14   EUTECTIC SOLIDIFICATION

Eutectic reactions, liquid $\rightarrow \alpha + \beta$, can result in several geometric configurations of α and β. When the volumes of both phases are nearly equal (say from 30% to 50%), the most common morphology is lamellar. This is true of Cu–Ag and Pb–Sn. If the amount of one phase is much less than the other (say <30%), the eutectic is likely to

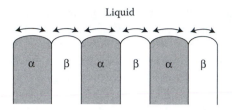

Liquid

FIGURE 2.23   Growing eutectic. The solute must partition between the two phases.

be rods of one phase surrounded by the other phase (e.g., NiAl–Cr and TaC–Ni). If the volume fraction of one phase is very low, that phase may form as isolated islands (graphite in Fe, and Si in Si–Al alloys).

As the eutectic front advances during solidification, the solutes must partition between the two phases as suggested in Figure 2.23. With increased rates of solidification, there is less time for diffusion; hence the widths of the lamellae decrease. Thus lamellae platelet width depends on freezing rate.

EXAMPLE PROBLEM 2.6

Determine the eutectic morphology (platelets, rods, and isolated spheres) that minimizes the total interphase area as a function of the volume fraction, f, of minor phase and the separation distance, λ, assuming simple geometric arrangements.

SOLUTION

For spheres in a simple cubic array, the volume fraction is $f = (4/3)\pi(r/\lambda)^3$ and the surface area per volume, A_v, is $4\pi r^2/\lambda^3$. Combining, $A_v = (4\pi)^{1/3}(3f)^{2/3}/\lambda$.

For a square array of rods, $f = \pi(r/\lambda)^2$ and $A_v = 2\pi r/\lambda^2$. Combining, $A_v = 2(f\pi)^{1/2}/\lambda$.

For parallel platelets, the surface area per volume, A_v, is $2/\lambda$, regardless of f. Figure 2.24 is a plot of $A_v/(A_{v\ \text{parallel plates}})$ for the three geometries as a function of f. According to this simple analysis, parallel plates have the least area for $f \geq 1/\pi = 31.8\%$, rods for $4\pi/81$ $(= 14.2\%) \leq f \leq 1/\pi$ (31.8%), and isolated spheres for $f \leq 4\pi/81 = 14.1\%$.

2.15   PERITECTIC FREEZING

Consider the freezing of an alloy of the peritectic composition, C_0, as shown in Figure 2.25a. The reaction, $\alpha + \text{liquid} \rightarrow \beta$ should start as soon as the peritectic temperature is reached. The β will form on the surface of preexisting α where there is contact between the α and liquid. As a film of β forms, it prevents direct contact between the liquid and α. Further reaction can occur only by diffusion of A or B atoms through the β so as the film thickens, the reaction becomes extremely slow. Usually peritectic reactions do not go to completion. Microstructures usually contain α-phase, even though the phase diagram predicts it should not exist. The term *surrounding* is used to describe this phenomenon.

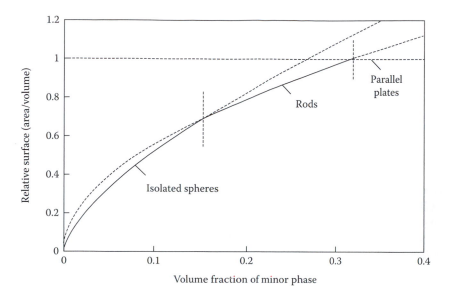

FIGURE 2.24 Relative amount of interphase area in eutectics composed of platelets, rods, and spheres. The morphology with the lowest interphase area has the least energy.

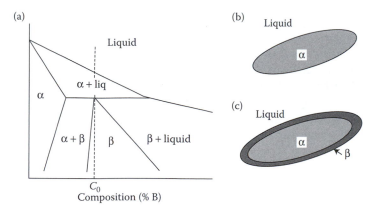

FIGURE 2.25 Portion of a phase diagram with a peritectic reaction (a). Particle just before the peritectic reaction (b) and the same region as b begins to form between the a and the liquid (c).

2.16 METAL GLASSES

If some alloy compositions are cooled fast enough, crystallization can be prevented. Instead a metallic glass will be formed. Most glass forming compositions for which much redistribution of elements by diffusion would be necessary to form the equilibrium crystalline phases. Therefore most compositions have many components. The compositions correspond to deep wells in the equilibrium phase diagram, so diffusion in the liquid is slow (Figure 2.26).

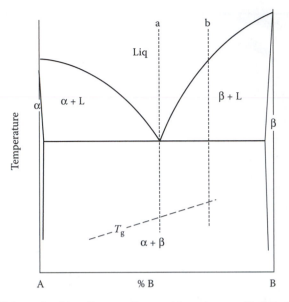

FIGURE 2.26   Schematic phase diagram. Composition a is more likely to form a glass than composition b, because of the much lower temperature at which crystallization starts.

Most of the early compositions included appreciable amounts of small atoms including P, Si, Be, and Ge. Cooling rates of 10^3 K/s to $\approx 10^5$ K/s were necessary to prevent crystallization. This limited alloys to thin ribbons or wires. More recently, magnesium-base, iron-base and zirconium–titanium-base alloys have been developed that do not require such rapid cooling. The first commercial alloy available in bulk form is Vitreloy 1, which contains 41.2 a/o Zr, 13.8 a/o Ti, 12.5 a/o Cu, 10 a/o Ni, and 22.5 a/o Be. Because the critical cooling rate is about 1 K/s, glassy parts can be made with dimensions of several centimeters.

The very high yield strengths typical of metallic glasses permit very high elastic strains and therefore storage of a large amount of elastic energy. This has lead to use as heads of golf clubs. Its properties are given in Table 2.4.

TABLE 2.4
Properties of Vitreloy 1

Density	5.9 Mg/m³
Young's modulus	95 GPa
Shear modulus	35 GPa
YS	1.9 GPa
K_{IC}	55 MPa√m
T_g	625 K
Endurance limit/UTS	≈0.03

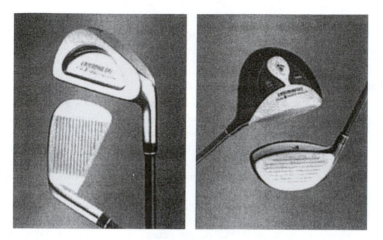

FIGURE 2.27 Golf club heads of vitreloy-1. Irons at the left and drivers at the right. (Courtesy of Otis Buchanan, Liquidmetal Technologies, Lake Forest, CA.)

The high ratio of yield strength to Young's modulus permits elastic strains of 1.9/95 = 2%. The fracture toughness is very high but the fatigue strength is very low. The ratio of endurance limit to yield strength of 0.03 is very much lower than the ratios of 0.3–0.5 typical of crystalline metals.

There are two principle uses: as soft magnetic materials (there are no barriers for domain wall movement) and applications based on the large amount of elastic energy that can be stored. Use of metallic glass in the heads of golf clubs (Figure 2.27) has permitted longer drives.

2.17 MISCELLANY

Ice worms are the channels of voids that form during freezing of ice cubes. They are the result of dissolved gas being released during solidification and forming a channel. In the 1950s, a researcher at the General Electric Research Laboratories noted that the diameter of these ice worms varied periodically along their length and became interested in developing a fundamental theory to explain the periodic variation in diameter. He felt that the ratio of the wavelength to diameter should be some simple constant. Then a technician pointed out to him that the temperature in a refrigerator cycles so that the rate of freezing should cycle too. The wavelength was found to correspond to the cycling frequency.

Old-time gold assayers tested a gold ore by heating it until a molten droplet of gold formed. As the molten gold droplets cooled, they would darken as they underwent substantial undercooling. Once nucleation occurred, however, the color would brighten again as the drop was heated by the release of latent heat. This color change is called recalescence.

In the past, there were two commercial grades of low-carbon steels, *killed* and *rimming*. Rimming steels were only partially deoxidized before casting into ingots. During solidification, large amounts of CO were released and caused a violent stirring action, sending out a spray of sparks. The stirring broke up any boundary layer, and as a consequence prevented dendrite formation and allowed center-to-surface segregation of carbon. The very low carbon content of the surface was preferred if coatings were to be applied.

PROBLEMS

1. The rate at which metals freeze is controlled by the rate at which heat can be extracted. Consider the freezing of aluminum. If the liquid–solid interface advances at 1mm/s, what is the thermal gradient (°C/mm) in the solid?

    Data for aluminum: melting point = 660°C, specific heat = 0.215 cal/ (g-°C), heat of fusion = 94.5 cal/g, atomic wt = 27, density = 2.7 mg/m³, thermal conductivity = 0.22 (W/mm²)/(°C/mm), and coefficient of linear expansion = 22.5×10^{-6}/°C.

2. An ingot of Al-5% Cu is directionally solidified. Assume that there is no diffusion in the solid and that there is perfect mixing in the liquid. Pure aluminum melts at 660°C. At the eutectic temperature of 548°C, the liquid composition is 33.2% Cu and the solid composition is 5.35% Cu. Assume that the liquidus and solidus are straight lines.

    a. Find the distribution coefficient expressed as $k = C_S/C_L$ where C_S and C_L are expressed as % Cu.

    b. Calculate the composition of the liquid when the solidification is 40% complete.

    c. What is the average composition of the solid, C_{Sav}, at this point. (Make sure that $0.4C_{Sav} + 0.6C_L = 5\%$.)

    d. What is the liquid–solid interface temperature at this point?

    e. How much eutectic will be formed?

3. Consider the freezing of an aluminum alloy containing 0.005% copper.

    a. What would be the composition of the first solid to freeze?

    b. What would be the average composition of the first half to freeze?

4. Consider the steady-state freezing of an aluminum alloy containing 0.55% Cu. In steady-state freezing the boundary layer is such that the solid freezing has the same composition as the alloy. Assume that the liquid–solid interface moves at a rate of 80 μm/s. The diffusion coefficient of copper in liquid aluminum is 3×10^{-9} m²/s.

    a. What is the interface temperature?

    b. What is the thickness of the boundary layer?

    c. What temperature gradient would be required to maintain plane-front growth?

5. The dependence of the dendrite arm spacing, λ, on the cooling rate, r, is given by $\lambda = kr^{-n}$. λ was found to be 100 μm at $r = 0.1$ K/s and 10 μm at $r = 60$ K/s.

    a. Find k and n.

    b. What rate of cooling would be required to produce a spacing of $\lambda = 1.0$mm?

6. At the melting point of aluminum and 1 atm partial pressure of hydrogen, the equilibrium solubility of hydrogen is 7×10^{-3} cm³/g of Al in the liquid and 4×10^{-4} cm²/g of Al in the solid. The solubilities, 4×10^{-4} cm³/g and 7×10^{-3} cm²/g, are expressed as the volumes measured at 20°C and 1 atm (STP), *not* the volumes at the melting point.

    a. Calculate the equilibrium solubilities in the liquid and solid at 0.1 atmosphere H_2. Express your answer in STP.

b. What volumes of H_2 would be liberated during the freezing per volume of aluminum, if the partial pressure of H_2 were 0.1 atmospheres? (The H_2 is liberated at a total pressure of 1 atmosphere and at the melting point of aluminum. Assume the perfect gas law. Your answer should equal the percent gas porosity if the gas is trapped interdendritically.)

7. Consider an aluminum-rich binary aluminum–silicon alloy. The melting temperature of aluminum is 660°C; the eutectic is at 577°C and 12.6 wt% Si. The maximum solubility of silicon in aluminum is 1.65% Si at 577°C. The liquidus and solidus can be approximated by straight lines. The diffusivity of silicon in liquid aluminum is 8×10^{-8} m²/s. Freezing occurs at a rate of 10 μm/s.

a. For an alloy of 0.05% Si, what is the interface temperature for steady-state freezing?

b. Find the thickness of the boundary layer.

c. What temperature gradient is necessary to maintain plane-front growth?

d. Repeat a, b, and c for an alloy containing 1% silicon.

8. Predict the morphology of each of the eutectics listed below. The compositions are from phase diagrams in the *Metals Handbook*, Vol. 8, 8th ed. Some of the densities are estimates.

System	Phase	Composition	Density (Mg/m³)
Bi/Cd	α Bi	0% Cd	9.8
	Eutectic	39.7% Cd	
	β Cd	100% Cd	7.9
Fe/C	γ	2.1% C	7.9
	Eutectic	4.3% C	
	Graphite	100% C	2.25
Cu/Al	θ	47% Al	6.6
	Eutectic	66.8% Al	
	α	94% Al	2.7
BiPb	BiPb$_2$	42% Bi	11.4
	Eutectic	56% Bi	
	Bi	100% Bi	9.8

9. The melting point of pure aluminum is 660°C and aluminum and silicon form a eutectic, the eutectic temperature is 577°C, the eutectic composition is 12% Si, and the maximum solubility of silicon in solid aluminum is 1.65%. Assume the phase diagram consists of straight lines.

If aluminum containing 0.15 wt% Si were solidified, what would be the composition of the first solid to form?

10. Some solutes raise the melting temperature, causing both the liquidus and solidus to increase with additional solute. In this case, the distribution coefficient $k > 1$.

a. Is the Scheil Equation 2.11 still valid?

b. Describe qualitatively how having $k > 1$ affects the segregation.

REFERENCES

T. F. Bower, P. Krishna, R. D. Pehlke, and K. T. Bilkey, in *Light Metals* 2001, Canadian Institute of Mining, Metallurgy and Petroleum, 2001.

M. C. Flemings, *Solidification Processing*, McGraw-Hill, 1974.

R. A. Flinn, *Fundamentals of Metal Casting*, Addison-Wesley, 1964.

B. R. T. Frost in *Progress in Metal Physics*, Vol. 5, Pergamon Press, 1954.

A. G. Guy, *Elements of Physical Metallurgy*, Addison-Wesley, 1959.

J. H. Hollomon and D. Turnbull, in *Solidification of Metals and Alloys*, AIME, 1951.

W. Kurz and D. J. Fisher, *Fundamentals of Solidification*. Trans Tech Publications, Switzerland, 1984.

Liquid Metals and Solidification: *A Seminar on Liquid Metals and Solidification*, 39th National Metal Congress and Exposition, Chicago, 1957, ASM, Cleveland, 1958.

W. G. Pfann, *Zone Melting*, Wiley, NY, 1958.

D. A. Porter and K. E. Easterling, *Phase Transformations in Metals & Alloys*, Chapman & Hall, 1981.

E. Scheil, *Z. Metallkunde*, 34, 70, 1942.

P. G. Shewman, *Transformations in Metals*, McGraw-Hill, 1969.

3 Diffusion

Diffusion is involved in phase changes and in treatments that alter surface compositions. This chapter treats Fick's first and second laws, several applications of Fick's second law, diffusion mechanisms, the Kirkendall effect, and Darken's treatment of interdiffusion. Diffusion in systems with more than one phase is also treated.

3.1 FICK'S FIRST LAW

This law states that in a solution with a concentration gradient there will be a net flux of solute atoms from regions of high solute concentration to regions of low solute concentration and that the net flux of solute is proportional to the concentration gradient. This can be expressed as

$$J = -D\frac{dc}{dx},$$ (3.1)

where J is the net flux of solute, D is the diffusivity (or diffusion coefficient), c is the concentration of solute, and x is the distance (see Figure 3.1).

The flux is the net amount of solute crossing an imaginary plane per area of the plane and per time. The flux may be expressed as solute atoms/(m²s), in which case the concentration, c, must be expressed as atoms/m³. Alternatively J and c may be expressed in terms of mass of solute; the units of J being (kg solute)/(m²s) and of c being (kg solute)/m³. The diffusivity has dimensions of m²/s and depends on the solvent, the solute, the concentration, and the temperature.

Fick's first law for mass transport by diffusion is analogous to the laws of thermal and electrical conduction. For heat conduction, $q = kdT/dx$, where dT/dx is a thermal gradient (°C/m), k is the thermal conductivity, J/(ms°C), and q is the flux, J/(m²s). Ohm's law, $I = E/R$, can be expressed in terms of a current density, $i = \sigma\varepsilon$, where i is the current density in coulombs/(m²s) is a flux, ε is the voltage gradient (V/m), and the proportionality constant is the conductivity, $\sigma = 1/\rho$, in (ohm-m)$^{-1}$.

Direct use of Fick's first law is limited to steady-state (or nearly steady-state) problems in which the variation of dc/dx over the concentration range of concern can be neglected.

3.2 FICK'S SECOND LAW

This expresses how the concentration at a point changes with time. Consider an element of area A and thickness dx in a concentration gradient (Figure 3.2).

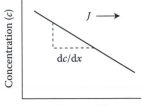

FIGURE 3.1   A concentration gradient. Note that with a negative gradient ($dc/dx < 0$) as shown here the flux, J, is positive.

According to Fick's first law, the flux into the element is $J_{in} = -Ddc/dx$ and the flux out of it is $J_{out} = -Ddc/dx - d(-Ddc/dx)/dx$.

The rate of change of the composition within the element is then $dc/dt = J_{in} - J_{out}$ or

$$\frac{dc}{dt} = \frac{\partial(Ddc/dx)}{\partial x}.$$

(3.2)

This is a general statement of Fick's second law which recognizes that the diffusivity may be a function of concentration and therefore of distance, x. In applications where the variation of D with distance and time can be neglected, Equation 3.2 can be simplified into a more useful form,

$$\frac{dc}{dt} = D\frac{d^2c}{dx^2}.$$

(3.3)

Rigorously, concentration should be expressed in atoms or mass per volume, but if density changes are neglected concentration may be expressed in atom% or wt%. Several specific solutions to Fick's second law are given below.

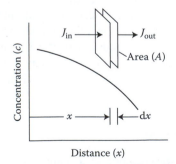

FIGURE 3.2   Rate of change of composition in an element of volume Adx equals the differences between the fluxes into and out of the element.

3.3 SOLUTIONS OF FICK'S SECOND LAW

3.3.1 ADDITION OF MATERIAL TO OR REMOVAL FROM A SURFACE

If the composition at the surface of a material is suddenly changed from its initial composition, c_0, to a new composition, c_s, and held at that level (Figure 3.3), the solution to Equation 3.3 is

$$c = c_s - (c_s - c_0)\, \mathrm{erf}\left[\frac{x}{2\sqrt{(Dt)}}\right], \tag{3.4}$$

where erf is the error function. Table 3.1 and Figure 3.4 show how the erf(y) depends on y. One application of this solution involves carburizing and decarburizing of steels.

Three straight lines make a rough approximation to the error function as shown in Figure 3.5.

For $y \geq 1$, erf(y) = 1; for $1 \geq y \geq -1$, erf(y) = y; and for $-1 \geq y$, erf(y) = -1.

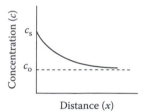

FIGURE 3.3 Solution to Fick's second law for a constant surface concentration, c_s.

TABLE 3.1
Values of the Error Function, $\mathrm{erf}(x) = 2/\pi^{1/2} \int_0^x \exp(-t^2)\, dt$

y	erf(y)	y	erf(y)	y	erf(y)	y	erf(y)
0.00	0.0000	0.05	0.0564	0.10	0.1125	0.15	0.1680
0.20	0.2227	0.25	0.2763	0.30	0.3286	0.35	0.3794
0.40	0.4284	0.45	0.4755	0.50	0.5205	0.55	0.5633
0.60	0.6039	0.65	0.6420	0.70	0.6778	0.75	0.7112
0.80	0.7421	0.85	0.7707	0.90	0.7970	0.95	0.8209
1.00	0.8427	1.10	0.8802	1.20	0.9103	1.30	0.9340
1.40	0.9523	1.50	0.9861	1.60	0.9763	1.70	0.9838
1.80	0.9891	1.90	0.9928	2.00	0.9953	2.20	0.9981
2.40	0.9993	2.60	0.9998	2.80	0.9999		

Note: erf($-y$) = $-$erf(y), i.e., erf(-0.20) = -0.2227. For small values of y, erf(y) $\approx 2y/\sqrt{\pi}$.

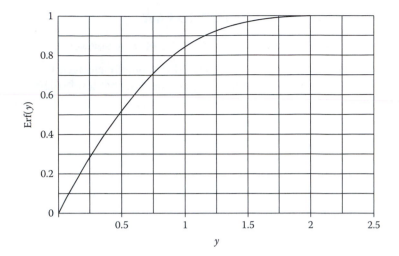

FIGURE 3.4　Dependence of erf(y) on y.

3.3.2　JUNCTION OF TWO SOLID SOLUTIONS

Another simple solution is for two blocks of differing initial concentrations, c_1 and c_2, that are welded together. In this case,

$$c = \frac{c_1 + c_2}{2} - \left[\frac{c_1 - c_2}{2}\right] \mathrm{erf}\left[\frac{x}{2\sqrt{(Dt)}}\right]. \tag{3.5}$$

Figure 3.6 illustrates this solution. Note that Equation 3.6 is similar to Equation 3.4, except that $(c_1 + c_2)/2$ replaces c_s and $(c_1 - c_2)/2$ replaces $c_s - c_o$.

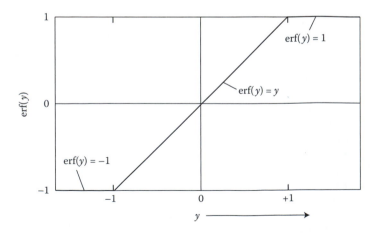

FIGURE 3.5　Approximation of the variation of erf(y) with y. For $y \leq -1$, erf(y) = -1; for $y \geq 1$, erf(y) = 1; for $-1 \leq y \leq 1$, erf(y) = y.

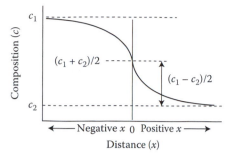

FIGURE 3.6   Solution to Fick's second law for two solutions with different concentrations.

3.3.3   Homogenization

Interdendritic segregation during solidification causes local composition variations that can be approximated by a sine wave of wavelength, $2x$ and amplitude, c_{omax}. Homogenization by diffusion decreases the amplitude c_{omax} to c_{max} as shown in Figure 3.7.

Defining c and c_o as the differences between local concentrations and the average composition, the extent of homogenization is described by

$$\frac{c}{c_o} = \exp\left[-\left(\frac{\pi}{X}\right)^2\right], \quad \text{where } X = \frac{x}{\sqrt{(Dt)}}. \tag{3.6}$$

All solutions to Fick's second law are of the form

$$f(\text{concentrations}) = \frac{x}{\sqrt{(Dt)}}, \tag{3.7}$$

where $f(\text{concentrations})$ is a function of c, (concentration at a specific point), c_o (initial concentration), c_s (surface concentration), and so on. In many problems, these concentrations are fixed so

$$\frac{x}{\sqrt{(Dt)}} = \text{constant}. \tag{3.8}$$

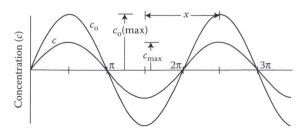

FIGURE 3.7   Sinusoidal concentration profile, c_o, resulting from interdendritic segregation, and the profile, c, after some homogenization.

3.4 MECHANISMS OF DIFFUSION

In interstitial solid solutions, diffusion occurs by interstitially dissolved atoms jumping from one interstitial site to another. For an atom to move from one interstitial site to another, it must pass through a position where its potential energy is at maximum. The difference between the potential energy in this position and that in the normal interstitial site is the activation energy for diffusion and must be provided by thermal fluctuations. The overall diffusion rate is governed by an Arrhenius-type rate equation;

$$D = D_o \exp\left(\frac{-E}{kT}\right), \tag{3.9}$$

where D_o is a constant for the diffusing system, k is Boltzmann's constant, T is the absolute temperature, and E is the activation energy (the energy for a single jump). Often this equation is written as

$$D = D_o \exp\left(\frac{-Q}{RT}\right), \tag{3.10}$$

where the activation energy, $Q = N_o E$, is for a mole of jumps. (N_o is Avogadro's number $= 6.02 \times 10^{23}$ jumps). Correspondingly, R is the gas constant ($=N_o k$). Interstitial solutes diffuse much faster than substitutional solutes. Table 3.2 shows that carbon, which dissolves interstitially, diffuses much faster than do nickel and manganese, which dissolve substitutionally. Also note that diffusion in bcc iron is much faster than that in fcc iron.

For self-diffusion and diffusion in substitutional solid solutions, the diffusion mechanism is not so obvious. For years many experts considered the possibility of interchange mechanisms (whereby two atoms exchanged places) and ring mechanisms (which involves a cooperative rotation of a ring of four, six, or more atoms). In mechanisms of these types, both species of atoms would diffuse at the same rate (Figure 3.8). Today we know that the dominant mechanism of diffusion is by movement

---

TABLE 3.2

Diffusion Constants for Several Systems

Solute	Solvent	Solvent Structure	D_o (m²/s)	Q (kJ/mol)
Carbon	α-iron	bcc	2.2×10^{-4}	122.5
Carbon	γ-iron[a]	fcc	0.2×10^{-4}	142
Nickel	γ-iron	fcc	0.8×10^{-4}	280
Manganese	γ-iron	fcc	0.35×10^{-4}	282

[a] The diffusion rate of carbon increases with carbon content.

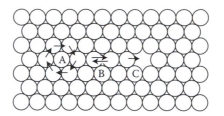

FIGURE 3.8 Schematic illustration of several mechanisms proposed for substitutional diffusion. A—ring interchange, B—simple interchange, and C—vacancy migration.

of vacancies (vacant lattice sites). Earlier it was pointed out that the equilibrium number of vacancies depends exponentially on temperature.

$$n_v = n_o \exp\left(\frac{-E_f}{kT}\right), \tag{3.11}$$

where n_v/n_o is the fraction of the lattice sites that are vacant and E_f is the energy to form a vacancy.

The rate a given vacant site will be filled by a substitutional atom moving into it is also dependent on thermal activation

$$\text{Rate} = \exp\left(\frac{-E_m}{kT}\right), \tag{3.12}$$

where E_m is the energy barrier to the movement of a vacancy by it being filled by an adjacent substitutional atom. The net rate of diffusion is proportional to the product of the number of vacancies and the rate at which they contribute to diffusion. Therefore, $D = D_o\exp(-E_f/kT) \exp(-E_m/kT)$, which simplifies to

$$D = D_o \exp\left(\frac{-E}{kT}\right), \tag{3.13}$$

where

$$E = E_f + E_m. \tag{3.14}$$

Of course this equation can also be expressed in an equivalent form in terms of Q, the activation energy per mole of diffusion jumps,

$$D = D_o \exp\left(\frac{-Q}{RT}\right). \tag{3.15}$$

In general, the activation energies for self-diffusion and diffusion of substitutional solutes are considerably higher than those interstitial diffusions and therefore the diffusion rates are much lower. Data for self-diffusion in several metals is given in Table 3.3. In comparing these data, several other trends are apparent. One is that the

TABLE 3.3
Self Diffusion Data

Metal	Crystal Structure	Q (kJ/mol)	D_o (m²/s)	T_m (K)	Q/RT_m
Cu	fcc	198	20×10^{-6}	1356	17.5
Ag	fcc	185	40	1234	18.0
Ni	fcc	281	130	1726	19.6
Au	fcc	198	9.1	1336	17.8
Pb	fcc	102	28	600	20.4
α–Fe	bcc	240	190	1809	15.9
Nb	bcc	441	1200	2741	19.4
Mo	bcc	461	180	2883	19.2
Mg	hcp	136	125	923	17.8

Source: From J. H. Hollomon and D. Turnbull, *Solidification of Metals and Alloys*, AIME, 1951. With permission.

activation energies increase with melting point. In fact, for most relatively close-packed metals (fcc, bcc, and hcp), Q/T_m is nearly the same.

With the use of radioactive isotopes, it has been possible to measure the rates of self-diffusion in solids. Self-diffusion can play a significant role in such diverse phenomena as sintering, creep, thermal etching, and grain boundary migration.

3.5   KIRKENDALL EFFECT

In early studies of diffusion it was assumed that in binary solid solutions both species of atoms diffuse in opposite directions at the same velocity. However, experiments by Smigelskas and Kirkendall showed that this could not be so. They studied diffusion in a diffusion couple formed by plating copper onto a brass bar containing 30% zinc. Molybdenum wires were wrapped around the bar before plating so that the initial interface could be located after diffusion. Examination of the couple after diffusion revealed an apparent movement of the wires. That is, the distance between the wires and center of the bar had decreased. Since the wires, themselves, cannot diffuse, the only reasonable interpretation is that the net flux of zinc past the wires in one direction was faster than the flux of copper in the opposite direction as sketched in Figure 3.9. This observation is inconsistent with all of the exchange mechanisms for diffusion and thereby provided strong supporting evidence for the vacancy mechanism of diffusion.

Porosity can be formed because of the different rates of diffusion of two species. Because zinc is diffusing faster than copper, there is a net flux of vacancies into the zinc-rich brass. Under some circumstances, these vacancies may diffuse to grain boundaries. In this case, there is a volume contraction as the vacancies disappear into the boundaries.

However, if volume contraction of the zinc-rich region is prevented by macroscopic constrains, porosity will result from the precipitation of the vacancies to form voids. Such porosity can be formed as concentration gradients formed by

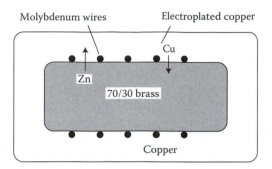

FIGURE 3.9 Diffusion couple in the Kirkendall experiments. Because zinc atoms diffused faster than copper atoms, the molybdenum wires appeared to move toward the center of the specimen.

interdendritic segregation are minimized during annealing. In brass ingots, the centers of the dendrites arms are copper rich and the interdendritic regions are zinc rich. When mechanical working of an ingot has reduced the dendritic spacing enough for homogenization during annealing, porosity will be formed in the zinc-rich regions.

3.6 SPECIAL DIFFUSION PATHS

Diffusion occurs rapidly along grain boundaries, dislocations, and free surfaces. Probably the most important of these is grain boundary diffusion. Data for self-diffusion in silver are given in Table 3.4. These data are based on experiments on both polycrystals and single crystals of silver.

The overall diffusivities measured are shown in Figure 3.10. It should be noted that the effects of grain boundary diffusion are observable only at low temperatures. In most practical diffusion problems, grain boundary diffusion can be neglected.

EXAMPLE PROBLEM 3.1

A. If the grain size of silver were doubled, by what factor would the net diffusion by grain boundary diffusivity change?
B. At what temperature would the net transport by grain boundary diffusion be the same as the grain boundary diffusional transport with the original grain size at 1000 K?

TABLE 3.4
Self-Diffusion in Silver

Path	D_o (m²/s)	Q (kJ/mol)
Lattice	90×10^{-6}	193
Grain boundary[a]	2.3×10^{-9}	110.9

[a] To express the diffusion in terms of a diffusivity, an effective width of the grain boundary path was assumed to be 30 nm.

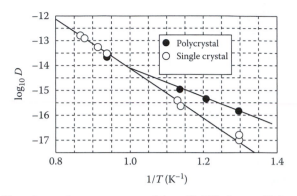

FIGURE 3.10 Experimental measurements of the self-diffusion coefficient in silver. Silver single crystals and polycrystals of a 35 mm grain size. Higher values of D for the polycrystal at low temperatures (high values of $1/T$) are a result of grain boundary diffusion. At high temperatures, lattice diffusion masks the contribution of grain boundary diffusion.

SOLUTION

A. The net transport by diffusion must be proportional to the product of the diffusivity and the cross-sectional area through which the diffusion occurs. Assuming a constant effective thickness, t, of the grain boundary, the cross section of grain boundary, for one grain of diameter, d, is $A_{gb} = \pi dt$. The cross-sectional area of the grain is $A_g = \pi d^2/4$. The relative amount of grain boundary area is $A_{gb}/A_g = t/4d$. Therefore, doubling the grain size will result in cutting the effective transport in half.

B. The temperature, T_2, at which the transport is half of that at T_1 is such that $D_2 = 0.5D_1$ or

$$D_2 = D_o \exp\left[\left(\frac{-Q}{RT_2}\right)\right] = 0.5D_2 = 0.5D_o \exp\left[\left(\frac{-Q}{RT_1}\right)\right].$$

$$\frac{\exp[(-Q/RT_1)]}{\exp[(-Q/RT_2)]} = 2 = \exp\left[\left(\frac{-Q}{R}\right)\left(\frac{1}{T_1} - \frac{1}{T_2}\right)\right],$$

$$\frac{1}{T_1} - \frac{1}{T_2} = \left(\frac{-R}{Q}\right)\ln 2,$$

$$\frac{1}{T_2} - \frac{1}{T_1} = \left(\frac{-R}{Q}\right)\ln 2.$$

Substituting $Q_{gb} = 110,900$ J/mol,

$$\frac{1}{T_2} = \frac{1}{T_1} + \left(\frac{8.314}{110,900}\right)\ln 2 = \frac{1}{T_1} + 5.2 \times 10^{-5}.$$

for $T_1 = 1000$, $T_2 = 950$ K.

3.7 DARKEN'S EQUATION

The solutions to Fick's second law (Equations 3.5 through 3.7) are based on a single diffusivity, D, whereas the Kirkendall experiments show that each specie has its own diffusivity. Darken has shown that Fick's second law should be written as

$$\frac{\partial N_A}{\partial t} = \frac{\partial}{\partial x}\left(\tilde{D}\,\frac{\partial N_A}{\partial x}\right), \tag{3.16}$$

where $\tilde{D}$ is the effective diffusivity and is related to the intrinsic diffusivities of the two species, D_A and D_B, by

$$\tilde{D} = N_B\,D_A + N_A\,D_B, \tag{3.17}$$

where N_A and N_B are the atomic fractions of the two species. It should be noted that for dilute solutions (say $N_B \to 0$), ~D approaches the diffusivity of the solute ($\tilde{D} \to D_B$). In this case, $\tilde{D}$ may be substituted for D in Equations 3.5 through 3.7.

Darken also showed that the velocity, v, of the markers in the Kirkendall experiments is given by

$$v = (D_A - D_B)\frac{\partial N_A}{\partial x}. \tag{3.18}$$

3.8 DIFFUSION IN SYSTEMS WITH MORE THAN ONE PHASE

In analyzing diffusion couples involving two or more phases, there are two key points:

1. Local equilibrium is maintained at interfaces. Therefore there are discontinuities in composition profiles at interfaces. The phase diagram gives the compositions that are in equilibrium with one another.
2. No net diffusion can occur in a two-phase microstructure because both phases are in equilibrium and there are no concentration gradients in the phases. These points will be illustrated by several examples.

Consider diffusion between two pure metals in a system that has an intermediate phase as illustrated in Figure 3.11. Interdiffusion between blocks of pure A and pure B at temperature, T, will result in the concentration profiles shown in Figure 3.12.

Note that a band of β will develop at the interface. The compositions at the $\alpha - \beta$ and $\beta - \gamma$ interfaces are those from the equilibrium diagram, so the concentration profile is discontinuous at the interfaces. No two-phase microstructure will develop.

Consider the decarburization of a steel having a carbon content of c_o when it is heated in the austenite (γ) region and held in air (Figure 3.13). At this temperature, the reaction $2C + O_2 \to 2CO$ effectively reduces the carbon concentration at the

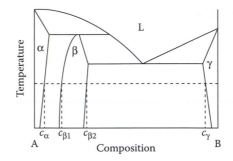

FIGURE 3.11   A–B phase diagram.

surface to zero. A layer of α forms at the surface and into the steel to a depth, x. The concentration profile near the surface is shown in Figure 3.14. The concentration gradient is $dc/dx = -c_\alpha/x$, where c_α is the carbon content of the α in equilibrium with the γ. Fick's first law gives the flux,

$$J = -D\frac{dc}{dx} = D\frac{C_\alpha}{x}. \tag{3.19}$$

As the interface advances a distance, dx (Figure 3.15), the amount of carbon that is removed in a time interval, dt, is approximately $(c_\gamma - c_\alpha)\,dx$; hence the flux is

$$J = (c_\gamma - c_\alpha)\frac{dx}{dt}. \tag{3.20}$$

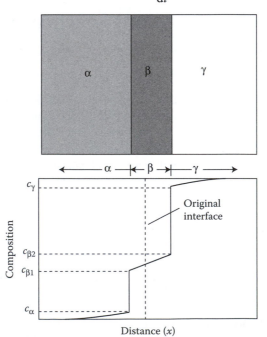

FIGURE 3.12   Microstructure of diffusion couple between A and B (above) and concentration profile (below).

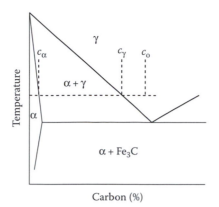

FIGURE 3.13   Iron–carbon diagram.

Equating the two expressions,

$$(c_\gamma - c_\alpha) \frac{dx}{dt} = D \frac{C_\alpha}{x},$$ (3.21)

hence $x\,dx = D[c_\alpha/(c_\gamma - c_\alpha)]\,dt$. Integrating gives $x^2 = [2Dc_\alpha/(c_\gamma - c_\alpha)]t$ or

$$x = \left[\frac{2Dtc_\alpha}{c_\gamma - c_\alpha} \right]^{1/2}.$$ (3.22)

Note that for fixed concentrations, $x/\sqrt{(Dt)}$ is constant or x is proportional to $\sqrt{(Dt)}$.

EXAMPLE PROBLEM 3.2

It is known that with a certain carburizing atmosphere, it takes 8 h at 900°C to obtain a carbon concentration of 0.75 at a depth of 0.020 in. Find the time to reach the same carbon concentration at a depth of 0.30 in at another temperature.

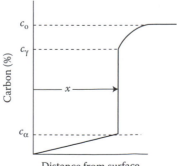

FIGURE 3.14   Carbon concentration profile.

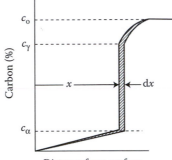

FIGURE 3.15   Change of concentration profile as the interface advances a distance, dx, in a time interval, dt.

SOLUTION

$x_2/\sqrt{(D_2 t_2)} = x_1/\sqrt{(D_1 t_1)}$. Letting $t_1 = 8$ h, $x_1 = 0.020$ in, $x_2 = 0.030$ in, and $D_2 = D_1$. Then $t_2 = t_1 (D_1/D_2)(x_2/x_1)^2$ h, that is, $(D_1/D_2)(0.03/0.02)^2 = 18(D_1/D_2)$.

EXAMPLE PROBLEM 3.3

A steel containing 0.20% C is to be carburized in an atmosphere that maintains a carbon concentration of 1.20% at the surface.

 A. After 10 h at 870°C, at what depth below the surface would you find a concentration of 0.40% C? (For diffusion of C in austenite, $D_o = 2.0 \times 10^{-5}$ m^2/s and $Q = 140 \times 10^3$ J/mol).
 B. How long would it take, still at 870°C, to double the depth (part A) at which the concentration is 0.40%?
 C. What carburizing time at 927°C gives the same results as 10 h at 870°C?

SOLUTION

 A. Using $c = c_s - (c_s - c_o)\,\mathrm{erf}[x/(2\sqrt{Dt})]$, with $c = 0.4$, $c_o = 0.2$, $c_s = 1.2$, so $(c - c_o)/(c_s - c_o) = 0.2$, or $\mathrm{erf}[x/(2\sqrt{Dt})] = 0.8$.
   Interpolating, $x/(2\sqrt{(Dt)}) = 0.90 + 0.05 - (0.8 - 0.7970)/(0.8209 - 0.7970) = 0.906$, so $x = 0.906 - 2\sqrt{(Dt)}$, where $D = 2.0 \times 10^{-5}\exp[-140,000/(8.31 - 1143)] = 7.939 \times 10^{-12}$ m^2/s and $t = 36,000$ s.
   $x = 0.906 \times 2(7.939 \times 10^{-12} \times 36,000)^{1/2} = 9.69 \times 10^{-4}$ m or about 1 mm.
 B. $x/\sqrt{(Dt)} = $ constant. For the same temperature, D is fixed, so $x_2/\sqrt{t_2} = x_1/\sqrt{t_1}$, or $t_2 = t_1(x_2/x_1)^2 = 10(2)^2 = 40$ h.
 C. For the "same carburizing results," the concentration profile must be the same. Therefore, $Dt = $ constant, or $D_2 t_2 = D_1 t_1$,

$$t_2 = t_1(D_1/D_2) = 10\ \text{h} \times \exp[(-Q/R)(1/T_1 - 1/T_2)] = 10\ \exp[(-140,000/8.31)/(1/1143 - 1/1200)] = 4.97\ \text{or 5 h.}$$

EXAMPLE PROBLEM 3.4

A steel containing 0.25% C was heated in air for 10 h at 700°C. Find the depth of the decarburized layer (i.e., the layer in which there is no Fe$_3$C). Given the solubility of C in α-Fe at 700°C is 0.016%. One may assume that the carbon concentration at the surface is negligible.

SOLUTION

At 700°C, the steel consists of two phases, α and Fe₃C. The concentration profile must appear as sketched in Figure 3.13. Near the surface, there is a decarburized layer containing only α and the concentration in the a must vary from 0% C at the outside surface to $C_\alpha = 0.16\%$ C, where it is in contact with Fe₃C (Figure 3.16).

An approximate solution can be obtained by using Fick's first law to make a mass balance as the interface moves a distance dx. The amount of carbon transported to the surface in a period, dt, is $(\bar{c} - c_\alpha)dx$ and this must equal the flux times dt, $-Jdt = D(dc/dx)dt$. Substituting $dc/dx = (c_\alpha - 0)/x$, $(\bar{C} - c_\alpha)dx = D(c_\alpha/x)dt$ and integrating, $x^2/2 = Dtc_\alpha/(\bar{c} - c_\alpha)$, $x = [2Dtc_\alpha/(\bar{c} - c_\alpha)]^{1/2}$.

Now substituting, $D = 2 \times 10^{-6}\ \exp[-84{,}400/(8.31 \times 973)] = 5.86 \times 10^{-11}\ m^2/s$, $\bar{c} = 0.25$, $c_\alpha = 0.016$, and $t = 36{,}00$ s, $x = 0.00057$ m or 0.6 mm.

3.9 MISCELLANY

In the 1930s, in the brass industry in Connecticut, porosity was found in the microstructure of brass that had been worked down to a diameter of about an inch and then

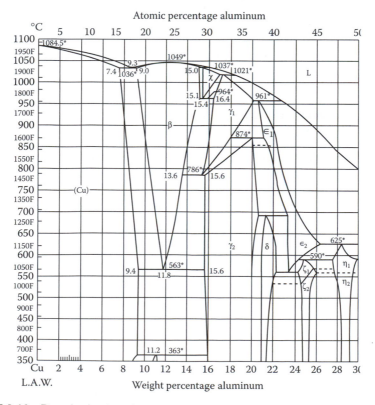

FIGURE 3.16 Decarburization of a steel heated in the α + Fe₃C phase region. (From *ASM Handbook*, Vol. 8, 8th ed., 1973. With permission.)

given a very high temperature anneal. It was felt that this porosity must be the result of a precipitate that had been etched out or perhaps had dissolved in the water during polishing. Because no one could isolate the precipitate, it was dubbed the *Ghost precipitate*. It was not until many years later that it was realized that this porosity was a result of the Kirkendall effect, with zinc diffusing faster than copper. It took reduction of the diameter to about 1 in before the distance between compositional variations caused by dendritic segregation to be reduced enough that diffusion during annealing would result in appreciable porosity.

PROBLEMS

1. A block of an alloy of Cu-6% Al was welded to a block of Cu-14% Al and heated to 700°C. Sketch the concentration profile after some diffusion occurred. The phase diagram is given in Figure 3.16.
2. Consider a piece of steel containing 0.20% C at 750°C in an atmosphere that reduces the carbon concentration at the surface to zero. The solubility of carbon in α-iron at 750°C is 0.019%.
   a. Sketch the concentration profile near the surface. (Plot % C versus distance from surface.)
   b. Find an appropriate diffusivity for C in iron containing 0.20% C at 750°C.
   c. Using Fick's first law, $J = -D dc/dx$, express the flux, J, in terms of the depth of the decarburized layer.
3. Note that in an incremental time period, dt, the decarburized depth increases by dx, such that $J dt = (c_o - c_\alpha)dx$, where $J = Dc_\alpha/x$.
   a. Find x as a function of time by integration.
   b. What would be the depth of the decarburized layer after 4 h?
4. a. If the grain size of silver were doubled, by what factor would the net diffusion by grain boundary diffusivity change?
   b. At what temperature would the net transport by grain boundary diffusion be the same as with the original grain size at 1000°C?
5. When iron is exposed to C at 850°C, it is found that after 2 h, the concentration of carbon is 0.65% at 1 mm below the surface. If iron were exposed to the same atmosphere at 900°C for 1.5 h, at what depth below the surface would the concentration of carbon be 0.65%?
6. a. Calculate the ratio of the diameter of the largest interstitial hole in a bcc crystal to the diameter of an interstitial atom that could just squeeze from one interstitial hole to another.
   b. Repeat for an fcc crystal.
   c. In view of your answers to A and B, how would you expect the diffusivity of an interstitial atom in an fcc crystal to compare with that in a bcc crystal at the same temperature?

REFERENCES

ASM Handbook, Vol. 8, 8th ed., 1973.
Diffusion in BCC Metals, ASM, 1965.
J. H. Hollomon and D. Turnbull, *Solidification of Metals and Alloys*, AIME, 1951.
P. G. Shewmon, *Diffusion in Solids*, McGraw-Hill, 1963.

4 Surface Tension and Surface Energy

There is characteristic surface energy associated with all interfaces and surfaces. The energy to create the surface is proportional to the area of surface. This energy depends on what is on each side of the interface. A surface energy may also be thought of as a surface tension. The dimensions of specific surface energy (energy/area as J/m^2) are equivalent to the units of surface tension (force/distance as N/m). For our purposes they will be considered equal.

4.1 DIRECT MEASUREMENTS OF SURFACE ENERGY

Most measurements have been indirect, comparing the energy of one surface to that of another. There have been relatively few direct measurements of surface energy. A classic experiment in some physics courses is the measurement of the surface tension of soapy water. This is illustrated in Figure 4.1. Measurement is made of the force, F, that must be applied to a soap film of fixed length, L, to keep it from contracting. A force balance gives $F = 2L\gamma$, so the surface tension is given by

$$\gamma = \frac{F}{2L}. \tag{4.1}$$

The reason that F is divided by $2L$ is that there are two liquid–vapor interfaces associated with a soap film.

Another way to measure the surface tension of a soap film is to measure the pressure, P, necessary to maintain a soap bubble of diameter, D (Figure 4.2).

From a simple force balance on an imaginary cut, $P\pi D^2/4 = 2\pi D\gamma$, or

$$\gamma = \frac{PD}{8}. \tag{4.2}$$

Buttner et al. (1951) measured the surface energy for gold (or more properly the energy of the interface between solid gold and its vapor). A weighted gold wire was suspended from the cover of an evacuated gold box (Figure 4.3); the system was heated to a high temperature and the rate of creep was measured.

The experiment was repeated with different weights. With low weights, the wire would contract because of its surface tension but with greater weights the wire would elongate. They determined the weight, W, at which the wire neither elongated nor

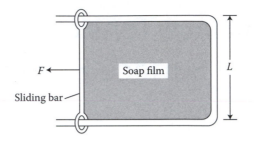

FIGURE 4.1   Measurement of surface tension of a soap film.

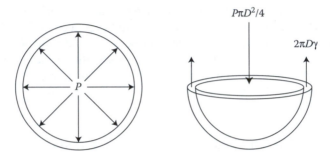

FIGURE 4.2   Force balance on a soap bubble.

contracted. One might be tempted to calculate the surface tension, γ, by equating the weight W to the longitudinal force, $\pi D\gamma$, where D is the wire diameter and πd is its circumference. However, this would neglect the fact that the surface tension is also acting to contract the circumference. Consider instead the change of energy of the system associated with an incremental change of length, dL. The surface area of the wire is $A = \pi DL$. As the wire elongates, its diameter also changes, so $dA = \pi D dL + \pi L dD$. Since the volume, $V = \pi D^2 L/4$ is constant, $dV = (\pi/4)(D^2 dL + 2LD dD) = 0$, or $dD = -(D/2L)dL$. Substituting this in the expression for the area, $dA = \pi[D dL - (D/2)dL]$,

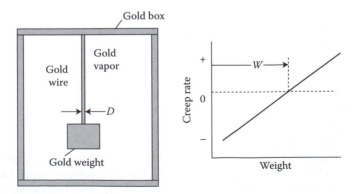

FIGURE 4.3   Measurement of the surface energy of gold.

so $dA = (\pi D/2)dL$. The increase of total surface energy is γdA and this must equal the external work, WdL. Therefore $\gamma(\pi D/2)dL = WdL$, or

$$\gamma = \frac{2W}{\pi D}. \tag{4.3}$$

Values of surface energy measured by this technique are 1.140 J/m² for Ag at 903°C, 1.400 J/m² for Au at 1204°C, and 1.650 J/m² for Cu at 1000°C. The solid–vapor surface energy should be relatively independent of the temperature, but it should be realized that the surface energy of a solid–vapor interface depends on its crystallographic orientation; hence these measurements must reflect an average value for many orientations.

4.2 MEASUREMENTS OF RELATIVE SURFACE ENERGIES

Most surface energies have been determined from other surface energies by measuring the angles at which surfaces meet. If there is equilibrium, the angles at which three surfaces meet depend on the relative energies of the three interfaces (Figure 4.4). This is because there must be equilibrium of forces. Each surface exerts a force per length on the junction that is equal to its surface tension. The force vectors must form a triangle, so the law of sines gives

$$\frac{\gamma_{23}}{\sin \theta_1} = \frac{\gamma_{31}}{\sin \theta_2} = \frac{\gamma_{12}}{\sin \theta_3}. \tag{4.4}$$

Often two of the angles and two of the surface energies are equal. For example, consider the intersection of a grain boundary with a free surface. If the temperature is high enough and the time is long enough, the surface will thermally etch (by vaporization or surface diffusion) until equilibrium angles are formed. From a balance of forces parallel to the grain boundary (Figure 4.5),

$$\gamma_{gb} = 2\gamma_{sv} \cos\left(\frac{\theta}{2}\right). \tag{4.5}$$

Similar relations can be used to find the twin boundary energy from the grain boundary energy and the boundary energy between two phases from the energy of a grain boundary in one of them.

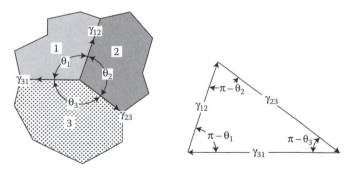

FIGURE 4.4 Relative surface energies.

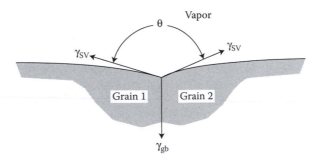

FIGURE 4.5   Intersection of a grain boundary with a free surface.

4.3   WETTING ANGLES

A liquid droplet resting on a solid surface (Figure 4.6) will form a characteristic angle, θ, with the surface, such that there is a force balance parallel to the surface;

$$\gamma_{SV} = \gamma_{LS} + \gamma_{LV} \cos \theta. \tag{4.6}$$

The liquid is said to be *nonwetting* if $\theta = 180°(\gamma_{SV} + \gamma_{LV} \geq \gamma_{SL})$. If $\gamma_{LS} + \gamma_{LV} < \gamma_{SV}$, there is no solution to Equation 4.6. This is a condition of complete wetting. The liquid will spread over the entire surface, replacing the solid–vapor interface with liquid–solid and liquid–vapor interfaces. Such wetting is desired in soldering and brazing. A flux may be required to clean the surface, in which case it replaces the vapor in the analysis.

4.4   RELATION OF SURFACE ENERGY TO BONDING

An approximate calculation of surface energy can be made by considering a hypothetical experiment in which new surfaces are formed by breaking of bonds. The energy of the two surfaces created should equal the energy required to break these bonds. One way to estimate this is to assume that the value of the latent heat of sublimation (solid $\rightarrow$ gas transformation), ΔH_S, expressed on a per-atom basis is the energy to break all of an atom's bonds. Then the surface energy can be found by

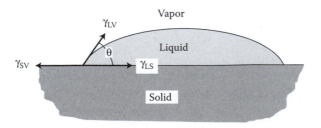

FIGURE 4.6   Force balance on a liquid drop on a solid surface.

calculating how many bonds must be broken/area to create the surface. Such calculations predict an orientation dependence of surface energy with more densely packed surfaces having the lower energies. This tends to be in accord with experimental observations of the prevalence of low-index surfaces when surface tension controls the surfaces present.

Another way of estimating surface energy is to envision the surface being formed by mechanical forces across them and calculating the work required to separate the two halves of a crystal.

$$\gamma_s = \left(\frac{1}{2}\right)\left(\frac{\text{work}}{\text{area}}\right) = \left(\frac{1}{2}\right)\int \sigma \, ds, \tag{4.7}$$

where σ is the stress that is required to create a separation, s, of the two surfaces. Here the lower and upper limits of integration should be $s = 0$ and $s = \infty$.

To proceed further along these lines, something must be assumed about the shape of the σ versus s curve. For example, let the curve be approximated by

$$\sigma = \left[\frac{E}{(\pi)}\right] \sin\left(\frac{\pi s}{s_o}\right) \tag{4.8}$$

from $(s/s_o) = 0$ to 1, so $\gamma_s = (1/2) \int \sigma \, ds = (1/2)[E/(\pi)] \int \sin(\pi s/s_o) \, ds$ (Figure 4.7).

Now integrating between the limits of $s = 0$ and $s = 1$, $\gamma_s = (1/2)[E(s_o/\pi)(\cos 0 - \cos \pi) = Es_o/(\pi^2)$. Evaluating and taking $s_o = d$ (the atomic diameter),

$$\gamma_s = \frac{Ed}{\pi^2}. \tag{4.9}$$

Evaluating this approximation for copper ($E = 128$ GPa and $d = 0.255$ nm) = 3.3 J/m^2. (This is a bit greater than the experimentally measured value.)

Because both surface energy and melting temperature are related to the bonding strength, the surface energy is closely related to the melting temperature as shown in Figure 4.8.

4.5 ANISOTROPY OF SURFACE ENERGY

The atoms in a surface have fewer near neighbors than those in the lattice. The energy of a free surface is largely attributable to the missing first near neighbors. Atoms in a {111} surface of an fcc crystal are missing three of the 12 neighbors they

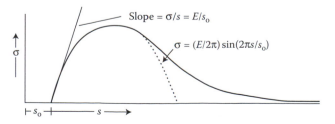

Slope = $\sigma/s = E/s_o$

$\sigma = (E/2\pi) \sin(2\pi s/s_o)$

FIGURE 4.7 Schematic plot of bonding strength as a function of atom separation.

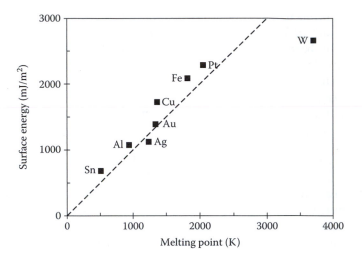

FIGURE 4.8 Correlation of surface energy with melting point.

should have. The surface energy per surface atom should be about $(1/4)U_b$, where U_b is the energy to break a bond. The atoms on a {100} surface of an fcc crystal are missing one third of their near neighbors, so the surface energy per surface atom should be about $(1/3)U_b$. Experimental measurements show that the surface energy per surface atom is about $0.15U_b$. This value reflects the fact that the entropy change of surface atoms is not negligible.

EXAMPLE PROBLEM 4.1

Estimate the ratios of the free surface energies of {111}, {100}, and {110} surfaces in fcc crystals.

SOLUTION

Each atom on a {111} surface is missing three near-neighbor bonds. For each atom, the surface area is $6r^2/\sqrt{3}$. The number of missing bonds per area = $3/[6r^2/\sqrt{3}] = 0.866/r^2$.

Each atom on a {100} surface is missing four near-neighbor bonds. For each atom, the surface area is $4r^2$. The number of missing bonds per area = $4/4r^2 = 1/r^2$.

Each atom on a {110} surface is missing five near-neighbor bonds. For each atom, the surface area is $8/\sqrt{2}r^2$. The number of missing bonds per area = $5/[8/\sqrt{2}r^2] = 0.8839/r^2$.

The ratio of $\gamma_{111} : \gamma_{100} : \gamma_{110} = 0.866 : 1.0 : 0.8839$.

EXAMPLE PROBLEM 4.2

Determine how the free surface energy of a two-dimensional square lattice depends on the orientation of the surface.

SOLUTION

Consider the square lattice in Figure 4.9 and let the area of the surface equal 1.

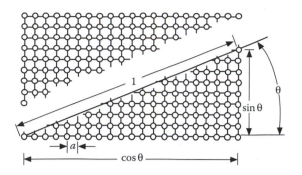

FIGURE 4.9 Two surfaces created by breaking bonds. The surface energy is proportional to the number of broken bonds per area.

For $0 = \theta = 90°$, the number of missing vertical bonds is $\sin \theta/a$ and the number of missing horizontal bonds is $\cos \theta/a$, so the total number of missing bonds is $(\sin \theta + \cos \theta)/a$. Breaking these bonds creates two surfaces of total length $= 2$, so the surface energy per area is

$$\gamma_\theta = U_b \frac{\cos \theta + |\sin \theta|}{2a},$$ (4.10)

where U_b is the energy per bond. This plots as a circle of radius $\sqrt{2}U_b/a$, centered at $x = y = U_b/f\sqrt{2}$. Symmetry indicates that in other quadrants γ_θ also plots as a circle of radius $\sqrt{2}U_b/a$. Figure 4.10 is a polar plot of γ.

The shape of a free crystal that minimizes the surface energy can be found by realizing that it will be bounded by surfaces of lowest energy. On a polar plot, these planes are normal to the shortest radii.

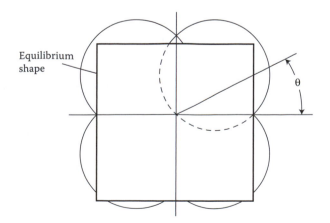

FIGURE 4.10 Polar plot of the surface energy of a two-dimensional square crystal as a function of orientation. The total surface energy of a crystal is a minimum when it is bounded by planes constructed perpendicular to the shortest normals.

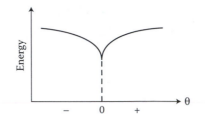

FIGURE 4.11   Dependence of surface energy on the angle of deviation from a low-index plane.

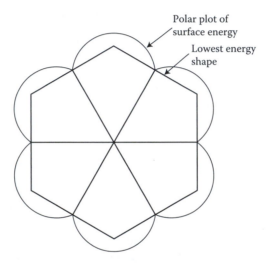

FIGURE 4.12   A γ-plot for a hypothetical crystal. The lowest energy shape for the crystal corresponds to planes normal to radii at the cusps.

Gibbs raised the question about the shape of solid that would minimize the total surface energy, $U_S = \int \gamma \, dA$. Wulff showed that the minimum energy configuration can be found by constructing planes perpendicular to radii at the σ_θ surface of the polar plot (Wulff planes). The equilibrium shape (particle shape with the lowest U_s) is bounded by portions of Wulff planes that can be reached from the origin without crossing any other Wulff planes.

Figure 4.11 is a schematic plot showing how the surface energy changes with θ for orientations near a low-index plane. Plots of surface energy versus orientation should show cusps at low-index orientations.

A two-dimensional Wulff plot for a three-dimensional crystal having both {100} and {111} faces is illustrated in Figure 4.12. The corresponding solid is sketched.

4.6   SEGREGATION TO SURFACES

There is a tendency for solutes to segregate to both free surfaces and grain boundaries and thereby lower the surface energy. The structure of grain boundaries in metals is such that there are places where atoms are too close and other places where they are too far apart. Substitutional solute atoms that are smaller than the solute will preferentially move to places where the atom spacing is crowded whereas large solutes will preferentially move to places where the atoms are too far apart. Interstitial solutes will preferentially segregate to open regions. In dilute solutions, the solute concentrations at or near grain boundaries are often many times the average concentration.

4.6.1 NOTE

The correlation of the free surface energy, γ with melting point does not apply with glass or thermoplastics. The reason is that in these materials, molecules at the surface are aligned so that any missing are the weakest ones and not typical of the bonding as a whole.

4.7 MISCELLANY

The reason that one can blow soap bubbles is not because soap raises or lowers the surface tension of water. It is because the surface tension depends on the concentration of soap at the surface. In a soap film, this tension must vary from place to place to balance the stresses due to gravity.

PROBLEMS

1. Estimate for a two-dimensional crystal, the relative energies of a {31} and {32} faces. Find γ_{31}/γ_{32}.
2. a. Consider a {111} face of an fcc crystal. For the surface atoms, what fraction of the near-neighbor bonds are missing?
   b. Consider a {100} face of an fcc crystal. For the surface atoms, what fraction of the near-neighbor bonds are missing?
   c. A reasonable first-order approximation is that the surface energy is proportional to the number of missing bonds per area. Using this assumption, estimate $\gamma_{\{111\}}/\gamma_{\{100\}}$.
3. For copper, what is the equilibrium angle between
   a. Free surface and a grain boundary?
   b. Grain boundary and a twin boundary?
4. evaluate $\gamma_\theta/(U_b/a^2)$ for a two-dimensional crystal with a square lattice over the range $-5° = \theta = +5°$ and plot $\gamma_\theta/(U_b/a^2)$ versus θ.
5. Calculate the pressure inside a droplet of copper having a diameter of 100 μm.
6. Estimate the free surface energy of lead and nickel. Lead melts at 327°C and nickel melts at 1455°C.

APPENDIX: DATA ON VARIOUS SYSTEMS

TABLE 4.1
Comparison between Interphase and Grain Boundary Energies

| System | Interface between | | Grain Boundary | $\gamma_{\alpha\beta}/\gamma_{gb}$ |
	Phase A	Phase B		
Cu/Zn	α(fcc)	β(bcc)	α/α	0.78
Cu/Zn	α(fcc)	β(bcc)	β/β	1.00
Cu/Al	α(fcc)	β(bcc)	α/α	0.71
Cu/Al	β(bcc)	γ(cmplx. cub.)	γ/γ	0.78
Cu/Si	α(fcc)	β(bcc)	α/α	0.53
Cu/Si	α(fcc)	β(bcc)	β/β	1.18
Fe/C	α(bcc)	Fe₃C	α/α	0.93
Fe/C	α(bcc)	γ(fcc)	α/α	0.71
Fe/C	α(bcc)	γ(fcc)	γ/γ	0.74

Source: From C. S. Smith, *Imperfections in Nearly Perfect Crystals*, Shockley, Ed., Wiley, 1952. p. 384. With permission.

TABLE 4.2
Grain Boundary and Free Surface Energies and Twin Boundaries[a]

Metal	γ_{gb} (J/m²)	γ_{gb}/γ_{sv}	γ_{tb} (mJ/m²)
Copper	0.62	0.36	21
γ-iron	0.76	0.40	
α-iron (4% Si)	0.76	0.55	
Aluminum	0.32	0.30	
Gold	0.38	0.27	
Tin	0.16	0.24	
Silver	0.40	0.35	8
304 stainless steel	0.84		19

Source: From R. Swalin, *Thermodynamics of Solids*, Wiley, 1962. With permission.

[a] γ_{gb} and γ_{gb}/γ_{sv}.

TABLE 4.3
Solid–Liquid Interfacial Energies and Comparison with Heat of Fusion (per Atom)

Metal	γ_{SL} (J/m²)	Ratio[a]
Hg	0.0244	0.53
Sn	0.0545	0.42
Pb	0.033	0.39
Ag	0.126	0.46
Au	0.132	0.44
Cu	0.177	0.44
Mn	0.206	0.48
Ni	0.255	0.44
Fe	0.204	0.45
Pt	0.240	0.46

Source: From J. H. Hollomon and D. Turnbull, Chapter 7, *Progress in Metal Physics 4*, Chalmers Ed., Pergamon, 1953. With permission.

[a] (surface energy/surface atom)/(latent heat of fusion/atom).

TABLE 4.4
Surface Energies of Pure Metals in mJ/m²

Metal	γ_{sf}	γ_{twin}	γ_{gb}	γ_{sv}
Ag	16	8	790	1140
Al	166	75	325	980
Au	32	15	364	1485
Cu	45	24	625	1725
Fe			780	1950
Ni	125	43	866	2280
Pd	180			
Pt	322	161	1000	3000
Rh	750			
Th	115			
Ir	300			
W				2800
Cd	175			
Mg	125			
Zn	140		340	

Source: From J. P. Hirth and J. Lothe, *Theory of Dislocations*, 3rd ed., Wiley, NY, 1982. With permission.

The terms γ_{sf}, γ_{twin}, γ_{gb}, and γ_{sv} are the surface energies of stacking faults, twin boundaries, grain boundaries, and the solid–vapor surfaces.

REFERENCES

J. H. Brophy, R. M. Rose, and J. Wulff, *The Structure and Properties of Materials, Vol. II, Thermodynamics of Structure*, Wiley, 1964.

F. H. Buttner, H. Udin, and J. Wulff, *J. Metals*, 3, 1209, 1951.

J. P. Hirth and J. Lothe, *Theory of Dislocations,* 3rd ed., Wiley, NY, 1982.

J. H. Hollomon and D. Turnbull, Chapter 7, *Progress in Metal Physics* 4, Chalmers, Ed., Pergamon, 1953.

Metal Interfaces, ASM, 1952.

D. A. Porter and K. E. Easterling, *Phase Transformations in Metals and Alloys*, 2nd ed., Chapman and Hall, 1992.

C. S. Smith, *Imperfections in Nearly Perfect Crystals*, Shockley, Ed., Wiley, 1952. p. 384.

R. Swalin, *Thermodynamics of Solids*, Wiley, 1962.

5 Solid Solutions

5.1 TYPES OF SOLID SOLUTIONS

There are two types of solid solutions in metals. *Interstitial* solid solutions are ones in which small solute atoms fit between the solvent atoms. The only atoms small enough to dissolve interstitially are hydrogen, boron, carbon, and nitrogen. Even these atoms, except hydrogen, are larger than the holes in metal lattices, so they cause a volume expansion. Only the transition elements (e.g., Fe, Ni, Ti, and Zr) have appreciable solubilities for C, N, and B.

In *substitutional* solid solutions, the solute atoms substitute for solvent atoms in lattice sites. The difference between the two types of solid solutions is illustrated schematically in Figure 5.1. Several factors govern the extent of solubility.

The factors governing solubility in substitutional solid solutions, often called the Hume-Rothery rules, are

1. Atoms of the two elements cannot differ greatly in size. If the atoms differ in size by 15% or less, the solubility can be appreciable (10% solute) if the other factors are favorable.
2. If there is strong chemical affinity between the elements, solubility will be very limited.
3. Solid solubility is affected by the relative valence of the elements. The ratio of number of valence electrons per atom, e/a, affects the solubility. Solutes that raise the e/a, generally have a higher solubility than those that lower the e/a. For example, the solid solubility of lithium in magnesium is 24.5 atom% whereas the solubility of magnesium in lithium is 70 atom%. Likewise, the solubility of aluminum in copper is much greater than the solubility of copper in aluminum.
4. For complete miscibility between two elements, they must have the same crystal structure. The same crystal structure is not necessary, however, for extensive solubility.

5.2 ELECTRON-TO-ATOM RATIO

The maximum solubility of many elements in copper and silver base alloys corresponds to an e/a ratio of 1.35–1.4. This corresponds closely to the concentration (1.36) at which the Fermi sphere approaches the first Brillouin zone. In calculating the e/a

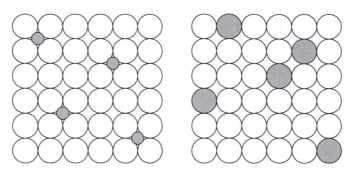

FIGURE 5.1   Schematic illustration of an interstitial solid solution (left) and a substitutional solid solution (right).

ratio, valences are assigned to elements according to Table 5.1. Table 5.2 shows the solubility limits of a number of metals in copper and silver. Figure 5.2 shows the solubility of several elements in silver.

EXAMPLE PROBLEM 5.1

Calculate the electron-to-atom ratio in a copper-base alloy containing 15% zinc and 3% silicon.

SOLUTION

Assume 100 g. 82 g of copper contribute (1 e/a)(82 g/63.5 g/mol) = 1.291 mol of e. Zinc contributes (2 e/a)(15 g/65.4 g/mol) = 0.459 mol of e. Silicon contributes (4e/a)(3g)/(28g/mol) = 0.428 mol of e. The total is 1.291 + 0.459 + 0.428 = 2.178 mol of e.

The total number of atoms is 82/63.5 + 15/65.4 + 3/28 = 1.628 mol of atoms. The e/a = 2.305/1.647 = 1.34. This would predict that the alloy would be a single fcc solid solution.

5.3   ENTROPY OF MIXING

It can be shown by statistical mechanics that the change of entropy, ΔS_m, when a random mixture is formed is given by

$$\Delta S_m = k \ln(p), \tag{5.1}$$

TABLE 5.1
Metal Valences

Metal	Cu	Ag	Au	Zn	Cd	Hg	Al	In	Ga	Sn	Si	Ge	Ni
Valence	1	1	1	2	2	2	3	3	3	4	4	4	0

TABLE 5.2

Electron-to-Atom Ratio at the Solubility Limits in Copper and Silver Alloys

Alloy	Electron-to-Atom Ratio at the Maximum Solubility in the α-Phase
Cu–Zn	1.38
Cu–Al	1.41
Cu–Ga	1.41
Cu–Si	1.42
Cu–Ge	1.36
Cu–Sn	1.27
Ag–Cd	1.42
Ag–Zn	1.38
Ag–In	1.35
Ag–Al	1.41
Ag–Ga	1.38
Ag–Sn	1.37

Source: From C. S. Barrett, *Structure of Metals*, McGraw-Hill, 1943, p. 266. With permission.

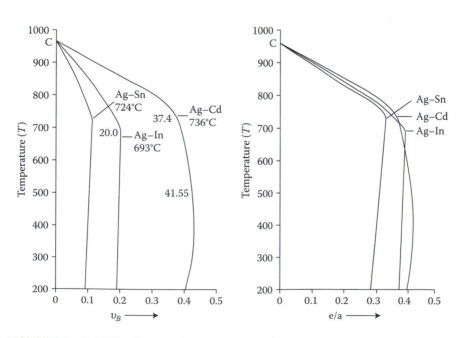

FIGURE 5.2 Solubility limits of Cd, In, and Sn in silver expressed as atom fraction (left) and as concentration of valence electrons (right). (From P. Haasen, *Physical Metallurgy*, 2nd ed., Cambridge, 1986, p. 127. With permission.)

where, p is the number of distinguishable arrangements of atoms and k is Boltzmann's constant. Let us consider a mole of atoms, where n_a is the number of atoms of element a, n_b is the number of atoms of element b, and n_o is the total number of atoms $(n_o = n_a + n_b = 6.02 \times 10^{23})$. Imagine a crystal with n_o possible locations for atoms (atom sites) and imagine filling these sites, one at a time, with n_a atoms of a. The first atom of a can be put in any of n_o sites, the second in any of $(n_o - 1)$ sites, the third in any of $(n - 2)$ sites, and so on until the last atom of a can be put in any of $(n_o - n_a + 1)$ sites. The total ways of filling the sites is then equal to $n_o \cdot (n_o - 1) \cdot (n_o - 2) \cdot \cdots (n_o - n_a + 1)$. However, this is not the number of distinguishable arrangements because it does not matter whether we put the third a atom in the third site or the first site or the 112th site. The number of distinguishable arrangements is

$$p = \left[\frac{n_o(n_o - 1)(n_o - 2) \cdots (n_o - n_a + 1)}{n_a!} \right] \tag{5.2}$$

If both the numerator and denominator are multiplied by $n_b!$, Equation 5.1 can be expressed as

$$p = \frac{n_o!}{n_a! n_b!}, \tag{5.3}$$

so

$$\Delta S_m = k \ln(p) = k[\ln(n_o!) - \ln(n_a!) - \ln(n_b!)]. \tag{5.4}$$

Using Stirling's approximation,

$$\ln(x!) = x \ln x - x,$$
$$\Delta S_m = k[n_o \ln(n_o) - n_a \ln(n_a) - n_b \ln(n_b) - n_o + n_a + n_b]. \tag{5.5}$$

Simplifying, $\Delta S_m = k[(n_a + n_b) \ln(n_o) - n_a \ln(n_a) - n_b \ln(n_b)]$, or $\Delta S_m = -k[n_o \ln(n_a/n_o) + n_o \ln(n_b/n_o)]$. Now recognizing that the mole fraction of element a is $N_a = n_a/n_o$, the mole fraction of element b is $(1 - N_a) = N_b = n_b/n_o$, and that the gas constant $R = kn_o$,

$$\Delta S_m = -R[N_a \ln N_a + N_b \ln N_b]. \tag{5.6}$$

Figure 5.3 shows how $\Delta S_m/R$ varies with composition.

5.4 ENTHALPY OF MIXING

The enthalpy may increase, decrease, or remain unchanged as a solution forms. The enthalpy of the two pure components before mixing is

$$H = X_a H_a^o + X_b H_b^o, \tag{5.7}$$

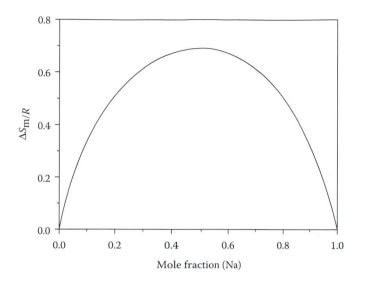

FIGURE 5.3 Variation of the entropy of mixing with composition according to Equation 5.6.

where H_a^o and H_b^o are the enthalpies of the pure components and $X_a = N_a/N_o$ and $X_b = N_b/N_o$ are the mole fractions of a and b. If there is no change of enthalpy on solution ($\Delta H_m = 0$) the solution is called an *ideal solution*. On the other hand, ΔH_m may be positive (positive deviation from ideality). In this case, cooling occurs when elements a and b are mixed (heat is absorbed to form the solution). This indicates that there is a preference for having a–a and b–b near neighbors instead of a–b near neighbors. On the other hand, if there is strong attraction between unlike atoms, there will be a negative deviation from ideality ($\Delta H_m < 0$) and heat will be released. A good approximation for ΔH_m is

$$\Delta H_m = X_a X_b \Omega, \tag{5.8}$$

where $\Omega = (zn_o/2)[E_{ab} - (E_{aa} + E_{bb})/2]$. In this expression, z is the coordination number; hence $(zn_o/2)$ is the total number of bonds, and E_{ab}, E_{aa}, and E_{bb} are the energies of the ab, aa, and ab bonds.

5.5 FREE ENERGY CHANGE ON MIXING

The change of Gibbs' free energy on mixing is then

$$\Delta G_m = \Delta H_m - T\Delta S_m = \Delta H_m + RT[X_a \ln X_a + X_b \ln X_b]. \tag{5.9}$$

Note that the second term is always negative since both X_a and X_b are both <1 and the natural log of a number <1 is negative. ΔH may be either negative or positive, so ΔG can also be either negative or positive. Substituting $\Delta H_m = X_a X_b \Omega$

$$\Delta G_m = -X_a X_b \Omega + RT[X_a \ln X_a + X_b \ln X_b]. \tag{5.10}$$

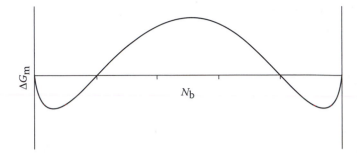

FIGURE 5.4 Variation of ΔG_m with composition for a positive value of W. The solubilities correspond to tangent points to both minima.

If Ω is negative, a and b are completely miscible. However, the solubility is limited if Ω is positive. Figure 5.4 is a plot of Equation 5.10 for a positive value of Ω. The solubilities correspond to the tangent points for a line tangent to both minima.

5.5.1 SOLUBILITY IN DILUTE SOLUTIONS

The solubility limit of b corresponds to $d(\Delta G_m)/dX_b = 0$. Expressing Equation 5.10 as $\Delta G_m = X_b(1 - X_b)\Omega + RT[(1 - X_b)\ln(1 - X_b) + X_b \ln X_b]$, and differentiating,

$$\frac{d(\Delta G_m)}{dX_b} = \Omega(1 - 2X_b) + RT[-(1 - X_b)\ln(1 - X_b) - \ln(1 - X_b) + X_b \ln X_b + \ln X_b] = 0$$

$$RT[(1 + X_b)\ln X_b - \ln(1 - X_b)(2 - X_b) = \Omega(1 - 2X_b) \tag{5.11}$$

EXAMPLE PROBLEM 5.2

Find the solubility of b in a at 860°C, if $\Omega = -30,000$ J/mol.

SOLUTION

Substituting $T = 860°C = 1133$ K and $R = 8.134$ into Equation 5.11,

$$\frac{[(1 + X_b)\ln X_b - \ln(1 - X_b)(2 - X_b)}{1 - 2X_b} = \Omega$$

Solving by trial and error, $X_b = 0.079$.

For a dilute solutions $X_b \to 0$, so $\ln(1 - X_b) \to 0$, $1 + X_b \to 1$, and $(1 - 2X_b) \to 1$ so $RT \ln X_b = -\Omega$. The solubility is given by

$$X_b = \exp\left[\frac{-\Omega}{RT}\right] \tag{5.12}$$

Equation 5.11 is a good approximation to the solvus is many systems. It can also be used to predict the equilibrium solubilities of vacancies and interstitial defects.

EXAMPLE PROBLEM 5.3

The solubility of carbon in a-iron is 0.0218% at 723°C. Estimate the solubility at 500°C.

SOLUTION

This is a dilute solution so Equation 5.12 is applicable. $\Omega = -RT \ln(X_b)$. Substituting $X_b = 0.000218(55.85/12) = 0.001015$ at 1000 K, $\Omega = 5.75 \times 10^{-4}$. At 500°C (773 K), $X_b = \exp[-5.75 \times 10^4/(8.314 \times 773)] = 1.30 \times 10^{-4}$ or 13 ppm. This corresponds to $1.30 \times 10^{-4}(12/55.85) \times 100 = 0.0028\%$.

5.6 EQUILIBRIUM VACANCY AND INTERSTITIAL CONCENTRATIONS

The energy to create a mole of vacancies in copper has been estimated to be 83,700 J/mol. Substituting in Equation 5.11, the equilibrium concentration of vacancies in copper is given by $N_v = \exp[-83,700/RT]$. Just below the melting point of copper (1086°C = 1356 K), this predicts $N_v = 6 \times 10^{-4}$. About one out of every 1600 lattice sites is vacant. This concentration is about the same for all metals just below their melting point. At room temperature the equilibrium concentration is negligible. For copper, $N_v = \exp[-83,700/293R] = 5 \times 10^{-16}$.

Where do the vacancies come from when a metal is heated? Where do they go when the metal is cooled? Edge dislocations and grain boundaries are the primary sources and sinks for vacancies. Free surfaces can also act as sources and sinks, but they are much less important. If cooling is rapid, not all of the vacancies can diffuse to these sinks, so there will be an excess of vacancies. Quenched metals therefore contain more vacancies than slowly cooled ones.

Vacancies in excess of the equilibrium number are produced by the intersection of dislocations during plastic deformation. They can also be produced by irradiation.

There is also an equilibrium number of interstitial defects. Again Equation 5.11 can be used to estimate the number but the energy to produce an interstitial is about four times as high as that to produce a vacancy. If $Q = 300,000$ J/mol and $T = 1356$ K is substituted into Equation 5.11, it is found that $N_v = 10^{-12}$; so even just below the melting point, the number of equilibrium interstitials is negligible. However, dislocation intersections and irradiation can create more than the equilibrium number of interstitials.

5.7 ORDERED VERSUS RANDOM SOLID SOLUTIONS

The positions of atoms in a substitutional solid solution may not be random. If the average aa and bb bond strength is greater than the aa bond strength, there will be clustering. a atoms will have more a near neighbors than would be expected in a random solution. On the other hand, if the average aa and bb bond strength is weaker than the ab bond strength, there will be *short-range ordering*. There will be more ab bonds than those in a random solution. The degree of ordering or clustering can be described by a parameter, s, defined as

$$s = \frac{N_{ab} - N_{ab(random)}}{N_{ab(max)} - N_{ab(random)}}, \tag{5.13}$$

where N_{ab} is the number of ab bonds, $N_{ab(max)}$ is the maximum possible number of ab bonds, and $N_{ab(random)}$ is the number that would be expected in a random solution. For a random solution, $s = 0$. For clustering, s is negative and a positive value of s indicates ordering, with $s = 1$ corresponding to perfect order.

EXAMPLE PROBLEM 5.4

Calculate the value of s for the two-dimensional solutions shown in Figure 5.5a and b.

SOLUTION

There are 50 a atoms and 50 b atoms and 180 bonds. $N_a = N_b = 50$, so $N_{ab(max)} = 180$, $N_{ab(random)} = 90$. In Figure 5.5a, $N_{ab} = 94$, so

$$s = \frac{N_{ab} - N_{ab(random)}}{N_{ab(max)} - N_{ab(random)}} = 0.044.$$

In Figure 5.5b, $N_{ab} = 131$, so $s = (N_{ab} - N_{ab(random)})/(N_{ab(max)} - N_{ab(random)}) = 0.456$.

For alloys having compositions that correspond to simple ratios of a and b, *long-range order* is possible. Figure 5.6 shows several ordered structures. Other compositions with these ordered structures are listed in Table 5.3. The ordered fields in the Cu–Au phase diagram are shown in Figure 5.7.

5.8   MISCELLANY

Hume-Rothery, Mallott, and Channel-Evans (1934) first proposed that solid solubility is very limited if the atomic diameters of solute and solvent vary by >15%. In the

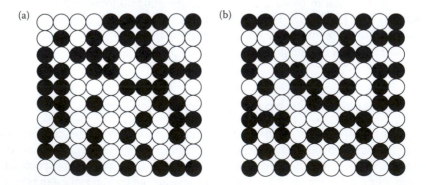

FIGURE 5.5   An a–b solid solution with $N_a = N_b = 50$.

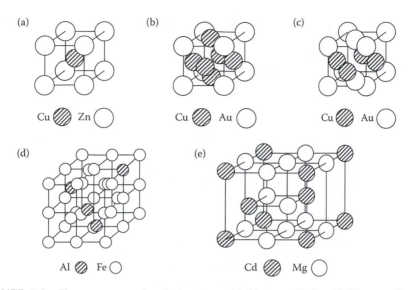

FIGURE 5.6 Five common ordered structures: (a) $L2_0$ type CuZn, (b) $L1_2$ type Cu_3Au, (c) $L1_0$ type CuAu, (d) DO_3 typeFe_3Al, and (e) DO_{19} typeMg_3Cd.

TABLE 5.3
Ordered Structures

$L2_0$	$L1_2$	$L1_0$	DO_3	DO_{19}
CuZn	Cu_3Au	CuAu	Fe_3Al	Mg_3Cd
FeCo	Au_3Cu	CoPt	Fe_3Si	Cd_3Mg
NiAl	Ni_3Mn	FePt	Fe_3Be	Ti_3Al
FeAl	Ni_3Al		Cu_3Al	Ni_3Sn
AgMg	Pt_3Fe			

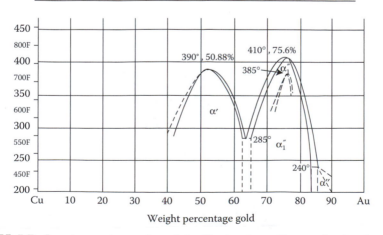

FIGURE 5.7 Low-temperature region of the Cu–Au phase diagram showing the ordered phase regions, Cu_3Au (α') and CuAu (α''). (From *ASM Handbook* Vol. 8, 8th ed., 1973. With permission.)

same year, Jones (1934) realized solid solutions in Cu, Ag, and Au become unstable at e/a ratios over 1.4 and related this solubility to the shape of the electron density curves. Solubilities in magnesium of Group II and III elements that have favorable size factors decreases with valence. However, it is not possible to fix an exact e/a ratio for the solubility limit.

PROBLEMS

1. The solubility of carbon in austenite is given as $C_\gamma = 1.165 \exp[-28,960/RT]$, where the units of 28,960 are cal/mol, C_γ is the atomic fraction carbon, and R should be expressed as cal/mol-K. Use this equation to predict the metastable solubility of carbon at 600°C. Express your answer as wt% carbon.

2. Consider an alloy of copper containing 20% Zn, 3% Al, and 3% Sn. Calculate the electron-to-atom ratio. Assuming the maximum solubility corresponds to e/a of 1.38, would this alloy be a single solid solution?

3. In ordered solutions there are different "phases" in the sense that different regions may be out of "phase" with each other. Using the term "phase" in this sense, how many different phases are there in
   a. CuZn?
   b. Cu_3Au?
   c. CuAu?

4. The equilibrium concentration of vacancies per volume can be approximated by $N_v = \exp[-E_f/RT]$. For copper, E_f has been estimated to be 85,000 J/mol.
   a. Calculate Nv for copper at 1000°C.
   b. Assume that all of these vacancies are retained when the copper is quenched to room temperature. What is the energy/mole associated with these vacancies?
   c. If all of the energy associated with these vacancies were converted to heat, what would be the increase of temperature, ΔT?
      You will have to find the appropriate data for copper.

5. Determine the short-range order parameter, s, for the two-dimensional solution illustrated in Figure 5.8.

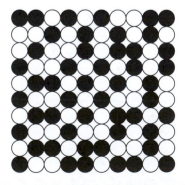

FIGURE 5.8   Two-dimensional solid solution of 50% white and 50% black.

6. Recall that for a solution of a and b, $\Delta G_m = \Delta H_m - T\Delta S_m$ and a good approximation for ΔH_m is $\Delta H_m = N_a N_b \Omega$, where $\Omega = (zn_o/2)[E_{ab} - (E_{ab} + E_{ab})/2]$. ΔS_m can be taken as $\Delta S_m = -R(N_a \ln N_a + N_b \ln N_b)$.
   A. Make a plot of ΔG_m versus N_b at $T = 800°C$. Assume that $\Omega = 27$ kJ/mol.
   B. Determine the solubility limits at 800°C.
   C. Determine the solubility limits at 600°C.
7. The solubility of Ca in Mg is 1.6% at 517°C. Estimate the solubility at 300°C.
8. The fcc crystal structure is more densely packed than the bcc structure. Yet the solubility of C in γ-Fe is much greater than in α-Fe. Calculate the size if the largest interstitial holes in both the fcc and bcc as fractions of the size of the iron atom.
9. Consider a copper-base alloy containing 8 g of Al and 92 g of Cu. What is the maximum amount of silicon grams that could be added and still retain a single fcc solid solution (e/a = 1.35)?

Metal	Cu	Al	Si
Atomic mass	63.5	27	28
Valence	1	3	4

REFERENCES

ASM Handbook Vol. 8, 8th ed., 1973.
C. S. Barrett, *Structure of Metals*, McGraw-Hill, 1943, p. 266.
W. Hume-Rothery, *Structure of Metals and Alloys*, The Institute of Metals, 1944.
W. Hume-Rothery, G. W. Mallott, and K. M. Channel-Evans, *Phil. Trans. Roy. Soc.*, A233, 1934.
P. Haasen, *Physical Metallurgy*, 2nd ed., Cambridge, 1986, p. 127.
H. Jones, *Proc Roy Soc*, 1934.
N. F. Mott and H. Jones, *The Theory of the Properties of Metals and Alloys*, 1936.
G. V. Raynor, in *Progress of Metal Physics*, Butterworth, 1949.
F. Seitz, *Modern Theory of Solids*, McGraw-Hill, 1940.

6 Intermediate Phases

Intermediate phases exist because they have lower free energies than terminal solid solutions. The reduction of free energy may be attributable to one of several phenomena: electron energy levels, covalent or ionic bonding, or geometric packing of atoms.

6.1 HUME-ROTHERY OR ELECTRON PHASES

The ratio of valence electrons to atoms, e/a, has a strong effect on the energies of different crystal structures. The terminal fcc solid solutions of copper, gold, and silver with other elements are stable up to an e/a ratio of about 1.38. This corresponds to the e/a ratio at which the Fermi sphere approaches the Brillouin zone. Other crystal structures are stable at higher electron-to-atom ratios. Table 6.1 shows the valences used in calculating the number of valence electrons. Figure 6.1 shows the composition ranges over which the β-bcc phase exists in several copper-base alloy systems. Instead of plotting the composition in terms of wt% or atom%, composition is plotted in terms of the e/a ratio. The phase boundaries for the various solutes are similar. The β-phase is centered about an e/a ratio of 1.5, which corresponds to CuZn, Cu_3Al, Cu_5Sn, and so on. However, the range of compositions is very wide and the atoms are arranged randomly; so the phase should be regarded as a solid solution rather than a compound. In some systems, a cubic manganese structure is found at e/a = 1.5. The β phase is very ductile.

Similar coincidence of e/a ratios is found for other intermediate phases. Table 6.2 is a list of a number of electron phases in alloy systems.

6.2 COVALENT COMPOUNDS

Other intermediate phases have stochiometric compositions with limited solubility. Among these are compounds with covalent bonding. Those having the zinc-blende structure are AlSb, GaSb, and ZnSb. Those with a nickel arsenide structure (Figure 6.2) include Mn_7Ge_4, Mn_2Sn, Fe_3Sn_2, Co_3Sn_2, CoSb, Ni_2Ga, Ni_3Sn_2, Cu_4In_3, and Cu_6Sn_5.

6.3 IONIC COMPOUNDS

These include Mg_2Si, Mg_2Sn, Mg_3As_2, and MgSe. Figure 6.3 is the Mg–Si phase diagram. The β-phase (Mg_2Si) appears as a line because its solubility for excess Mg or Si is extremely low.

TABLE 6.1

Valences

Valence	Metals
1	Cu, Au, Ag
2	Zn, Cd, Mg, Be, Hg
3	Al, Ga, In
4	Sn, Si, Ge, Pb
5	P, As, Sb, Bi
0	Fe, Co, Ni, Ru, Rh, Pd, Pt, Ir, Os

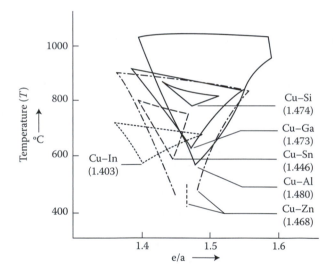

FIGURE 6.1   Electron-to-atom ratios over which the bcc β-phase exists in several copper-base alloy systems. (From T. B. Massalski and H. W. King, *Progress in Materials Science*, 10, 1, 1961. With permission.)

6.4   INTERSTITIAL COMPOUNDS

These are compounds of transition metals with carbon, nitrogen, and hydrogen. In interstitial compounds, the radius of the interstitial atom must be <2/3 that of the transition metal atom. Bonding has a strong covalent character. These compounds have very high melting points (Table 6.3).

6.5   LAVES PHASES

These are compounds with a composition AB_2 that arise because of the dense packing that can be achieved if ratio of the diameters of the B atoms to the A atoms is about 1.2. In these phases, the A atoms have 12 B nearest neighbors and 4 A nearest neighbors. Each B atom has 6 A and 6 B near neighbors. This amounts to an average

TABLE 6.2

Electron Phases in Various Alloy Systems

	Electrons per Atom				
	3/2			21/13	7/4
bcc Structure	Cubic Manganese Structure	HCP Structure		γ Brass Structure	HCP Structure
CuBe	Cu_5Si	Cu_3Ga		Cu_5Zn_8	$CuZn_3$
CuZn	AgHg	Cu_5Ge		Cu_5Cd_8	Cu_3Sn
Cu_3Al	Ag_3Al	AgZn		Cu_5Hg_8	Ag_5Al_3
Cu_3Ga	Au_3Al	Ag_3Al		Cu_9Al_4	
Cu_3In	$CoZn_3$	Ag_3Ga		$Cu_{31}Si_8$	
Cu_5Sn		Ag_3In		Ag_9In_4	
AgMg		Ag_5Sn		Mn_5Zn_{21}	
Ag_3Al		Ag_7Sb		Fe_5Zn_{21}	
FeAl		Au_3In		$Na_{31}Pb_8$	
CoAl		Au_5Sn			
PdIn					

Source: From W. Hume-Rothery and Raynor, *Structure of Metals*, p. 197. With permission.

coordination number of $(2 \times 12 + 16)/3 = 13.33$. There are three crystal structures: cubic $MgCu_2$, hexagonal $MgZn_2$, and hexagonal $MgNi_2$. They are all composed of tetrahedral of A atoms surrounded by B atoms. The $MgCu_2$ unit cell has eight formula units per cell. Figure 6.4 illustrated the structure of these compounds.

The packing in $MgZn_2$ and $MgNi_2$ is similar except that the tetrahedral are arranged in a hexagonal packing. Figure 6.4 shows the geometric arrangements of the tetrahedral. The essential feature of these three phases is that the ratio of atomic radii is very close to $r_A/r_B = \sqrt{(1.5)} = 1.225$. Table 6.4 lists some of the intermetallic phases of the Laves types.

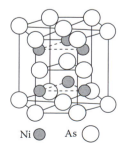

Ni ● As ○

FIGURE 6.2 Nickel arsenide structure. Note that the As atoms form tetrahedral around the Ni atoms.

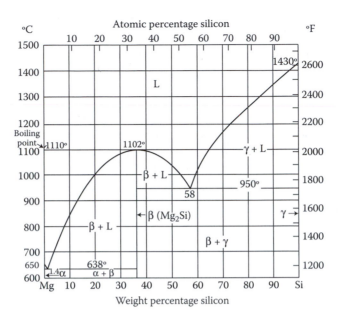

FIGURE 6.3 Magnesium–silicon phase diagram. Note that the β-phase (Mg_2Si) appears as a line. (From *Metals Handbook*, 7th ed., ASM p. 1226. With permission.)

TABLE 6.3
Melting Points of Some Interstitial Compounds

Compound	Melting Point (K)	Compound	Melting Point (K)
TiC	3425	NbC	3775
TiN	3215	NbN	2475
VC	2475	HfC	4175
ZrC	3800	TaC	4175
ZrN	3250	TaN	3360

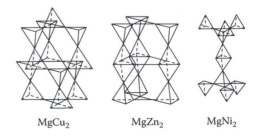

FIGURE 6.4 Arrangement of tetrahedral in the three Laves phases. (From P. Haasen, *Physical Metallurgy*, 2nd ed., Cambridge University Press, 1986, p. 140. With permission.)

TABLE 6.4
Some Intermetallic Phases of the
Laves Types

MgCu$_2$ Type	MgZn$_2$ Type	MgNi$_2$ Type
AgBe$_2$	BaMg$_2$	ReBe$_2$
BiAu$_2$	CuAg$_2$	FeB$_2$
CuAl$_2$	CuBe$_2$	MoBe$_2$
CeAl$_2$	KNa$_2$	TaCo$_2$
CeFe$_2$	MoBe$_2$	WBe$_2$
GdFe$_2$	NbFe$_2$	ZrFe$_2$
KBi$_2$	ReBe$_2$	
ZnAl$_2$	TaFe$_2$	
NaAu$_2$	UNi$_2$	

Source: From W. Hume-Rothery and Raynor, *Structure of Metals*, p. 228. With permission.

6.6 MISCELLANY

Originally metallurgy was a branch of chemistry. The break between metallurgy and chemistry came as a result of the failure of Dalton's laws of simple proportions when applied to many intermetallic phases. The compositions of the electron phases do not correspond to simple ratios of species; these phases have large solubilities for excess atoms of both species and the positions in the atom species were not fixed in the lattice.

PROBLEMS

1. a. If an alloy of 50 atom % Zn and 50 atom % Al were added to copper, at what composition would be the center of the β-field? (i.e., at what atom% Zn would the e/a ratio equal 1.5?)
   b. What would be the corresponding weight % Zn and Al?
2. Fe$_3$C has an orthorhombic unit cell with dimensions, $a = 0.509$ nm, $b = 0.647$ nm, and $c = 0.452$ nm. Its density is approximately the same as iron (7.9 mg/m^3). How many iron and carbon atoms are there in each unit cell?

REFERENCES

C. S. Barrett, *Structure of Metals*, Oxford, 1980.
P. Haasen, *Physical Metallurgy*, 2nd ed., Cambridge University Press, 1986, p. 140.
W. Hume-Rothery, *Structure of Metals and Alloys*, The Institute of Metals, 1944.
W. Hume-Rothery and Raynor, *Structure of Metals*, p. 197.
W. Hume-Rothery and Raynor, *Structure of Metals*, p. 228.
T. B. Massalski and H. W. King, *Progress in Materials Science*, 10, 1, 1961.
Metals Handbook, 7th ed., ASM p. 1226, 1973.
N. F. Mott and H. Jones, *The Theory of the Properties of Metals and Alloys*, 1936.
F. Seitz, *Modern Theory of Solids*, McGraw-Hill, 1940.

7 Phase Diagrams

Phase diagrams indicate how the equilibrium between phases depends on overall composition, temperature, and pressure. In dealing with liquids and solids, we do not usually consider pressure as a variable because changes of pressure usually have only minor effects on equilibrium. Engineering materials usually contain many components. However, for simplicity, we will begin with a review of binary phase diagrams, which indicate the equilibrium between two components. For most students this should be a review.

7.1   REVIEW OF BINARY PHASE DIAGRAMS

Binary phase diagrams show the equilibrium as a function of overall composition and temperature at a fixed atmospheric pressure (Figure 7.1). There are two simple ways to view a phase diagram. A phase diagram may be viewed as a map showing the phase compositions that are permissible and those that are not. Single-phase regions represent conditions (composition and temperature) that result in solutions (liquid solutions or solid solutions). Two-phase regions may be regarded as forbidden regions—combinations of temperature and composition that cannot exist in a very small region. If the temperature and overall composition fall in a two-phase region, equilibrium can be obtained only if the material breaks up into two different phases (solutions), one richer and one leaner in solute and the overall composition. The composition of each phase may be read by drawing a horizontal line through the overall composition and noting the intersections with the boundaries of the two-phase field.

Alternatively, binary phase diagrams may be regarded as plots of solubility limits. Each boundary of a single-phase region represents the maximum solubility of that phase for the component at that end of the diagram.

Some lines have names. For example, the *solidus* is the line or lines representing the lowest temperature at which there is any liquid. It is also the highest temperature at which the material is 100% solid. In Figure 7.2, a and d are the solidus lines. The *liquidus* is the lowest temperature at which the alloy is 100% liquid (or the highest temperature at which any solid can exist). Lines b and c form the liquidus in Figure 7.2. A line representing the maximum solid solubility (e and f) is called the *solvus*.

7.2   INVARIANT REACTIONS

Reactions that occur at a constant temperature between three phases are called *invariant* reactions. Several examples are given in Figure 7.3.

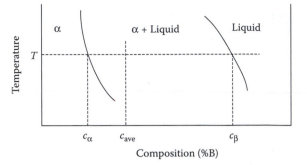

FIGURE 7.1 If the overall composition, c_{ave}, and temperature, T, lie in the two-phase α + liquid field, the alloy will be composed of α-phase of composition c_α and liquid of composition c_L.

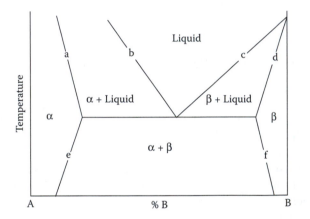

FIGURE 7.2 Lines a and e represent the maximum solubility element B in the α-phase. Line c is the maximum solubility of B in the liquid phase and line b is the maximum solubility of A in the liquid phase. Lines d and f represent the solubility of A in the β-phase.

7.3 LEVER LAW

In a two-phase region, a mass balance gives the relative amounts of the two phases

$$f_\alpha = \frac{c_\beta - c_{av}}{c_\beta - c_\alpha} \quad \text{and} \quad f_\beta = \frac{c_\alpha - c_{av}}{c_\alpha - c_\beta}, \tag{7.1}$$

where c_α and c_β are the compositions of the α- and β-phases and c_{av} is the overall composition. If the compositions, c_α, c_β, and c_{av} are in wt%, f_α and f_β are the weight fractions of the alloy that are in the form of α- and β-phases.

7.4 GIBBS' PHASE RULE

Gibbs' phase rule compares, for equilibrium, the number of variables necessary to describe the system and the number of equations relating these variables. The difference

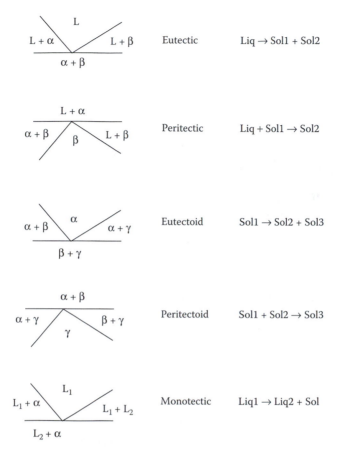

	Eutectic	Liq → Sol1 + Sol2
	Peritectic	Liq + Sol1 → Sol2
	Eutectoid	Sol1 → Sol2 + Sol3
	Peritectoid	Sol1 + Sol2 → Sol3
	Monotectic	Liq1 → Liq2 + Sol

FIGURE 7.3 Several invariant reactions.

is the number of independent variables (degrees of freedom), or variables that can be changed independently without affecting the number of phases in equilibrium.

7.4.1 EQUALITIES

Equilibrium implies that there is no driving force for a change of composition. Therefore, the vapor pressure of each element must be the same in all phases.[*] If it were not, there would be a driving force for compositional change. Suppose, for example, that the equilibrium vapor pressure of zinc were higher for the α-phase than for the β-phase. This would mean that Zn vapor evaporating at the α-phase diffuse to and condense at β-phase. It does not matter that the vapor pressures may

---

[*] Some authors describe equilibrium in terms of partial free energy or chemical activities instead of vapor pressure. Since these quantities are uniquely related to vapor pressure, the result is the same.

be extremely small so that the rate of compositional change would be small. Equilibrium therefore implies that

$$p_A^\alpha = p_A^\beta = p_A^\gamma = \cdots,$$

$$p_B^\alpha = p_B^\beta = p_B^\gamma = \cdots,$$ (7.2)

$$p_C^\alpha = p_C^\beta = p_C^\gamma = \cdots,$$

Here the subscripts refer to the components (elements or compounds) and the superscripts refer to the phases. For example, p_A^β is the equilibrium vapor pressure of element A in phase β. In this set of equations there are $P - 1$ equalities (equal signs) in each line (where P is the number of phases, α, β, γ, etc.), and there are C lines (where C is the number of components, A, B, C, etc.), so the total number of equalities is $C(P - 1)$.

7.4.2 VARIABLES

The variables in the system are the composition of *each phase* and the environmental variables. To express the composition of each phase we need $C - 1$ independent terms. For example, to describe the composition of the α-phase we would need to fix $C_A^\alpha, C_B^\alpha, \ldots, C_{C-1}^\alpha$.

The reason that the number of compositional variables for the α-phase is $C - 1$ rather than C is that once the percent (or fraction) of all but one of the components has been established, the amount of the last one is fixed since the percentages must sum to 100% (or fractions to 1). Since there are P phases and $C - 1$ compositional variables for each phase, the total number of compositional variables is $P(C - 1)$.

The usual environmental variables are temperature and pressure. However, one can imagine a system where one wanted to change some other variable (e.g., magnetic field) or often we do not want to consider pressure as a variable because we want to describe equilibrium at a fixed total pressure (e.g., atmospheric); so to be general we will designate the number of environmental variables as E. Most phase diagrams are at constant pressure, so the only environmental variable is temperature. In this case, $E = 1$. (If both variations in both temperature and pressure are considered, $E = 2$.)

7.4.3 DEGREES OF FREEDOM

The degrees of freedom are the number of variables that can be changed independently without changing which phases are in equilibrium. The number of degrees of freedom, F, is the number of variables not fixed by the equalities, so it equals the number of variables, $P(C - 1) + E$, minus the number of equalities, $C(P - 1)$.

$$F = P(C - 1) + E - C(P - 1) \quad \text{or}$$
$$F = C + E - P.$$ (7.3)

This is the Gibbs' phase rule. For constant pressure, but variable temperature, it can be written as $F = C - P + 1$. (One simple way of remembering the phase rule is to write it as $P + F = C + E$, which might also be a short-hand way of saying a **P**olice **F**orce equals **C**ops plus **E**xecutives.)

EXAMPLE PROBLEM 7.1

Consider the following systems:

1. A binary Pb–Sn alloy with 40% Sn at about 200°C. Pressure fixed at atmospheric, so $E = 1$. The phase diagram (Figure 7.4) shows that there are two phases present, liquid and α, so $P = 2$. Since this is a binary alloy, $C = 2$ (Pb and Sn). Substituting in the phase rule, $F = 2 + 1 - 2 = 1$, so there is one degree of freedom. We can change the temperature and still have the same two phases in equilibrium, but their compositions will automatically adjust. Or else the composition of one phase might be changed independently, requiring that both the temperature and the composition of the other phase change accordingly.

EXAMPLE PROBLEM 7.2

Consider the same alloy, but at a higher temperature where there is only one phase (liquid). Now $P = 1$, $C = 2$, and $E = 1$; so from Equation 7.3, $F = 2$. We can change both the temperature and the composition of the liquid without altering the equilibrium (maintaining a single liquid phase).

EXAMPLE PROBLEM 7.3

Consider the same alloy at a temperature where three phases (α, β, and liquid) are at equilibrium. $E = 1$, $C = 2$, and now $P = 3$, so the phase rule predicts that $F = 0$. This means that neither the composition of the α-phase or the β-phase nor the temperature can be changed while still maintaining equilibrium between these three phases. That is why the eutectic is called an "invariant." Note however that if we varied the pressure on the system by a sufficient amount, the eutectic temperature and compositions would be altered. This is in accord with considering

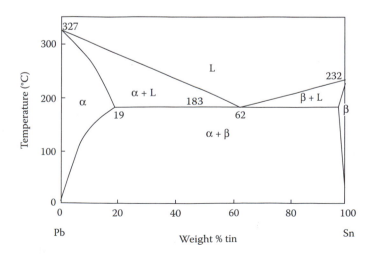

FIGURE 7.4 Lead–tin phase diagram.

both temperature and pressure as variables, so $E = 2$. In this case, $F = 1$, which means that the eutectic equilibrium can be maintained by independently varying the pressure, the temperature, or the composition of one of the phases and allowing the other variables to change accordingly.

EXAMPLE PROBLEM 7.4

Consider pure water (H_2O). In this case $C = 1$. Let pressure be a variable in addition to temperature, so $E = 2$. It is possible for three phases, pure water vapor, liquid water, and ice, to coexist ($P = 3$) but the phase rule says that $F = 1 + 2 - 3 = 0$, so there are no degrees of freedom. The three phases can coexist only at one temperature and pressure which is called the *triple point*.

7.5   TERNARY PHASE DIAGRAMS

In systems involving three components, the composition may be plotted on a triangular section (Figure 7.5). Pure components are represented at the corners and the grid lines show the amount of each component. Confusion about the labeling of lines will be cleared up by realizing that all of the lines parallel to AB are lines on which the % C is constant. Those nearest C have the greatest amount of C.

To represent temperature, a third dimension is needed. Figure 7.6 is a sketch of a three-dimensional ternary diagram in which temperature is the vertical coordinate.

Ternary equilibrium can be represented in two dimensions by two types of sections, isothermal or vertical sections. A typical isothermal section is indicated in Figure 7.7. There are single-phase regions, two-phase regions, and three-phase regions. As in binary diagrams, there are always two-phase regions between single-phase regions. Three-phase regions are triangular and meet single-phase regions only at their corners.

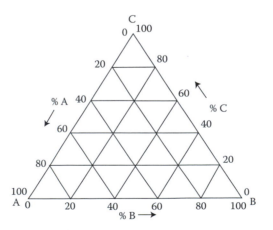

FIGURE 7.5   Triangular grid for representing the compositions in a three-component system.

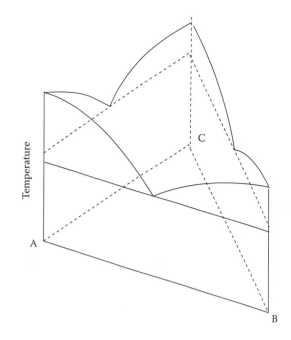

FIGURE 7.6 Three-dimensional model of a ternary phase diagram.

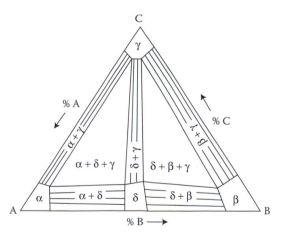

FIGURE 7.7 Isothermal section of a hypothetical ternary phase diagram showing one-, two-, and three-phase regions.

7.5.1 Two-Phase Regions

In the two-phase regions, tie lines are needed to indicate the compositions of the two phases that are in equilibrium. Often they are not shown, but usually they can be approximated by using a little judgment. The lever law may be used to find the relative amounts of the two phases at opposite ends of a tie line going through the overall composition. (In applying the lever law, the composition may be expressed in terms of any of the three components but greatest accuracy will be obtained by using the component that differs the most between the two phases.)

EXAMPLE PROBLEM 7.5

Consider an overall composition of 50% A, 10% B, and 30% C lying in a two-phase region consisting of α with a composition $c_\alpha = 92\%$ A, 6.5% B, and 11% C; and γ with a composition $c_\gamma = 31\%$ A, 14.6% B, and 54.4% C. The fraction of the alloy present as α can be calculated from the % A as $f_\alpha = (31-60)/(31-82) = 56.9\%$ or from the % C as $f_\alpha = (54.4-30)/(54.4-11.5) = 56.9\%$, or from the % B as $f_\alpha = (14.6-10)/(14.6-6.5) = 56.8\%$. Note that a small error in reading the compositions of the phases (e.g., if c_γ is misread as $c_\gamma = 31.2\%$ A, 14.8% B, and 54.0% C, the calculated value of f_α would be $f_\alpha = 56.7\%$ using the % A, and $f_\alpha = 56.5\%$ using the % C, but the error is larger using the % B, $f_\alpha = 57.8$).

7.5.2 THREE-PHASE REGIONS

These are triangular. The compositions of the phases are at the corners. The relative amounts of the three phases can be found from a modified lever law, $f_a = (c_{av} - c_a)/(c'_a - c_a)$, where c'_a is found by extrapolating a line drawn through c_α and c_{av} to the opposite side of the triangle as shown in Figure 7.8.

Much can be deduced about the progression of phases formed during solidification from a representation of the liquidus on triangular coordinates. Figure 7.9 shows the liquidus surface of the Ag–Cu–Zn system during freezing that the composition of the liquid moves away from the solid phase that is forming and down the temperature gradient. Once a eutectic valley is reached, the liquid composition will follow the eutectic valley toward lower temperatures.

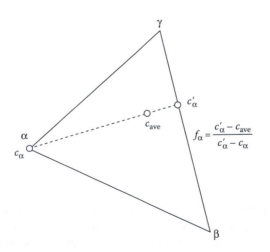

FIGURE 7.8 Use of the lever law in a three-phase region.

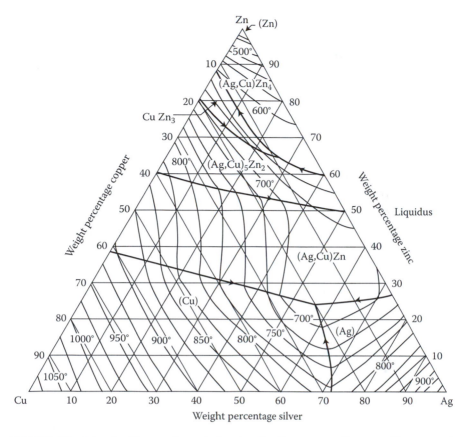

FIGURE 7.9 The liquidus surface of the Ag–Cu–Zn ternary diagram. (From *Metals Handbook*, 8th ed., Vol. 8, ASM, p. 380, 1973. With permission.)

7.5.3 VERTICAL SECTIONS

Vertical cuts through a ternary are sometimes made at a constant % of one of the components and sometimes at a constant ratio of two components (Figure 7.10).

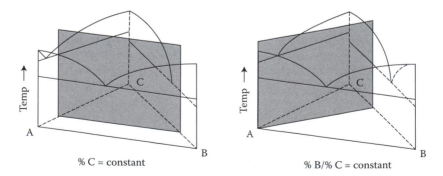

FIGURE 7.10 Pseudobinary sections of a ternary.

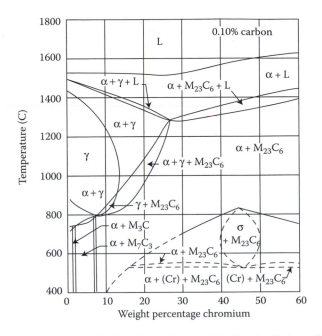

FIGURE 7.11   Section at 0.10% C of the iron-rich end of the Fe–Cr–C phase diagram. (From *Metals Handbook*, 8th ed., Vol. 8, ASM, p. 403, 1973. With permission.)

These are sometimes called pseudobinaries, but they differ from binary phase diagrams in several respects. One is that the tie lines between compositions in equilibrium with one another do not in general lie in the plane of the cut, so it is impossible to read the compositions of the phases in equilibrium or to use the lever law.

The section of the Fe–Cr–C diagram at 0.10% C is given in Figure 7.11. Note that Figure 7.11 shows an $\alpha + \gamma$ two-phase region but no α-single-phase region. The α-phase cannot dissolve 0.1% C, so there is not a single-phase α-region on the 0.1% carbon section.

7.6   MISCELLANY

Without a doubt, Josiah Willard Gibbs was the greatest American scientist of the nineteenth century and probably of all times. He was a theorist in an America that was preoccupied by practical inventions. Gibbs was born in 1839 in New Haven, Connecticut. He studied at Yale University, where in 1863 he received the first doctorate in engineering given in North America. After leaving Yale, he studied at various European Universities and then returned to live in New Haven with his sister. In 1871, he was appointed professor of Mathematical Physics without salary. He worked by himself in virtual obscurity. His work was almost unknown and certainly unappreciated in the United States because he published his work in an obscure journal, Transactions of the Connecticut Academy of Science. His work, however, was well known to the Maxwell, Helmholtz, and other leading European

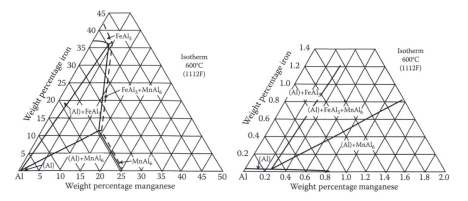

FIGURE 7.12 The Al–Fe–Mn phase diagram. The aluminum-rich corner is enlarged at the right. (From *Metals Handbook*, 8th ed., Vol. 8, ASM, p. 392, 1973. With permission.)

scientists because he sent them reprints. The story is told that when Maxwell came to an international exposition in Philadelphia, he asked "Where is Gibbs?" The reply was "Who is Gibbs?"

PROBLEMS

1. A portion of the isothermal section of Al–Fe–Mn phase diagram at 600°C is shown in Figure 7.12. Assuming equilibrium, list the phases present, give their compositions, and calculate the relative amounts of them for (a) 0.20% Fe, 0.80% Mn and (b) 0.80% Fe, 1.00% Mn.

2. Four distinct phases were observed in the microstructure of a ternary alloy at room temperature. Discuss this observation briefly in terms of the phase rule. What is the most likely explanation?

3. Consider the freezing of a ternary eutectic. The pressure and temperature are constant. The liquid simultaneously freezes to three solid phases, so there are four phases present. One student applies the phase rule and concludes that there are zero degrees of freedom. Another student says that this is wrong because the amounts of the phases are not constant. Who is right? Discuss briefly.

4. Considering both temperature and pressure to be variables, what is the largest number of phases that can coexist at equilibrium in a binary alloy?

5. Consider a binary alloy of copper and silver. At a temperature near 850°C and an alloy composition near 80% Cu–20% Ag, the phase diagram says that there should be a copper-rich solid and a liquid. Considering the pressure to be fixed at a pressure of one atmosphere, how many degrees of freedom are there? What is (are) the independent variable(s).

REFERENCE

Metals Handbook, 8th ed., Vol. 8, ASM, p. 380; 392; 403, 1973.

8 Dislocations

8.1 THE NATURE OF DISLOCATIONS

Dislocations are line imperfections in a crystal. The lattice around a dislocation is distorted, so the atoms are displaced from their normal lattice sites. The lattice distortion is greatest near a dislocation and decreases with distance from it.

One special form of a dislocation is an *edge* dislocation that is sketched in Figure 8.1. The geometry of an edge dislocation can be visualized by cutting into a perfect crystal and then inserting an extra half plane of atoms in the cut as shown in Figure 8.2. The dislocation is the bottom edge of this extra half plane. An alternative way of visualizing dislocations is illustrated in Figure 8.3. An edge dislocation is created by shearing the top half of the crystal by one atomic distance perpendicular to the end of the cut (Figure 8.3b). This produces an extra half plane of atoms, the edge of which is the center of the dislocation. The other extreme form of a dislocation is the *screw dislocation*. This can be visualized by cutting into a perfect crystal and then shearing half of it by one atomic distance in a direction parallel to the end of the cut (Figure 8.3b). The end of the cut is the dislocation. Around it, the planes are connected in a manner similar to the levels of a spiral parking ramp.

In both cases, the dislocation is a boundary between slipped and unslipped regions. The direction of slip that occurs when an edge dislocation moves is perpendicular to the dislocation. In contrast, movement of a screw dislocation causes slip in the direction parallel to itself. The edge and screw are extreme cases. A dislocation may be neither parallel nor perpendicular to the slip direction.

A dislocation may be characterized by a *Burgers vector*. Consider an atom-to-atom circuit (Figure 8.4) that would close on itself if made in a perfect crystal. If this same circuit is constructed so that it goes around a dislocation, it will not close. The closure failure is the Burgers vector, denoted by **b**. The Burgers vector can be considered a *slip vector* because its direction is the slip direction and its magnitude is the magnitude of the slip displacement caused by the movement of the dislocation.

As a dislocation wanders through a crystal, its orientation changes from place to place, but its Burgers vector is the same everywhere (Figure 8.5). If the dislocation branches into two dislocations, the sum of the Burgers vectors of the branches equals its Burgers vector.

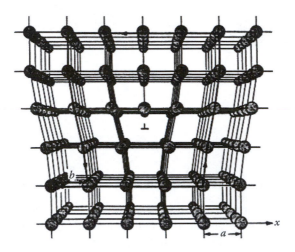

FIGURE 8.1   An edge dislocation is the edge of an extra half plane of atoms. (From A. G. Guy, *Elements of Physical Metallurgy*, Addison-Wesley, 1951. With permission.)

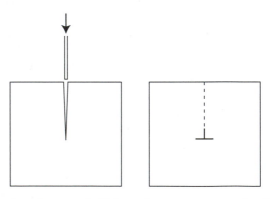

FIGURE 8.2   Insertion of an extra half plane of atoms creates an edge dislocation.

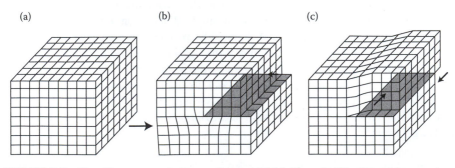

FIGURE 8.3   Consider a cut made in a perfect crystal (a). If one half is sheared by one atom distance perpendicular to the end of the cut, an edge dislocation results (b). If one half is sheared by one atom distance parallel to the end of the cut, a screw dislocation results (c).

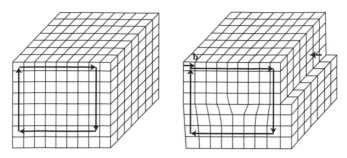

FIGURE 8.4 The Burgers vector of a dislocation can be determined by drawing a clockwise circuit that would close, if it were drawn in a perfect crystal. If the circuit is drawn around a dislocation, the closure failure is the Burgers vector.

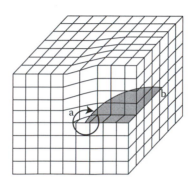

FIGURE 8.5 A dislocation may wander through a crystal, but everywhere it has the same Burgers vector. Where it is parallel to its Burgers vector, it is a screw (a); where it is perpendicular to its Burgers vector, it is an edge (b). Elsewhere it has mixed character.

8.2 ENERGY OF A SCREW DISLOCATION

The energy associated with a screw dislocation is the energy required to distort elastically the lattice surrounding the dislocation. The distortion is severe near the dislocation but decreases with distance. Consider an element at a distance, r, from the center of a screw dislocation of length, L (Figure 8.6a). The volume of the element is $2\pi rLdr$. Imagine unwrapping this element, so that it lies flat (Figure 8.6b).

$$\gamma = \frac{\mathbf{b}}{2\pi r},\qquad(8.1)$$

where $\mathbf{b}$ is the Burgers vector of the screw dislocation. The energy/volume required for this elastic distortion is

$$w = \frac{1}{2}\tau\gamma,\qquad(8.2)$$

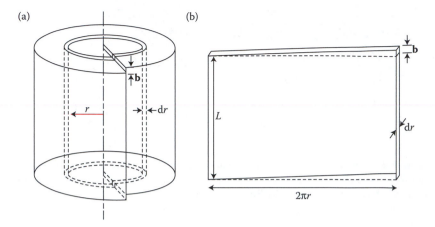

FIGURE 8.6   (a) Screw dislocation in a cylindrical crystal and (b) flattened element.

where τ is the shear stress necessary to cause the shear strain, γ. According to Hooke's law,

$$\tau = G\gamma, \tag{8.3}$$

where G is the shear modulus. Combining Equations 8.1 through 8.3, the energy/volume is

$$w = \frac{G\gamma^2}{2} = \left(\frac{1}{2}\right)\frac{G\mathbf{b}^2}{(2\pi r)^2}. \tag{8.4}$$

The total elastic energy, dU, associated with the element is its energy/volume times its volume,

$$dU = w(2\pi r L\, dr) = \left(\frac{G\mathbf{b}^2 L}{4\pi}\right)\frac{dr}{r}. \tag{8.5}$$

The total energy of the dislocation is obtained by integrating Equation 8.5.

$$U = \int\left(\frac{G\mathbf{b}^2 L}{4\pi}\right)\int\left(\frac{dr}{r}\right) = \left(\frac{G\mathbf{b}^2 L}{4\pi}\right)\ln\left(\frac{r_1}{r_0}\right). \tag{8.6}$$

EXAMPLE PROBLEM 8.1

Equation 8.4 gives the energy/volume as a function of r. Calculate the distance from the core of a screw dislocation at which the energy/volume equals the heat of vaporization, H_v, and express this distance in terms of r/b. Evaluate this critical value of r/b for copper. $\rho = 8.93$ mg/m³, $G = 77$ GPa, $b = 0.255$ nm, $H_v = 4.73$ MJ/kg.

SOLUTION

Solving Equation 8.4 for r/b, $(r/b)^2 = G/(8\pi^2 U_v) = (77 \times 10^9$ Pa$)/(8\pi^2 \times 8.93 \times 10^3$ kg/m³ $\times 4.73 \times 10^6$ J/kg$) = 0.23$, $r/b = 0.15$.

For copper, $r = 0.15 \times 0.255 = 0.04$ nm.

Clearly Equation 8.4 cannot hold for values of r less than this because it predicts that material within this radius would vaporize.

The problem, now, is to decide what are *appropriate* values for the lower limit, r_0, and the upper limit, r_1, of the integral. We might be tempted to let r_0 be 0 and r_1 be ∞, but both would cause the value of U in Equation 8.6 to be infinite. Consider first the lower limit, r_0. Equation 8.1 predicts an infinite strain at $r = 0$, which corresponds to an infinite energy/volume at the core of the dislocation, $(r_0 = 0)$. This is clearly unreasonable. The energy/volume cannot possibly be higher than the heat of vaporization. The discrepancy is a result of the breakdown of Hooke's law at the atomic level and at very high strains. The stress, τ, is to be proportional to γ only for small strains. A lower limit, r_0, can be chosen so that the neglected energy of the core $(0 < r < r_0)$ is equal to the overestimation of the integral for $r > r_0$ (Figure 8.7). A value of $r_0 = \mathbf{b}/4$ has been suggested.

Now consider the upper limit, r_1. This cannot be any larger than the radius of the crystal. A reasonable value for r_1 is half of the distance between dislocations because the stress fields of dislocations tend to annihilate one another. It is reasonable to approximate r_1 by $10^5 b$ that corresponds to a dislocation density of 10^{10} dislocations/m² which is typical of an annealed metal. This value is convenient because $\ln(10^5 b/0.25b) = \ln (4 \times 10^5) = 12.9 \approx 4\pi$. With this approximation, Equation 8.6 simplifies to

$$U_L \approx G\mathbf{b}^2. \tag{8.7}$$

The process of choosing the limits of integration so that the expression for the energy/length simplifies to Equation 8.7 may seem arbitrary. Yet the results, although approximate, are reasonable. The derivation of the energy of an edge dislocation is more complicated because the stress field around an edge dislocation is more complex. The end result is that for an edge dislocation,

$$U_L \approx \frac{G\mathbf{b}^2}{1 - \upsilon} \tag{8.8}$$

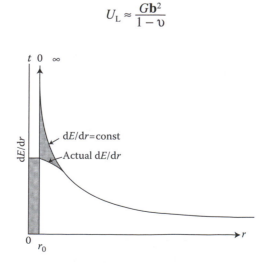

FIGURE 8.7 Energy density near the core of a dislocation.

where υ is Poisson's ratio. Thus the energy of an edge dislocation is greater than that of a screw by a factor of $1/(1 - \upsilon)$ or about 1.5.

There are two important features of Equations 8.7 and 8.8. One is that the energy of a dislocation is proportional to its length. This means that a dislocation exerts a line tension or a contractile force parallel to itself. The units of energy/length are J/m, which is the same as the units of force, N.

The second important feature is that the energy of a dislocation is proportional to $\mathbf{b}^2$. The energetics of reactions between parallel dislocations are controlled by this.

8.3 BURGERS VECTOR NOTATION

There is a simple notation system for describing the magnitude and direction of the Burgers vector of a dislocation. The direction is indicated by direction indices and the magnitude by a scalar preceding the direction. For example, $\mathbf{b} = (a/3)[\bar{2}11]$ in a cubic crystal means that the Burgers vector has components of $2a/3$, $-a/3$, and $a/3$ along the [100], [010], and [001] directions, respectively, where a is the lattice parameter. The magnitude of $(a/3)[\bar{2}11]$ is

$$|\mathbf{b}| = \left(\frac{a}{3}\right)[2^2 + (-1)^2 + 1^2]^{1/2} = \left(\frac{a}{3}\right)\sqrt{6}.$$

A dislocation corresponding to a full slip displacement in an fcc crystal would have a Burgers vector $(a/2)\langle 1\,1\,0\rangle$. In this case, the magnitude is $(a/2)\sqrt{2}$. Sometimes the lattice parameter, a, is omitted from the notation, being understood.

EXAMPLE PROBLEM 8.2

Using proper notation, write the Burgers vector for a dislocation corresponding to the shortest repeat distance in a bcc crystal.

SOLUTION

The shortest repeat distance is half of the body diagonal of the unit cell. $\mathbf{b} = (a/2)<1\,1\,1>$.

8.4 REACTIONS BETWEEN PARALLEL DISLOCATIONS AND FRANK'S RULE

Two parallel dislocations may combine and form a third dislocation. If they do, the Burgers vector of the product dislocation will be the vector sum of the Burgers vectors of reacting dislocations. That is, if $\mathbf{b}_1 + \mathbf{b}_2 \rightarrow \mathbf{b}_3$, then $\mathbf{b}_1 + \mathbf{b}_2 = \mathbf{b}_3$. The reaction should be energetically favorable if it lowers the energy of the system. Frank's rule states that since the energy of a dislocation is proportional to $\mathbf{b}^2$, the reaction is favorable if $\mathbf{b}_1^2 + \mathbf{b}_2^2 > \mathbf{b}_3^2$.

EXAMPLE PROBLEM 8.3

Consider the reaction of two parallel dislocations of Burgers vectors $\mathbf{b}_1 = (a/2)[101]$ and $\mathbf{b}_2 = a[0\bar{1}1]$. If they reacted, what would be the Burgers vector of the product dislocation? Would the reaction be energetically favorable?

SOLUTION

$$\mathbf{b}_3 = \mathbf{b}_1 + \mathbf{b}_2 = (a/2)[101] + a[0\overline{1}1] = (a/2)[1+0, 0+0+\overline{1}, 1+1] = (a/2)[1\overline{1}2].$$

$$\mathbf{b}_1^2 + \mathbf{b}_2^2 = \left(\frac{a^2}{2}\right)(1^2 + 1^2 + 0^2) + a^2\,[0^2 + (-1)^2 + 1^2] = 2.5a^2.$$

$$\mathbf{b}_3^2 = \left(\frac{a}{2}\right)^2\,[1^2 + (-1)^2 + 2^2] = 1.5a^2.$$

$\mathbf{b}_1^2 + \mathbf{b}_2^2 > \mathbf{b}_3^2$, so the reaction is energetically favorable.

Because it is energetically favorable for dislocations with large Burgers vectors to dissociate into dislocations with smaller Burgers vectors, the dislocations in crystals tend to have small Burgers vectors.

8.5 STRESS FIELDS AROUND DISLOCATIONS

Around a dislocation, the atoms are displaced from their normal positions. These displacements are equivalent to the displacements caused by elastic strains arising from external stresses. In fact, the lattice displacements or strains can be completely described by these stresses and Hooke's law can be used to find them.

It should be noted that the equations given below are based on the assumption of isotropic elasticity. The coordinate system is shown in Figure 8.8. For a screw dislocation with a Burgers vector, $\mathbf{b}$, parallel to the z axis,

$$\tau_{z\theta} = \frac{-Gb}{2\pi r} \quad \text{and} \quad \tau_{r\theta} = \tau_{rz} = \sigma_r = \sigma_\theta = \sigma_z = 0, \tag{8.9}$$

where G is the shear modulus and r is the radial distance from the dislocation. This stress field will cause a force per unit length, f_L, on another parallel dislocation equal to the dot product of the stress and its Burgers vector. The minus sign indicates that the repulsive force is inversely proportional to the distance between the dislocations. Equations 8.9 indicate that a screw dislocation creates no hydrostatic tension or

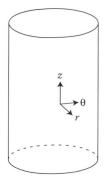

FIGURE 8.8 Coordinate system.

compression because $\sigma_H = (\sigma_r + \sigma_\theta + \sigma_z)/3 = 0$. Therefore, there should be no dilatation (volume strain) associated with a screw dislocation. However, real crystals are elastically anisotropic and with anisotropic elasticity there may be a small dilatation.

For an edge dislocation that lies parallel to z and with its Burgers vector parallel to x,

$$\tau_{xy} = Dx\frac{x^2 - y^2}{(x^2 + y^2)^2}, \tag{8.10a}$$

$$\sigma_x = -Dy\frac{3x^2 - y^2}{(x^2 + y^2)^2}, \tag{8.10b}$$

$$\sigma_y = Dy\frac{x^2 - y^2}{(x^2 + y^2)^2}, \tag{8.10c}$$

$$\sigma_z = \upsilon(\sigma_x - \sigma_y) = \frac{-2D\upsilon y}{x^2 + y^2}, \quad \tau_{yz} = \tau_{zx} = 0, \tag{8.10d}$$

where $D = Gb/[2\pi(1 - \upsilon)]$.

One of the important features of these equations is that there is a hydrostatic stress, $\sigma_H = (\sigma_x + \sigma_x + \sigma_x)/3$, around an edge dislocation. Combining Equations 8.10b–d,

$$\sigma_H = -\left(\frac{2}{3}\right)\frac{Dy(1 + \upsilon)}{x^2 + y^2} \quad \text{or}$$

$$\sigma_H = -\frac{A(y)}{x^2 + y^2}. \tag{8.11}$$

where, $A = Gb(1 + \upsilon)/[3\pi(1 - \upsilon)]$. Figure 8.9 shows how the hydrostatic stress varies near an edge dislocation. There is hydrostatic compression (negative σ_H) above the edge dislocation (positive y) and hydrostatic tension below it. The dilatation causes interactions between edge dislocations and solute atoms.

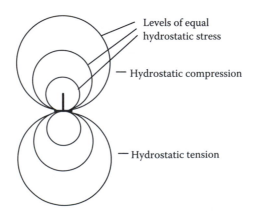

FIGURE 8.9 Contours of hydrostatic stress around an edge dislocation. Note that the level of hydrostatic stress increases near the dislocation.

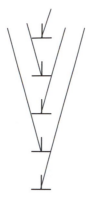

FIGURE 8.10 Low-angle tilt boundary formed by edge dislocations arranged so that the compressive stress field of one is partially annihilated by the tensile stress field of the one above.

In substitutional solutions, solute atoms that are larger than the solvent atoms are attracted to the region just below the edge dislocation where their larger size helps relieve the hydrostatic tension. Similarly, substitutional solute atoms that are smaller than the solvent atoms are attracted to the region just above the edge. In either case, edge dislocation will attract solute atoms. In interstitial solid solutions, all solute atoms are attracted to the region just below the edge dislocation where they help relieve the tension. It is this attraction of edge dislocations in iron for carbon and nitrogen that is responsible for the yield point effect and strain aging phenomenon in low-carbon steel.

Edge dislocations of like signs (same Burgers vectors) tend to form walls with one dislocation directly over another as shown in Figure 8.10. The hydrostatic tension caused by one dislocation is partially annihilated by the hydrostatic compression of its neighbor. This relatively low energy and therefore stable configuration forms a low-angle grain boundary.

8.6 FORCES ON DISLOCATIONS

A stress in a crystal causes a force on a dislocation. Consider a length, L, of dislocation of Burgers vector, $\mathbf{b}$, on a plane as shown in Figure 8.11. Let there be a shear stress, τ, acting on that plane. The force on the dislocation, per unit length, f_L, is

$$f_L = -\tau \mathbf{b}. \tag{8.12}$$

Note that here, the dot product is possible because once the plane of the stress is fixed, it can be treated as a vector (force). The stress, τ, may result from the stress field of another dislocation. Thus two screw dislocations exert an attractive force on each other of

$$f_L = -\frac{G\mathbf{b}_1 \cdot \mathbf{b}_2}{2\pi r}, \tag{8.13}$$

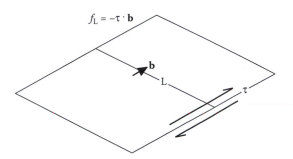

FIGURE 8.11 The force per length on a dislocation, $f_L = -\tau b$.

where **b** is the Burgers vector of the dislocation of concern. The minus sign means that they repel one another if the dot product is positive. An equivalent statement is that two dislocations repel each other if Frank's rule predicts that their combination would result in an energy increase. If the angle between $\mathbf{b}_1$ and $\mathbf{b}_2$ is greater than 90°, $|\mathbf{b}_1 + \mathbf{b}_2| > |\mathbf{b}_1| + |\mathbf{b}_2|$.

The interaction of two parallel edge dislocations is somewhat more complex. The stress field from one dislocation on the $x - y$ plane is given by Equation 8.10a, $\tau_{xy} = Dx(x^2 - y^2)/(x^2 + y^2)^2$, where $D = Gb/[2\pi(1 - \upsilon)]$. The mutual force on that plane is

$$f_L = -\left\{\frac{Gb_1 \cdot b_2}{2\pi(1 - \upsilon)}\right\} \frac{x^2 - y^2}{(x^2 + y^2)^2}. \tag{8.14}$$

This means that if $\mathbf{b}_1 \cdot \mathbf{b}_2$ is positive, (dislocations with like sign) there is mutual repulsion in the region $x > y$ and attraction in the region $x < y$. Again this is equivalent to saying that there is mutual repulsion if Frank's rule predicts that a reaction would cause an increase of energy and mutual attraction if it would cause a decrease of energy. Figure 8.12 shows the regions of attraction and repulsion. The stress, τ_{xy}, is zero at $x = 0$, $x = y$, and $x = \infty$.

Between $x = 0$ and $x = y$, τ_{xy} is negative, indicating that the stress field would cause an edge dislocation of the same sign to be attracted. Therefore, edge dislocations of the same sign tend to line up one above the other. For $x > y$, the stress, τ_{xy}, is positive, which indicates that the stress field would tend to repel another edge dislocation of the same sign. Figure 8.13 shows how the stress field varies. The attraction of like-sign edge dislocations of parallel planes leads to the formation of the low-angle grain boundaries during the recovery stage of annealing.

8.7 PARTIAL DISLOCATIONS IN fcc CRYSTALS

In fcc crystals, slip occurs on {111} planes and in ⟨110⟩ directions (Figure 8.14). Consider the specific case of $(111)[1\bar{1}0]$ slip. The Burgers vector corresponding to displacements of one atom diameter is $(a/2)[1\bar{1}0]$. A dislocation with this Burgers vector can dissociate into two partial dislocations,

$$\left(\frac{a}{2}\right)[1\bar{1}0] \rightarrow \left(\frac{a}{6}\right)[\bar{2}11] + \left(\frac{a}{6}\right)[\bar{1}2\bar{1}]. \tag{8.15}$$

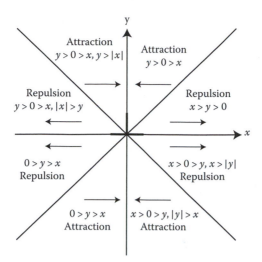

FIGURE 8.12 Stresses around an edge dislocation either attract or repel another parallel dislocation having the same Burgers vector, depending on how the two are positioned relative to one-another.

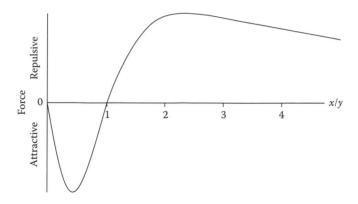

FIGURE 8.13 Magnitude of the force on an edge dislocation caused by the stress field of another edge dislocation of like sign at a location x, y.

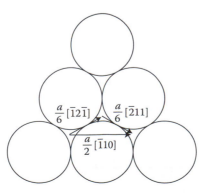

FIGURE 8.14 Burgers vectors for partial slip dislocations in an fcc crystal.

We can check that this reaction is vectorially correct by noting that $\mathbf{b}_1 =$ $(-a/2, a/2, 0)$ does equal $\mathbf{b}_2 = (-a/3, a/6, a/6) + \mathbf{b}_3 = (-a/6, a/3, -a/6)$. Figure 8.14 is a geometrical representation of these two partials.

EXAMPLE PROBLEM 8.4

When an $(a/2)[1\bar{1}0]$ dissociates into $(a/6)[\bar{2}11] + (a/6)[\bar{1}2\bar{1}]$, on which {111} plane must the partial dislocations lie?

SOLUTION

The normal to the plane must be at 90° to both Burgers vectors, so it can be found by taking the cross product, $[\bar{2}11] \times [\bar{1}2\bar{1}] = -3, -3, -3$ or (111).

An alternative way is to try all of the {111} possibilities and see which has a zero dot product with both $[\bar{2}11]$ and $[\bar{1}2\bar{1}]$.

EXAMPLE PROBLEM 8.5

Into which other pair of $(a/6)\langle 211 \rangle$ partials could the $(a/2)[\bar{1}10]$ dissociate? On what plane would they lie?

SOLUTION

By inspection, $(a/2)[\bar{1}10] \rightarrow (a/6)[\bar{2}1\bar{1}] + (a/6)[\bar{1}21]$. The $(a/6)[\bar{2}1\bar{1}]$ and $(a/6)[\bar{1}21]$ partials must lie in a $(11\bar{1})$. Note that $[\bar{2}1\bar{1}]\cdot[11\bar{1}] = 0$ and $[\bar{1}21]\cdot[11\bar{1}] = 0$.

8.8   STACKING FAULTS

If a single $(a/6)\langle 112 \rangle$ partial dislocation passes through an fcc crystal, it leaves behind a region in which the sequence of stacking of the close-packed {111} planes does not correspond to the normal fcc lattice. The third {111} plane is over neither the first nor the second {111} plane. Figure 8.15 shows that a $(a/6)\langle 112 \rangle$ partial dislocation changes the position of the third plane, so it is directly over the first. The normal stacking order in fcc and hcp lattices is shown in Figure 8.16. In Figure 8.17, the

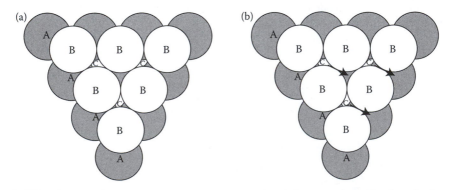

FIGURE 8.15   (a) Normal stacking of {111} planes in an fcc crystal can be described as ABCABC. (b) When a $(a/6)<112>$ partial passes through the crystal, the stacking sequence is changed to ABABC.

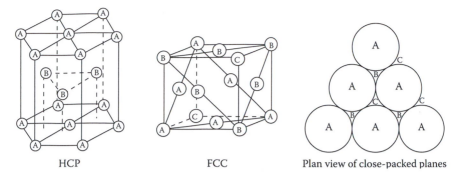

HCP   FCC   Plan view of close-packed planes

FIGURE 8.16 Stacking of close-packed planes in fcc and hcp crystals.

stacking order near a stacking fault in an fcc crystal is compared with the stacking in fcc and hcp lattices and near a twin boundary in an fcc crystal. There is a surface energy or surface tension associated with a stacking fault.

The packing sequence near a stacking fault in an fcc crystal is similar to the packing sequence in the hcp lattice. Since this is not the equilibrium structure of fcc, the stacking fault raises the energy and the increase of energy depends directly on the area of the fault. The stacking fault energy (SFE), γ_{SF}, is the energy/area of fault and can be regarded as a surface tension pulling the partial dislocations together. A stacking fault has twice as many incorrect second-nearest neighbors as a twin boundary. The similarity of the packing sequences at a twin boundary and at a stacking fault is clear in Figure 8.17. For most metals, the SFE is about twice the twin boundary energy as shown in Figure 8.18. The frequency of annealing twins is much higher in fcc metals of low SFE (Ag, brass, Al–bronze, and γ-stainless) than in those with higher SFE (Cu and Au). Annealing twins in the microstructures of aluminum alloys are rare. Table 8.1 lists the values of the SFE for a few fcc metals. The values of γ_{SF} for brass (Cu–Zn), aluminum–bronze (Cu–Al), and austenitic stainless steel are still lower than the value for Ag. In copper-base alloys, the SFE decreases as the electron-to-atom ratio increases.

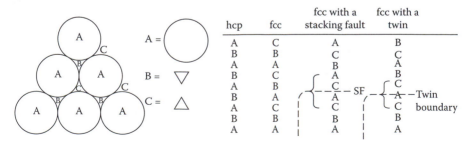

FIGURE 8.17 Stacking of close packed planes.

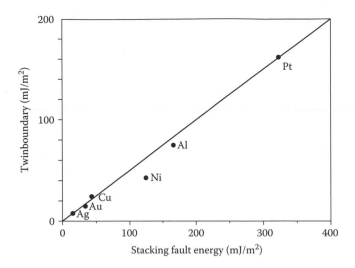

FIGURE 8.18 Relationship of SFE and twin boundary energy. The line represents $\gamma_{\text{stacking fault}} = 2\gamma_{\text{twin}}$.

The stacking fault between two $(a/6)\langle 112 \rangle$ partial dislocations is illustrated in Figure 8.19. The mutual repulsion of two partials tends to drive them away from each other. However, balancing this mutual repulsion is the attraction that results from the surface tension of the stacking fault between them (Figure 8.20).

EXAMPLE PROBLEM 8.6

Use Equation 8.13 to make an approximate calculation of equilibrium separation of the partial dislocations. Neglect the fact that both partials cannot be pure screws.

Solution

Taking $\mathbf{b}_1 = (a/6)[\bar{2}10]$ and $\mathbf{b}_2 = (a/6)[\bar{1}2\bar{1}]$, $\mathbf{b}_1 \cdot \mathbf{b}_2 = (a^2/36)(\bar{2} \cdot \bar{1} + 1 \cdot 2 + \bar{1} \cdot 1) = (a^2/12)$. Substituting into Equation 8.13, $f_L = -Ga^2/24\pi r$. The attractive force per length tending to pull the partials together is the SFE, g. At equilibrium, $\gamma + f_L = 0$.

$$r = \frac{Ga^2}{24\pi\gamma}. \tag{8.16}$$

TABLE 8.1

Stacking Fault Energies of Several fcc Metals (in mJ/m²)

Ag	Al	Au	Cu	Ni	Pd	Pt	Rh	Ir
16	166	32	45	125	180	322	750	300

Source: Adapted from listing in J.P. Hirth and J. Lothe, *Theory of Dislocations*, 2nd ed., Wiley, NY, 1982.

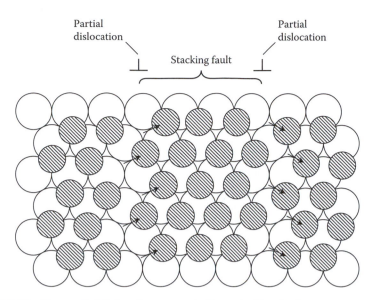

FIGURE 8.19   Two partial dislocations separated by a stacking fault.

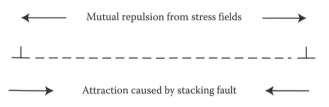

FIGURE 8.20   Equilibrium spacing between two partial dislocations corresponds to the separation at which the mutual repulsion and attraction balance.

Hirth and Lothe (1982) give a more exact relation for the force between two parallel dislocations that takes into account the angles between the Burgers vectors and the dislocation lines.

8.9   CROSS SLIP

A dislocation can move conservatively only on a plane that contains both the line of the dislocation and its Burgers vector. This limits it motion to a single plane unless the line of the dislocation and its Burgers vector are parallel. In this case, it is a screw dislocation and can change plane to avoid obstacles, a process called cross slip. Figure 8.21 illustrates cross slip.

However, dislocations that are dissociated into partials must recombine before they can dissociate onto the cross-slip plane as shown in Figure 8.22. If the SFE is very low, the partials will be widely separated and require a greater stress to cause recombination than if the SFE is high. Hence, alloys like brass with a very low SFE

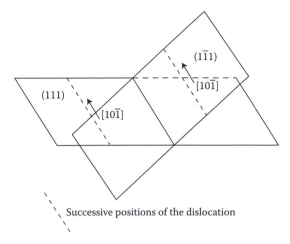

FIGURE 8.21 Cross slip. A screw dislocation with a Burgers vector parallel to $[10\bar{1}]$ can slip on either a (111) or a $(1\bar{1}1)$ plane.

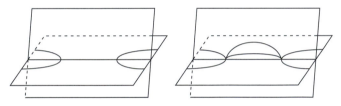

FIGURE 8.22 Cross slip of a dissociated screw dislocation. It must first recombine and then dissociate onto the cross-slip plane.

cross slip much less than high stacking fault metals like aluminum. As a consequence, slip lines in metals of high stacking fault, like aluminum, appear wavy, whereas slip lines in metals of low stacking fault are very straight. Figure 8.23 illustrates the difference.

8.10　MISCELLANY

8.10.1　POSTULATION OF DISLOCATIONS

Calculations of the strength of crystals by Frenkel (1926) showed the great discrepancy between theory and experimental observation. It was not until 1934 that three men, Taylor (1934), Orowan (1934), and Polanyi (1934) more-or-less independently proposed edge dislocations to explain the difference between σ_{theor} and σ_{exper}. The screw dislocation was postulated later by Burgers (1939). Orowan's interest in why crystals were as weak as they are was stimulated by an accident during research for his bachelor's degree at the Technische Universität Berlin. He dropped his only zinc single crystal on the floor. Rather than quit, he straightened the crystal and then tested it in creep. He later commented that reflection on the nature of the results contributed to his postulation of dislocations.

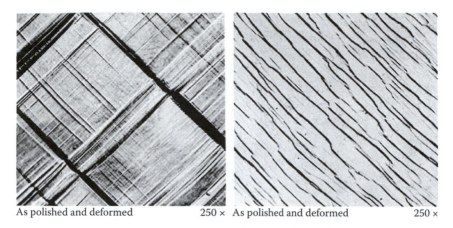

As polished and deformed   250 × As polished and deformed   250 ×

FIGURE 8.23 Straight parallel slip lines in an alloy of low SFE (left). Wavy slip lines in aluminum resulting from frequent cross slip (right). (From G. Y. Chin, *Metals Handbook*, Vol. 8, 8th ed. With permission.)

Without dislocations, crystal growth would require periodically the nucleation of a new plane of atoms. Theoretical physical chemists had come to the conclusion that any reasonable model based on nucleation of a new plane would predict crystal growth rates that were unreasonably slow. The postulation of screw dislocations removed this dilemma because nucleation of new planes would not be required.

8.10.2 DISCOVERY OF THE STRENGTH OF WHISKERS

As part of an investigation, in 1952 at the Bell Telephone Laboratories, of the failure of some electrical condensers, Treuting found whiskers of tin (about 2×10^{-6} m diameter) growing from the condenser walls. What distinguished this investigation from earlier findings of metal whiskers is that these whiskers were tested in bending (Herring and Galt, 1952). The surprising finding was that these whiskers could be bent to a strain of 2–3% without plastic deformation. This meant that the yield strength was more than 2.5% of the Young's modulus. For tin, $E = 16.8 \times 10^6$ psi, so $Y > 330,000$ psi. Thus overnight, tin became the strongest material known to man. Once the Herring–Galt observation was reported, many others began testing whiskers of other metals such as copper and iron with similar results and tin lost this distinction (Figure 8.24).

PROBLEMS

1. A crystal of aluminum contains 10^{12} m of dislocations/m³.
   a. Calculate the total amount of energy associated with dislocations/m³. Assume that half of the dislocations are edges and half are screws.
   b. If all of this energy could be released as heat, what would be the temperature rise?
   Data for aluminum: atomic diameter = 0.286 nm, crystal structure = fcc, density = 2.70 Mg/m³, atomic mass = 27 g/mol, $C = 0.90$ J/g°C, $G = 70$ GPa, and $\upsilon = 0.3$.

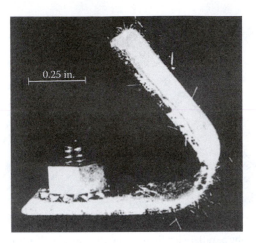

FIGURE 8.24 Tin whiskers growing on tin-plated steel. (From W. C. Ellis, D. F. Gibbons, and R. G. Treuting, in *Growth and Perfection of Crystals*, Doremus, Roberts, and Turnbull, eds, Wiley, 1958. With permission.)

2. Calculate the average spacing between dislocations in a 1/2° tilt boundary in aluminum. Look up any required data.

3. On what {110} planes of bcc iron can a dislocation with a Burgers vector $(a/2)[11\bar{1}]$ move?

4. A single crystal of aluminum was stretched in tension. Early in the test the specimen was removed from the testing machine and examined at high magnification. The distance between slip lines was found to be 100 μm and the average offset at each slip line was approximately 500 nm. Assume for simplicity that both the slip direction and the slip plane normal are oriented at 45° to the tensile axis as shown in Figure 8.25.

   a. On the average, how many dislocations must have emerged from the crystal at each observable slip line?

   b. What was the shear strain on the slip system, calculated over the whole crystal?

   c. What tensile strain (measured along the tensile axis) must have occurred? (i.e., What was the percentage elongation when the test was stopped?)

5. Consider the reactions between parallel dislocations given below. In each case, write the Burgers vector of the product dislocation and determine whether the reaction is energetically favorable.

   a. $(a/2)[1\bar{1}0] + (a/2)[110] \rightarrow$

   b. $(a/2)[101] + (a/2)[01\bar{1}] \rightarrow$

   c. $(a/2)[1\bar{1}0] + (a/2)[101] \rightarrow$

6. Consider the dislocation dissociation reaction $(a/2)[110] \rightarrow (a/6)[21\bar{1}] + (a/6)[121]$ in an fcc crystal. Assume that the energy/length of a dislocation is given by $E_L = Gb^2$ and neglect any dependence of the energy on the edge versus screw nature of the dislocation. Assume that this reaction occurs and the partial dislocations move very far apart.

   a. Express the total decrease in energy/length of the original $(a/2)[110]$ dislocation in terms of a and G.

   b. On which {111} plane must these dislocations lie?

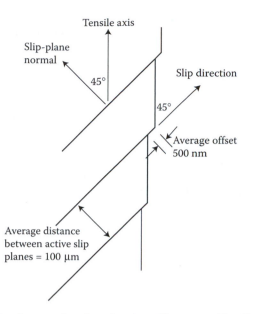

Tensile axis

Slip-plane normal

45°

Slip direction

45°

Average offset
500 nm

Average distance
between active slip
planes = 100 μm

FIGURE 8.25 Profile of a crystal surface showing offsets caused by slip.

7. A dislocation in an fcc crystal with a Burgers vector, $\mathbf{b} = (a/2)[0\ 1\ 1]$, dissociates on the $(1\bar{1}1)$ plane into two partial dislocations of the $(a/6)\langle 211\rangle$ type.
   a. Give the specific indices of the two $(a/6)\langle 211\rangle$ partial dislocations.
   b. Onto what other plane of the $\{111\}$ family could the $\mathbf{b} = (a/2)[011]$ dislocation have dissociated?
   c. Give the specific indices of the two $(a/6)\langle 211\rangle$ partials that would be formed if the $(a/2)[011]$ dislocation had dissociated on the plane in B.
8. Consider a crystal under a shear stress, τ, that has a circular dislocation loop of diameter, d, so that the region inside the circle has slipped relative to the material outside the circle. The presence of this dislocation increases the energy of the crystal by the dislocation energy/length times the length of the dislocation. The slip that occurs because of the formation of the dislocation under the stress, τ, lowers the energy of the system by τAb (τA is the shear force and b is the distance the force works through). If the diameter of the loop is small, the energy will be reduced if the loop shrinks. If the loop is large enough the loop will spontaneously expand. Find the diameter of critical size loop in terms of b, G, and τ. For simplicity take by $E_L = Gb^2$.
9. Referring to Figure 8.17, find the ratio of the wrong second-nearest neighbors across a stacking fault to the number across a twin boundary. If the surface energies are proportional to the number of wrong second-nearest neighbors, what is γ_{SF}/γ_{TB}?

REFERENCES

J. M. Burgers, *Proc. Kon. Ned. Akad. Wetenschap.*, 42, 378, 1939.
A. H. Cottrell, *Dislocations and Plastic Flow in Crystals*, Oxford, 1953.

G. Y. Chin, *Metals Handbook*, Vol. 8, 8th ed.

W. C. Ellis, D. F. Gibbons, and R. G. Treuting, in *Growth and Perfection of Crystals*, Doremus, Roberts, and Turnbull eds, Wiley, 1958.

J. Frenkel, *Z. Phys.* 37, 572, 1926.

A. G. Guy, *Elements of Physical Metallurgy*, Addison-Wesley, 1951.

C. Herring and J. K. Galt, *Phys. Rev.* 85, 1060, 1952.

J. P. Hirth and J. Lothe, *Theory of Dislocations*, 2nd ed., Wiley, NY, 1982.

D. Hull and D. J. Bacon, *Introduction to Dislocations*, 3rd ed., Butterworth, Heinemann, 1997.

E. Orowan, *Z. Phys.* 89, 604, 634, 1934.

M. Polanyi, *Z. Phys.* 89, 660, 1934.

W. T. Read, *Dislocations in Crystals*, McGraw-Hill, 1953.

G. I. Taylor, *Proc. Roy. Soc. A* 145, 362–387, 1934.

J. Weertman and J. R. Weertman, *Elementary Dislocation Theory*, Oxford, 1992.

9 Annealing

9.1 GENERAL

Annealing is heating of metal after it has been cold worked to soften it. Most of the energy expended in cold work is released as heat during the deformation. However, a small percent is stored by dislocations and vacancies (Figure 9.1). The stored energy is the driving force for the changes during annealing. There are three stages of annealing. In order of increasing time and temperature, they are

1. Recovery—often a small drop in hardness, rearrangement of dislocations to form subgrains. Otherwise overall grain shape and orientation remain unchanged.
2. Recrystallization—replacement of cold-worked grains by new ones. There are new orientations, a new grain size, and a new grain shape (but not necessarily equiaxed). Recrystallization causes the major hardness decrease.
3. Grain growth—growth of recrystallized grains at the expense of other recrystallized grains.

9.2 RECOVERY

The energy release during recovery is largely due to annealing out of point defects and rearrangement of dislocations. Most of the increase of electrical resistivity during cold work is attributable to vacancies. These anneal out during recovery, so the electrical resistivity drops (Figure 9.2) before any major hardness changes occur. During recovery, residual stresses are relieved and this decreases the energy stored as elastic strains. The changes during recovery cause no changes in microstructure that would be observable under a light microscope. Figure 9.3 shows the energy release and the changes of resistivity and hardness with increasing annealing temperatures.

There is also a rearrangement of dislocations into lower energy configurations such as low-angle tilt boundaries (Figure 8.10). In single crystals, these boundaries lead to a condition called *polygonization*, in which a bent crystal becomes facetted (Figure 9.4).

In polycrystals, these low-angle boundaries form subgrains that differ in orientation by only a degree or two. Figure 9.5 shows the subgrains formed in iron.

There may be subgrain coalescence by rotation as illustrated in Figure 9.6.

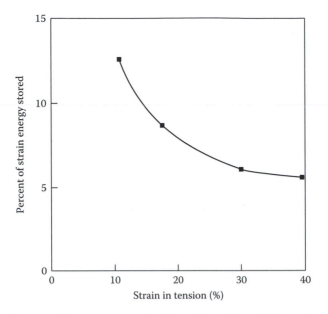

FIGURE 9.1   Percent of the energy expended in cold working of high-purity copper that is stored as lattice defects. (Data from P. Gordon, *Trans. AIME*, 203, 1043, 1955.)

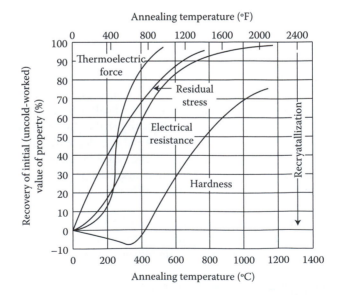

FIGURE 9.2   Property changes in tungsten during recovery. Note that the electrical conductivity improves before any major change in hardness. (From A. Guy and J. Hren, *Elements of Physical Metallurgy*, p. 439, 3rd ed., Addison-Wesley, 1974. With permission.)

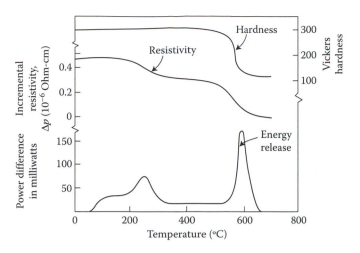

FIGURE 9.3 Annealing of cold-worked nickel. Energy release during annealing (bottom) and changes in resistivity and hardness (top). The first energy release peak is associated with recovery and the second with recrystallization.

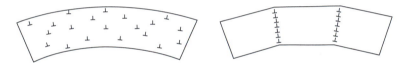

FIGURE 9.4 Alignment of low-angle grain boundaries in a bent single crystal causes polygonization.

FIGURE 9.5 Subgrain formation in iron during recovery. (From W. C. Leslie, J. T. Michalak, and F. W. Auk, AIME Conf. Series, *Iron and its Dilute Solutions*, p. 161. Interscience, 1963. With permission.)

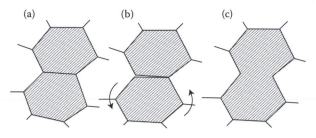

FIGURE 9.6 Subgrain coalescence by rotation. (From W. C. Leslie, J. T. Michalak, and F. W. Auk, AIME Conf. Series, *Iron and its Dilute Solutions*, p. 161. Interscience, 1963. With permission.)

In summary, during recovery, there is a large decrease in the number of point defects as they diffuse to edge dislocations and grain boundaries. Dislocations rearrange themselves into lower energy configurations leading to the formation of subgrains. Enough creep occurs to reduce elastic strains. There are major changes in electrical properties, some softening, and a reduction of residual stresses, but no changes of microstructure.

9.3 RECRYSTALLIZATION

During recrystallization, cold-worked grains having a high dislocation density are replaced by new grains having a much lower dislocation density. It is the decrease of energy associated with dislocations that is the driving force for recrystallization. Figure 9.7 shows the progress of recrystallization at 310°C of aluminum that had been cold-worked 5%.

Figure 9.8 is a schematic plot of the fraction of the microstructure that has been recrystallized for a constant temperature anneal. The rate of course depends on the temperature as shown in Figure 9.9 for high-purity copper.

It is common to define a recrystallization temperature as the temperature at which the microstructure will be 50% recrystallized in 30 min. As a rule of thumb, the recrystallization temperature for a metal of commercial purity is in the range of 1/3–1/2 of its absolute melting point (Figure 9.10). However, the recrystallization

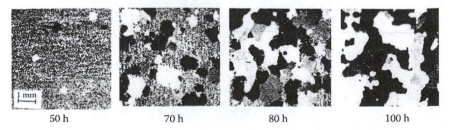

| 50 h | 70 h | 80 h | 100 h |

FIGURE 9.7 Progress of recrystallization at 310°C of aluminum that had been cold-worked 5%. (From W. C. Leslie, J. T. Michalak, and F. W. Auk, AIME Conf. Series, *Iron and its Dilute Solutions*, p. 161. Interscience, 1963. With permission.)

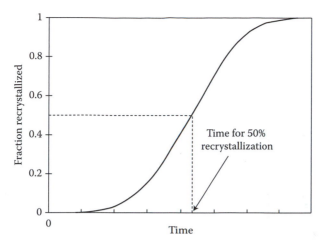

FIGURE 9.8 Schematic plot of how the recrystallized fraction of the microstructure changes with time for an isothermal anneal. (From A. Guy and J. Hren, *Elements of Physical Metallurgy*, p. 439, 3rd ed., Addison-Wesley, 1974. With permission.)

temperature is affected by purity, the amount of cold work, and the prior grain size. Solid solution impurities tend to segregate at grain boundaries and therefore slow their motion, thus raising the recrystallization temperature.

The effects of purity, cold work, and prior grain size can be understood in terms of how the nucleation of new grains and their growth are affected. Increasing the amount of cold work tends to increase both the nucleation rate and the rate of growth of new grains, and therefore lowers the recrystallization temperature (Figure 9.11).

Impurities that are present as second phases also retard grain boundary movement thereby raising the recrystallization temperature. The grain size before cold working affects the number of sites for nucleation. With finer grain sizes, there are more nucleation sites and therefore a lower recrystallization temperature (Figure 9.12).

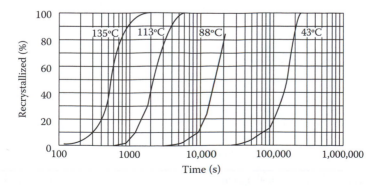

FIGURE 9.9 Isothermal recrystallization of 99.999% pure copper cold-worked 98%.

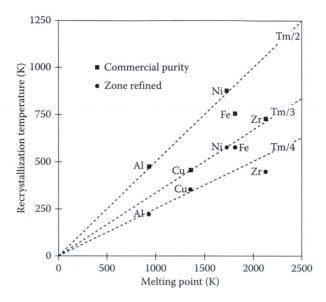

FIGURE 9.10 For most commercial metals the recrystallization temperature is between 1/3 and 1/2 of the absolute melting point. For zone-refined metal, the ratio of recrystallization temperature to melting point is nearer 1/4. (Data from O. Dimitrov, et al. in *Recrystallization of Metallic Materials*, Riederer-Verlag, GMBH, 1978.)

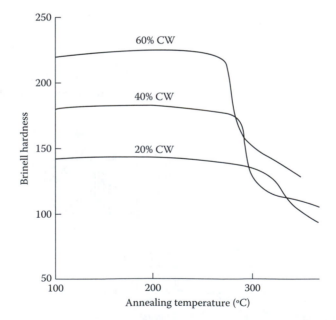

FIGURE 9.11 Effect of annealing temperature on cold-worked brass. Note that the sharp drop in hardness associated with recrystallization occurs at lower temperatures for material that is more heavily cold worked.

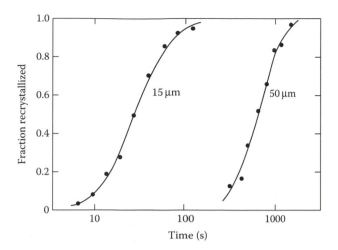

FIGURE 9.12 Kinetics of recrystallization of copper, cold-rolled 93% at 225°C. The material with the finer grain size recrystallizes much faster.

These same variables also affect the grain size that results from recrystallization. Increasing the amount of cold work increases both the nucleation rate and the growth rate. However, the nucleation rate is increased more than the growth rate; hence increased cold work decreases the grain size that results from recrystallization. Figure 9.13 shows this effect. Below about 5% cold work, recrystallization did not occur.

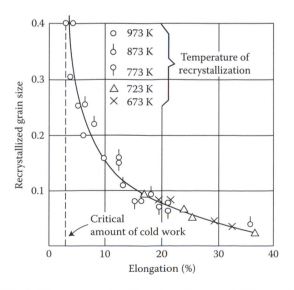

FIGURE 9.13 Effect of the amount of cold work on the recrystallized grain size of copper. Below about 3% cold work, recrystallization did not occur. Note that the temperature at which recrystallization did occur had little effect on the recrystallized grain size. (From W. B. Hutchinson, S. Jonsson, and L. Ryde, *Scripta Met.* 23, 671, 1989. With permission.)

Impurities slow down growth, thereby allowing more time for nucleation, so the recrystallized grain size is smaller. The increased nucleation sites with finer prior grain size results in a finer recrystallized grain size.

In summary

1. The recrystallization temperature goes down with increased purity, increased cold work, and finer prior grain size.
2. The recrystallized grain size decreases with increased cold work and finer prior grain size and increases with purity.

The kinetics of recrystallization follow an Arrhenius behavior. The time for a fixed amount of recrystallization (e.g., $f = 50\%$) is given by

$$t = A \exp\left(\frac{Q}{RT}\right), \tag{9.1}$$

where A is a constant and the activation energy, Q, is very nearly equal to that for self-diffusion.

Figure 9.14 is a plot of the time for 50% recrystallization of the copper in Figure 9.10 as a function of reciprocal temperature.

EXAMPLE PROBLEM 9.1

Using Figure 9.14, find the activation energy for recrystallization of copper.

SOLUTION

Since $t = A \exp(Q/RT)$, $t_2/t_1 = A \exp[(Q/R)(1/T_2 - 1/T_1)$,
$Q = R \ln(t_2/t_1)/(1/T_2 - 1/T_1)$. Taking points at the extreme ends of the line in Figure 9.15,
$t_2 = 40,000$ min at $1/T_2 = 0.0032$ and $t_1 = 8$ min at $1/T_1 = 0.0024$. Substituting,
$Q = 88,500$ J/mol.

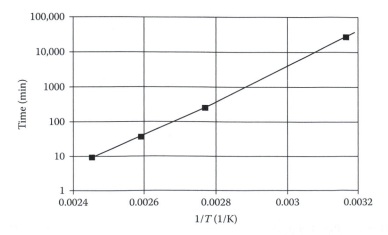

FIGURE 9.14 Arrhenius plot of the data in Figure 9.9. The slope equals Q/R. (From J. S. Smart and A. A. Smith, *Trans. AIME*, 152, 103, 1943. With permission.)

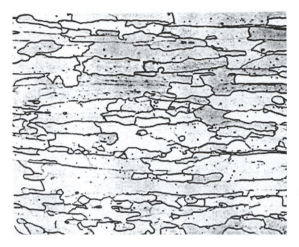

FIGURE 9.15   Microstructure of cold-rolled recrystallized brass. These very recrystallized grains are not equiaxed.

Two fables must be dispelled: (1) Recrystallized grain structures are rarely if ever equiaxed. Alignment of second-phase particles can restrict growth in some direction. Figure 9.15 shows the microstructure of cold-worked and recrystallized Mo–0.5Ti. (2) The orientations of recrystallized grains are not random. Figure 9.16

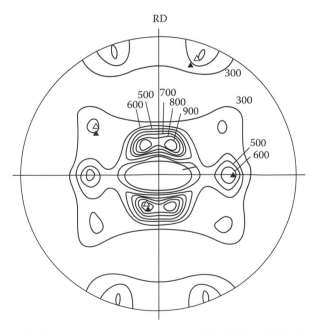

FIGURE 9.16   Pole figure of cold-rolled and recrystallized Mo–0.5Ti. Note that the strong crystallographic texture. (From *Metals Handbook*, Vol. 7, 8th ed., p. 199, ASM, 1972. With permission.)

is a pole figure of rolled sheet of brass that has been recrystallized. It shows a very strong texture.

9.4 AVRAMI KINETICS

Avrami kinetics can be applied to recrystallization. The Johnson–Mehl equation assumes a constant growth rate, G, in three dimensions and constant nucleation rate, $\dot{N}$. The fraction recrystallized, f, is then

$$f = 1 - \exp\left[-\frac{\pi}{3}\dot{N}G^3t^4\right]. \tag{9.2}$$

The volume of a spherical particle is $v = (4/3)\pi r^3$. If growth rate, G, is constant and starts at $t = \tau$, $v = (4/3)\pi[G(t - t)]^3$. If the faction transformed, f, is small, $f = \int (4/3) \pi\dot{N}[G(t - t)]^3 \, dt = (\pi/3)\dot{N}G^3t^4$. However, if f is large, both nucleation and growth have reduced volume, $1/(1 - f)$, in which they can occur. Inclusion of this leads to the Johnson–Mehl equation, $f = 1 - \exp[-(\pi/3)\dot{N}G^3t^4]$, so plot of $\ln\{\ln(1 - f)\}$ versus $\ln t$ should be a straight line of slope 4. However, often the slope is $\neq 4$ because the nucleation rate is not constant or growth is restricted to one or two dimensions instead of three.

Avrami generalized this equation to

$$f = 1 - \exp(-bt^n) \tag{9.3}$$

to account for these possibilities. Figure 9.17 illustrates schematically nucleation site saturation. If nucleation occurs only at grain boundaries (or grain corners) all of the possible nucleation sites will be used up quickly, so nucleation may stop. In this case, the contribution of nucleation to the Avrami exponent will be zero instead of one.

EXAMPLE PROBLEM 9.2

Find the value of n in the Avrami expression $f = 1 - \exp(-bt^n)$ that best fits the data for 135° in Figure 9.10.

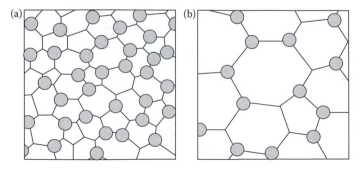

FIGURE 9.17 With a fine grain size, (a), there are many nucleation sites at the junctions of grain boundaries. However, with a large grain size, (b), there are fewer of these sites and they may become saturated. (From P. A. Beck and H. Hu in *Recrystallization, Grain Growth and Textures*, Margolin Ed., p. 393, ASM, 1966. With permission.)

Rearranging Equation 9.3, $\exp(-bt^n) = 1 - f$, $-bt^n = \ln(1 - f)$, Comparing two conditions, $(t_2/t_1)^n = \ln(1 - f_2)/\ln(1 - f_1)$, $n \ln(t_2/t_1) = \ln[\ln(1 - f_2)/\ln(1 - f_1)]$; so

$$n = \frac{\ln[\ln(1 - f_2)/\ln(1 - f_1)]}{\ln(t_2/t_1)}.$$

Taking two well-separated points on the line, $\ln(1 - f_2) = 0.91$ at $t_2 = 1000$ and $\ln(1 - f_1) = 0.10$ at $t_1 = 300$

$$n = \frac{\ln(\ln 0.91/\ln 0.90)}{\ln(1000/300)} = 3.02.$$

9.5 GRAIN GROWTH

In the grain growth stage of annealing, recrystallized grains grow at the expense of other recrystallized grains. The driving force for grain growth is the reduction of energy associated with grain boundaries. Important principles include

1. Boundaries tend to move toward center of curvature.
2. The dihedral angles between three grain boundaries average 120°.

Grain boundaries migrate toward their centers of curvature because by straightening they reduce their surface area. This can also be understood from a microscopic viewpoint. Atoms tend to move to positions with more correct near neighbors. This is equivalent to the boundary moving toward its center of curvature as shown in Figure 9.18.

The consequence of these two principles can be understood best in terms of a two-dimensional model of grains as illustrated schematically in Figure 9.19. If all of the grains were the same size, they would have six neighbors, so the 120° requirement

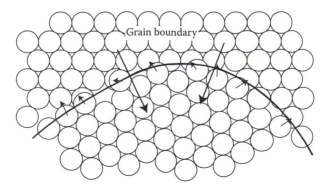

FIGURE 9.18 Movement of atoms to positions where they have a greater number of correct near neighbors. This is equivalent to the boundary moving toward its center of curvature.

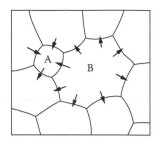

FIGURE 9.19 Large grains tend to be outwardly concave and therefore grow at the expense of their neighbors. Small grains tend to be inwardly concave and therefore shrink.

results in straight grain boundaries. A grain that is larger than its neighbors has more than six neighbors, so its boundaries outwardly concave.

Consequently, it will grow at the expense of its neighbors. A grain that is smaller than its neighbors has less than six neighbors, so its boundaries inwardly concave. It will shrink until it disappears. The result will be fewer grains, so the average grain size will increase.

To first approximation one would expect that the rate of migration would be proportional to curvature $(1/r)$ and that the average curvature of the grain boundaries would be proportional to the average grain diameter, D. In this case $dD/dt = a/D$, so

$$D^2 - D_0^2 = At. \tag{9.4}$$

The constant, A, ought to have temperature dependence of the Arrhenius form, so

$$D^2 - D_0^2 = Kt \exp\left(\frac{-Q}{RT}\right). \tag{9.5}$$

Figure 9.20 shows this temperature dependence. However, very often the kinetics is not so well behaved.

Empirically, the grain growth may be described by

$$D = kt^n, \tag{9.6}$$

where the exponent, n is not necessarily 1/2. Figure 9.21 shows this behavior. One cause of lower exponents is impurities in solid solution as shown in Figure 9.22.

Figure 9.23 shows that grain boundary migration rate depends on solute and misorientation. This orientation dependence is a major factor in the formation of recrystallization textures.

Another reason is that second-phase particles tend to pin grain boundaries. Consider a grain boundary that is pinned by a spherical inclusion as shown in Figure 9.24. The contact length between the boundary and the inclusion is $2\pi r \cos \theta$, where r is the radius of the inclusion and θ is the contact angle. This creates a drag force normal to the inclusion of $2\pi r \gamma \cos \theta$, where γ is the contact angle. This creates a drag

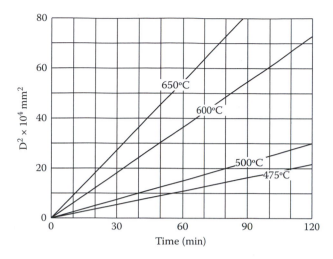

FIGURE 9.20 Grain growth of a 65% Cu–35% zinc brass. Note that D^2 is proportional to time as predicted by Equation 9.6. (Data from P. Feltham and G. J. Copley, *Acta Met.*, 6, 1958.)

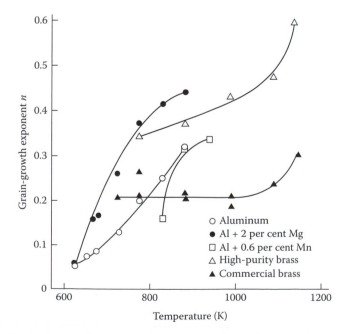

FIGURE 9.21 Grain growth exponent for several metals as a function of temperature. Note the deviation from Equation 9.7.

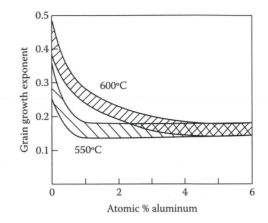

FIGURE 9.22 Dependence of the grain growth exponent on the amount of aluminum in solid solution in copper. (Data from S. Weinig and E. S. Machlin, *Trans. AIME*, 209, 844, 1957.)

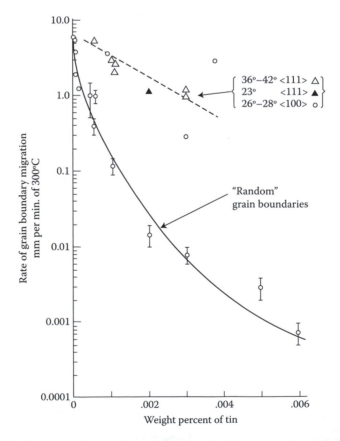

FIGURE 9.23 Rate of grain boundary migration in lead depends on the angle of misorientation as well as the amount of impurities. Special grain boundaries with certain angular relationships migrate much faster than normal grain boundaries. (From R. L. Fulman, *Metal Interfaces*, ASM, 1952, p. 187. With permission.)

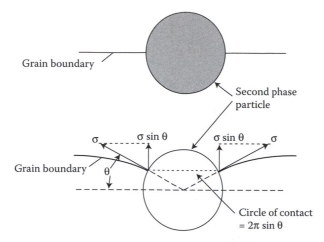

FIGURE 9.24 Drag of an inclusion on a grain boundary equals $\pi r\gamma$, where r is the particle radius and γ is the grain boundary surface energy.

force normal to the inclusion of $2\pi r\gamma \cos\theta$, where γ is the grain boundary surface energy. The component of this normal to the grain boundary is $F = 2\pi r\gamma \cos\theta \sin\theta$, which has a maximum value at $\theta = 45°$ of

$$F_{max} = \pi r\gamma_{gb}. \tag{9.7}$$

9.6 LIMITING GRAIN SIZE

Zener showed that second-phase particles limit the grain size during grain growth. Growth stops after a critical limiting size has been reached. The volume of a small spherical particle is $(4/3)\pi r^3$, so the number of particles/volume is $n_v = f/[(4/3)\pi r^3]$, where f is the volume fraction of second phase. The probability of a grain boundary being intersected by a given particle depends on its diameter, $D = 2r$, so the number of intersections/area is $n_A = 2rf/[(4/3)\pi r^3] = 3f/(2\pi r^2)$. Since the drag per particle equals $\pi r g_{gb}$ (Equation 9.7), the total drag force on a boundary per area should be $3r\gamma f/(2\pi r^2) = 3\gamma f/2r$. If it is assumed that driving force for growth force/area = $2\gamma/D$ (i.e., that the radius of curvature = the grain radius), $2\gamma/D = 3\gamma f/2r$. Growth will stagnate when the driving force for grain growth equals the drag force, so $3\gamma f/2r = 2\gamma/D_{max}$ or the limiting grain diameter is $\bar{D}$:

$$\bar{D}_{max} = \frac{(4/3)r}{f}. \tag{9.8}$$

The conclusion is that limiting grain size is proportional to r/f. This implies that the terminal grain size is smaller for a greater fraction of second phase and with smaller second-phase particles.

When metals are heated, grooves tend to form at grain boundaries at free surfaces as shown in Figure 9.25. These grooves can act to prevent or slow grain boundary movement in sheet metal, so the limiting grain size may be smaller than indicated by Equation 9.8.

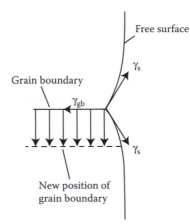

FIGURE 9.25 At high temperatures, grooves form at free surfaces to establish equilibrium between the grain boundary tensions. Grain boundary movement away from these grooves results in a greater length of boundary, so grooves retard grain boundary migration. (From J. W. Rutter and K. T. Aust, *Acta Met.* 13, 181, 1965. With permission.)

In wire, grain growth stops when grains occupy full diameter of grain, so the limiting grain size is roughly 1.5 times as long as the wire diameter. This is called a *bamboo* structure (Figure 9.26). In tungsten wires, grain boundaries are very brittle, so this must be avoided. To prevent this, fine particles of ThO_2 are incorporated into the microstructure.

The presence of inclusions can make a material very resistant to grain growth. Such materials are called *inherently fine grained*. However, the resistance to grain growth may breakdown at sufficiently high temperatures as shown in Figure 9.27. The result is *discontinuous grain growth* or *exaggerated grain growth* as illustrated in Figure 9.28. Dissolution of the second-phase particles or overcoming of their resistance allows a few grains to grow at the expense of others. This is sometimes

FIGURE 9.26 Bamboo structure of a wire. The grains occupy the full wire diameter.

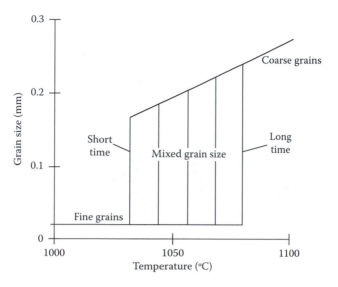

FIGURE 9.27 Inherently fine-grain steels (ones containing many fine particles) resist grain growth up to about 1050°C and then undergo exaggerated grain growth as the particle restrain is lost. Coarse grain steels (without many fine inclusions) undergo normal grain growth.

called *secondary recrystallization* and can lead to the formation of still new crystallographic textures.

The crystallographic texture is very important in silicon steels used for transformers. The recrystalization is (111)[110]. This means that the surfaces of the sheets are parallel to a {111} plane and the rolling direction is parallel to a ⟨110⟩ direction.

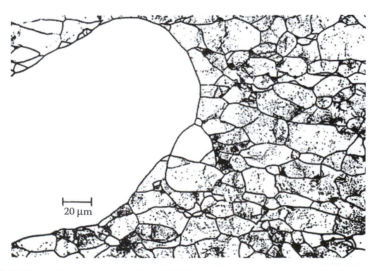

FIGURE 9.28 A large grain in an Fe–3% Si steel growing into finer-grained material. This is called *secondary recrystallization*, *exaggerated*, or *discontinuous grain growth*.

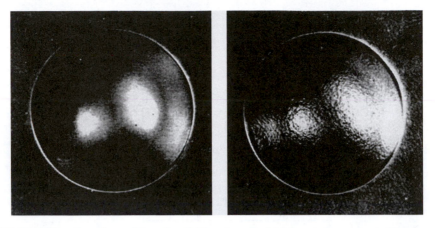

FIGURE 9.29 Cups formed from fine (left) and coarse (right) grain brass sheets. The orange peel on the coarse grain brass is apparent.

Secondary recrystallization produces a (110)[001] texture which is desired because $\langle 100 \rangle$ is the direction of easy magnetization.

There is a phenomenon known a tertiary recrystallization. The free energy of a surface is affected by the atmosphere. In experiments on silicon steel, the direction of boundary motion between grains with {100} and {110} surfaces could be reversed by changing the O_2 potential of the atmosphere.

There are a number of reasons for interest in grain size. One is the *orange peel* effect. Figure 9.29 shows the orange peel on a brass cup. When a metal is deformed, the surface becomes roughened on the scale of the grain size. If the grain size is large in a sheet that is stamped into automobile or appliance skin, this roughening cannot be covered with paint and will leave an undesirable appearance.

Coarse grains impart creep resistance at high temperatures. At low temperatures, fine grains result in higher hardness and toughness.

9.6.1 SUMMARY OF GRAIN GROWTH

Grain growth is the growth of already recrystallized grains at the expense of other recrystallized grains. The growth rate is retarded by impurities in solid solution and those present as second-phase particles. There is a limiting grain size that decreases with more particles. If the particles dissolve or otherwise cease to prevent grain growth, there may be a sudden increase of grain size.

9.7 CHAPTER SUMMARY

1. The driving forces for recovery are a decrease in vacancy concentration, energy reduction accompanying the realignment of dislocations into subgrain boundaries, and a reduction of stored elastic energy as stresses are relieved. During recovery subgrains formation, residual stresses are relieved

and there is a recovery of most of the original electrical conductivity. There may be some reduction of hardness.

2. The driving force for recrystallization is the reduction of energy as old grains with many dislocations are replaced by new grains with far fewer dislocations. New crystallographic textures are formed. There is a major drop in hardness.
3. The driving force for grain growth is the reduction of grain boundary surface energy. New textures are formed. Growth is limited by inclusions.

9.8 MISCELLANY

Benjamin Thompson (Count Rumford) as minister of war of the Kingdom of Bavaria noted a large increase of temperature during the boring of bronze cannons. He measured the temperature rise and with the known heat capacity of the bronze; he calculated the heat generated by machining. By equating this to the mechanical work done in machining, he was able to deduce the mechanical equivalent of heat. His value was a little too low, mainly because some of the plastic work done in machining is stored as dislocations in the chips.

PROBLEMS

1. Figure 9.30 shows the grains near a bullet hole in a sheet of tin. Explain why the grain size increases with distance from the hole and then suddenly decreases.
2. Assume that grains have the shape of Kelvin tetrakaidecahedra (See Appendix A1 for a description of a Kelvin tetrakaidecahedra). If, during recrystallization, new grains are nucleated at every corner and only at corners, what is the ratio of the number of recrystallized grains to old grains? What is the ratio of new grain diameter to old grain diameter?

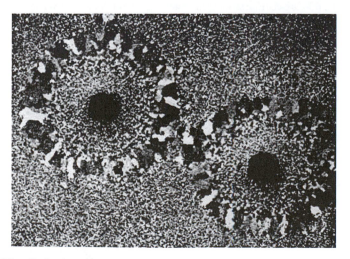

FIGURE 9.30 Grains in a tin sheet near a bullet hole revealed by etching several days after the bullet was shot. (From A. G. Guy and J. J. Hren, *Elements of Physical Metallurgy*, p. 439, 3rd ed., Addison-Wesley, 1974. With permission.)

3. It is desired to produce as large a grain diameter in aluminum as possible. Grains >1 cm diameter are required. What procedures would you recommend?

4. Two lots of a certain metal have been cold-rolled 20% and 60%. After annealing both at 400°C, the lot cold-rolled 60% is softer than the lot cold-rolled 20%. Explain.

   After annealing both at 500°C, the lot cold-rolled 20% is softer than the lot cold-rolled 60%. Explain.

5. Consider the annealing of copper in Figure 9.9.

   a. What is the recrystallization temperature (defined as the temperature for 50% recrystallization in 1/2 h)?

   b. Compare this with the general rule that the recrystallization temperature is between 1/2 and 1/3 of the melting point. Offer an explanation for any deviation from this rule.

   c. What is the activation energy for recrystallization?

   d. How long would it take for the copper to recrystallize at 25°C?

6. Figure 9.31 shows measurements of the heat released during recrystallization of high-purity copper. The area under the curve represents the energy/mol.

   a. Find the energy/mol in cal/mol and convert this to J/mol.

   b. If all of this heat went into heating the copper, by how much would the temperature rise?

   Data for copper: atomic weight = 63.54, density = 8.93 mg/m³, heat capacity = 386 J/kg K, and atomic diameter = 0.256 nm.

7. Examine the data plotted in Figure 9.9 and find the exponent in the Avrami equation that best fits the data.

8. Estimate the limiting grain size in a 1 cm thick plate of aluminum that contains a 1% volume fraction of spherical precipitates of a 500 nm diameter.

9. The surface energy of grain boundaries in copper has been measured to be 600 ergs/cm². Estimate the total grain boundary energy/volume (J/m³) in copper with a grain size of 500 μm.

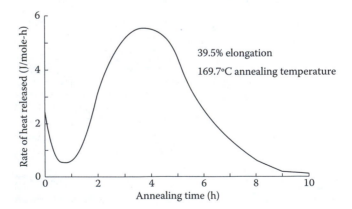

FIGURE 9.31 Heat released during an isothermal anneal of cold-worked copper. (From A. G. Guy and J. J. Hren, *Elements of Physical Metallurgy*, p. 439, 3rd ed., Addison-Wesley, 1974. With permission.)

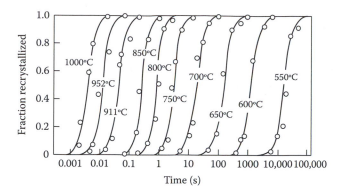

FIGURE 9.32 Recrystallization kinetics of Fe–3.5% Si after 60% cold work. (From J. Czochralski, in *Modern Metallkunde*, Berlin, 1924. With permission.)

10. Shown above in Figure 9.32 are data on the recrystallization kinetics of Fe-3.25% Si.
    a. Make an appropriate plot to determine whether this data can be represented by an Arrhenius equation.
    b. If does, determine the activation energy.
    c. Compare this with the activation energy for self-diffusion in iron.
    d. What is the recrystallization temperature (defined as the temperature for 50% recrystallization in 30 min)?
    e. Predict the time required for 50% recrystallization at 350°C.
11. Data on the recrystallization of copper with a 50 mm grain size at 225°C is given in Figure 9.12. Determine the Avrami exponent.
12. a. Using the data in Figure 9.10, determine the Avrami exponent.
    b. Use the times for 50% recrystallization to determine the activation energy for recrystallization of this copper.

REFERENCES

P. A. Beck and H. Hu in *Recrystallization, Grain Growth and Textures,* Margolin Ed., p. 393, ASM, 1966.

J. Czochralski, in *Modern Metallkunde*, Berlin, 1924.

B. F. Decker and D. Harker, *Trans. AIME*, 188, 1950.

O. Dimitrov, et al. in *Recrystallization of Metallic Materials*, Riederer-Verlag, GMBH, 1978.

P. Feltham and G. J. Copley, *Acta Met.*, 6, 1958.

R. L. Fulman, *Metal Interfaces*, p. 187, ASM, 1952.

P. Gordon, *Trans AIME*, 203, 1043, 1955.

A. G. Guy and J. J. Hren, *Elements of Physical Metallurgy*, p. 439, 3rd ed., Addison-Wesley, 1974.

F. J. Humphreys and M. Hatherly, *Recrystallization and Related Annealing Phenomena*, Pergamon, 1995.

W. C. Leslie, J. T. Michalak, and F. W. Auk, AIME Conf. Series, *Iron and its Dilute Solutions*, p. 161. Interscience, 1963.

Metals Handbook, Vol. 7, 8th ed., p. 199, ASM, 1972.

Recrystallization, Grain Growth and Textures, ASM, 1966.

Recrystallization, '90, TMS, 1990.

R. E. Reed-Hill and R. Abbaschian, *Physical Metallurgy Principles*, 3rd ed., PWS Publishing, 1994.

J. W. Rutter and K. T. *Aust, Acta Met*. 13, 181, 1965.

J. S. Smart and A. A. Smith, *Trans. AIME*, 152, 103, 1943.

G. R. Speich and R. M. Fisher, *Recrystallization, Grain Growth and Textures*, ASM, 1966.

S. Weinig and E. S. Machlin, *Trans. AIME*, 209, 844, 1957.

10 Phase Transformations

10.1 CLASSIFICATION OF SOLID–STATE PHASE TRANSFORMATIONS

Phase changes in solids may be classified as to whether nucleation and growth are required and as to whether compositional changes occur. Reactions that require nucleation and growth with composition change include

1. Precipitation reactions, $\alpha' \to \alpha + \beta$, where α' is a supersaturated solid solution.
2. Eutectoid transformation, $\alpha \to \beta + \gamma$.

Reactions that involve nucleation and growth without composition change are

3. Order-disorder, $\alpha \to \alpha'$.
4. Massive. This is the transformation of an alloy from one crystal structure to another, $\alpha \to \beta$. Usually the higher-temperature phase has greater symmetry.

Reactions that occur by nucleation and shear without compositional change are called

5. Martensitic transformations. There is no composition change and no growth.

Finally, phase changes can occur by growth without any nucleation stage. These reactions are

6. Spinodal, $\alpha \to \alpha' + \alpha''$. This involves the gradual separation of one phase into two phases.

10.2 MARTENSITIC TRANSFORMATIONS

The term *martensitic transformation* applies to phase transformations that occur *without* diffusion, by shearing of the lattice. The resulting martensite has the same composition as the phase from which it is formed. Lens-shaped regions shear, with the speed of sound, so that the atoms are in a new crystallographic arrangement. The composition is unchanged (Figure 10.1). If a spherical region underwent such a shear, its shape would be very different than that of the surrounding matrix, causing severe elastic strains in both the martensite and the surrounding matrix. This volume

137

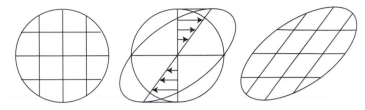

FIGURE 10.1 Schematic showing that shearing of a region creates a new crystal structure without compositional change.

of elastically strained material is greatly reduced if the shape of the martensite is lenticular (Figure 10.2).

While martensitic transformation is best known in steels, many other systems have martensitic transformations. Some examples are given in Table 10.1.

1. The martensite reaction in Fe–C alloys differs from that in most other systems in two important respects. In most of the systems, martensite is not very hard. The hardness of Fe–C martensite can be attributed to the martensite being a supersaturated solid solution of carbon in ferrite with so much carbon that the ferrite is distorted into a tetragonal lattice.

2. Most martensitic transformations are reversible (in the geometric sense) so that on heating the martensite transforms to the phase from which it forms by a reverse shear. The transformation is not reversible in a thermodynamic sense. (The temperature at which the reverse transformation occurs on heating is higher than the temperature for the martensite formation on cooling.) The result is a hysteresis on a plot of percent transformation (or some property) versus temperature. [The reason that there is no reversibility with iron–carbon martensites is that on heating iron–carbon martensites temper (decompose) to ferrite and iron carbide before the A_s temperature is reached.]

A typical plot showing this hysteresis in iron–nickel alloys is shown in Figure 10.3. On cooling martensite starts to form at $M_s = -30°C$, on heating the austenite starts forming at $A_s = 390°C$. Because martensite involves shear, applied stresses alter the transformation temperatures. The M_d and A_d are the temperatures of transformation while the material is being deformed. The temperature at which the free energies of the two phases are equal lies between the M_d and A_d temperatures.

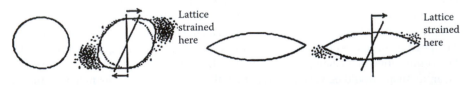

FIGURE 10.2 Volume of material that is elastically strained by the martensitic shear is much smaller if the shape of the martensite is lenticular rather than spherical.

TABLE 10.1
Systems with Martensitic Transformations

Material	Structure Change[a]
ZrO$_2$	tetr ll → mono
Fe–Ni alloys	fcc → bct
Co	fcc → hcp
Cu–23% Sn	cub → ortho
Cu–40% Zn	ordered bcc → fcc
Au–34% Cd	cub → ortho
Cu–13% Al	cub → ortho
In–30% Th	fcc → bct
Ti	bcc → hcp

[a] cub = cubic; tetr = tetragonal; hex = hexagonal; rhomb = rhombahedral; ortho = orthorhombic; mono = monoclinic, tric = triclinic.

There is a habit to martensitic transformations. Some (*hkl*) plane and some [*uvw*] direction in the martensite are parallel to some (*hkl*) plane and [*uvw*] direction, respectively, in the austenite. The transformations are very complex and we will not go into the geometric details. However, some insight can be gained by looking at the very simple picture presented by Bain (Figure 10.4). One fcc unit cell of austenite transforms into two unit cells of martensite. The dimension, *c*, of the martensite is less than the dimension, a_o, of the austenite and the dimension, *a*, of the martensite is larger than the dimension, a_o, of the austenite. If this were the total distortion,

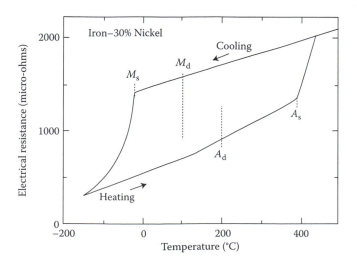

FIGURE 10.3 Hysteresis associated with the martensitic transformation in iron–nickel alloys. (After L. Kaufman and M. Cohen, *Trans. AIME*, 206, 1393, 1956.)

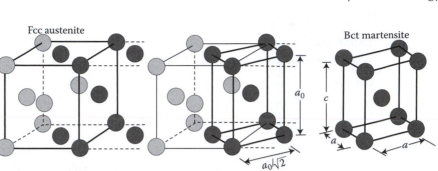

FIGURE 10.4 Relation between the unit cells of martensite and the austenite from which it forms.

$(100)_{mart}$ would be parallel to $(110)_{aust}$ and $[100]_{mart}$ would be parallel to $[110]_{aust}$, but the orientation relationships are more complicated because there are shears, which tend to minimize the distortions. The habit changes with carbon content because the lattice parameters and their changes are also carbon dependent. For 0.5–1.4% C, the relationships have been reported as $(111)_{aust}$ parallel to $(101)_{mart}$ and $[1\bar{1}10]_{aust}$ parallel to $[11\bar{1}]_{mart}$.

The shear strain associated with the transformation in Fe–C alloys is ~0.2. (The exact number depends on % C.) The shear causes a misfit between the martensite and the austenite, which is minimized when the martensite forms in the shape of discs or platelets (Figure 10.2). In some cases the martensite itself deforms, by twinning and/ or slip, so as to minimize the shear misfit.

There is usually a volume change associated with a martensitic transformation. Consequently, an increase of pressure will tend to favor the denser phase and shift the M_s and A_s correspondingly.

Martensite in iron–carbon steels has a body-centered tetragonal lattice. The tetragonality (ratio of c/a) depends on the carbon content. If a plot of c/a versus % C is extrapolated to 0% C, $c/a \to 1$ (bcc). The hardness of the martensite depends solely on % C (c/a) and a plot of hardness versus % C extrapolates to the hardness of ferrite as % C $\to$ 0. During tempering, carbides precipitate and there is a corresponding decrease of carbon dissolved in the martensite. Tempering usually causes a net softening because the loss of solid solution hardening has more of an effect than the precipitation hardening. High-speed tool steels are an exception.

> The geometric relations in martensitic transformations I must confess are such a mess, that they defy explanations.

10.3 SPINODAL DECOMPOSITION

Spinodal reactions may occur on cooling where there is a miscibility gap so that a single solid solution decomposes (breaks up) into two solid solutions as indicated in Figure 10.5. The corresponding free energy versus composition curves are shown in Figure 10.6. Consider alloy 1 cooled suddenly from the single-phase region (temperature 3) to temperature 1. This alloy can spontaneously segregate, forming some

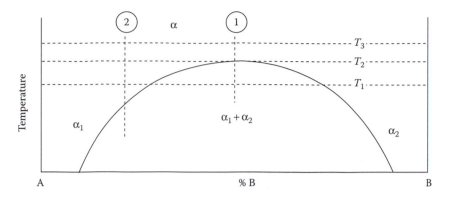

FIGURE 10.5 Phase diagram with a miscibility gap.

regions richer in B than the overall alloy and other regions, which are leaner in B. Such segregation will lower the free energy of the system (Figure 10.7) even though no discrete phase boundary is initially necessary. Because no new boundary area is required, there is no activation barrier to be overcome. Thus there is no nucleation stage in the usual sense of a thermally activated process. The single solid solution can start to break up into two phases without nucleation. Diffusion is, however, necessary. However, for alloy $\partial^2 G/\partial c^2 > 0$, such spinodal decomposition is not possible and nucleation of α_1 and α_2 is required.

Hence, spinodal reactions are only possible for composition/temperature combinations for which $\partial^2 G/\partial c^2$ is negative. Sometimes this region is shown by dotted lines on the phase diagram (Figure 10.8).

Figure 10.9 shows how the local composition changes during spinodal decomposition and compares it with the changes during precipitation by nucleation and growth. It is interesting to note that spinodal decomposition requires uphill diffusion. During spinodal decomposition, the boundary between the two phases sharpens rather than moves in contrast to growth, which involves the movement of an already sharp boundary between the two phases.

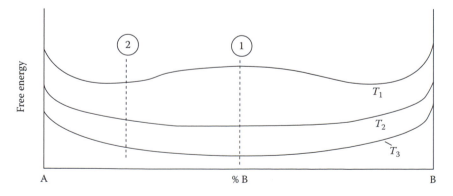

FIGURE 10.6 Free energy versus composition curves for the system in Figure 10.5 at three temperatures.

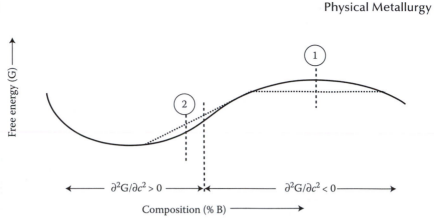

FIGURE 10.7 If alloy 1, for which $\partial^2 G/\partial c^2 < 0$, segregates into regions rich in B and lean in B, the free energy of the system is lowered. On the other hand, if alloy 2, for which $\partial^2 G/\partial c^2 > 0$, segregates, the free energy of the system increases.

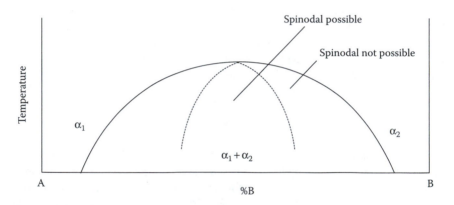

FIGURE 10.8 Phase diagram corresponding to Figure 10.7 showing regions in which spinodal reactions are not possible.

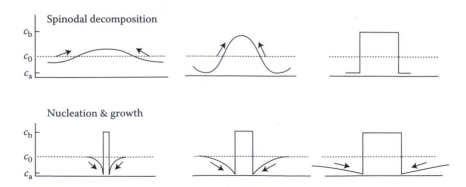

FIGURE 10.9 Comparison of spinodal decomposition and precipitation by nucleation and growth.

10.4   NUCLEATION KINETICS

There is an energy barrier to nucleation. During freezing, for homogeneous nucleation, $\Delta G = \Delta G_v \cdot (\text{vol}) + \gamma_{ls} \cdot (\text{surf area})$. For spherical particles this reduces to $\Delta G = \Delta G_v (4/3)\pi r^3 + \gamma_{ls} 4\pi r^2$. For a solid-state transformation, $\alpha \to \beta\beta$; however, there are usually additional terms. If there is a volume change (and there usually is), there will be a strain energy term, and, if nucleation occurs on a prior α–α grain boundary (it often does), there will be a term reflecting the decrease of grain boundary energy. Now the energy change in forming a small particle is

$$\Delta G = \Delta G_v (\text{vol})(\downarrow) + \gamma_{\alpha\beta}(\alpha\beta \, \text{surf area})(\uparrow) + \gamma_{gb}(\Delta \, \text{grain boundary area})(\downarrow)$$

$$+ \, \text{strain enegry term}\,(\uparrow) \tag{10.1}$$

Figure 10.10 illustrates these terms.

As with freezing, the ΔG_v term is roughly proportional to the degree of undercooling, so nucleation rates increase exponentially with undercooling.

10.5   GROWTH

In most cases, growth is controlled by diffusion.* Examples of diffusion-controlled growth include the transformation of austenite to pearlite. Here carbon must diffuse

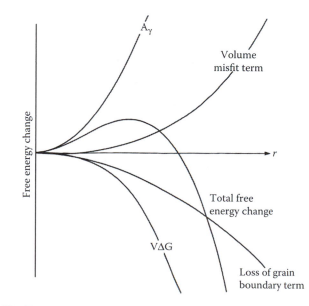

FIGURE 10.10   Energy terms in Equation 10.1 that are involved with nucleation in the solid state.

---

* This is in contrast to freezing, where the rate of freezing is usually controlled by rate of heat removal. One of the main differences is that the latent heat of freezing is very much greater than the latent heat of solid-state reactions.

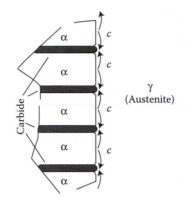

FIGURE 10.11 As a pearlite colony grows into austenite, the carbon atoms must diffuse to the growing carbide plates.

away from the growing ferrite platelets to the growing carbide platelets (Figure 10.11). Another example is the precipitation of a second phase, as in age hardening. Even if there is no net change of local composition as in the $\gamma-\alpha$ transformation in pure iron or during recrystallization, there is local reordering of atoms into a new crystal structure or a new crystal orientation, and this process is similar to diffusion. In any case, the rates of growth depend on the diffusivity and are therefore exponentially dependent on temperature.

10.6 OVERALL KINETICS

Since the transformation requires both nucleation and growth, the rate is fastest where neither is small. We will shortly see that the overall rate depends on $\dot{N}^a G^b$, where a and b depend on the details of the transformation. This leads to the familiar "C" curve of isothermal kinetics (Figure 10.12). Transformation at high temperatures with a high G and low $\dot{N}$ produces a course structure with few very large particles (or grains). On the other hand, transformation at very low temperatures produces fine microstructures with many small particles (or grains). Even where there is no net change of

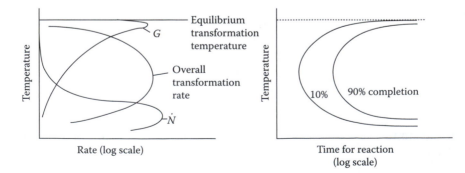

FIGURE 10.12 Temperature dependence of $\dot{N}$ and G, and the overall transformation rate (left) and the corresponding isothermal transformation curve (right).

local composition, as is the γ–α transformation in pure iron or in recrystallization, there is local reordering of atoms into a new crystal structure or crystal orientation, and this is similar to diffusion. In any case, the rates of growth depend on the diffusivity and are therefore exponentially dependent on temperature.

10.7 PRECIPITATION FROM A SUPERSATURATED SOLID SOLUTION

Figure 10.13 shows a typical phase diagram in which the solid solubility of a second phase, β, decreases with decreasing temperature. If an alloy of composition, c_o, is heated to temperature just below the eutectic temperature, the second phase will dissolve. If it is later held at a lower temperature, T_2, β can precipitate. The corresponding free energy composition diagram is shown in Figure 10.14.

Precipitation of β is governed by the rate of nucleation of β particles, $\dot{N}$, and their growth rate, $G = dr/dt$. The rate of nucleation increases as the temperature is lowered because the supersaturation increases.

10.8 PRECIPITATION HARDENING KINETICS

Figure 10.15 shows schematically the phase diagram of a precipitation-hardenable alloy. If after solution treatment the alloy of composition, c_{av}, is quenched to temperature and held, particles of β will precipitate out of α. The α-phase at the α–β boundary will have the composition c_α given by the phase diagram, while the composition of the α-phase remote from the particle will remain c_{av}. Thus there will be a concentration gradient dc/dx in the α-phase allowing diffusion of element B through the growing β and the growth rate of β is proportional to the diffusion flux, $J = -Ddc/dx$. However, as growth continues, the gradient near the β must decrease and so the rate of growth also decreases. In fact G is proportional to $t^{-1/2}$, so the contribution of each dimension of growth to the Avrami exponent is 1/2.

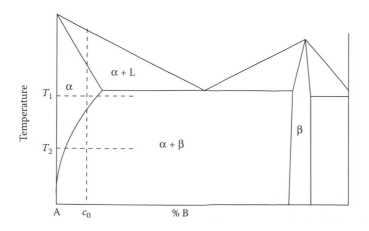

FIGURE 10.13 Schematic phase diagram to illustrate the precipitation of β in α.

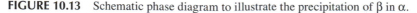

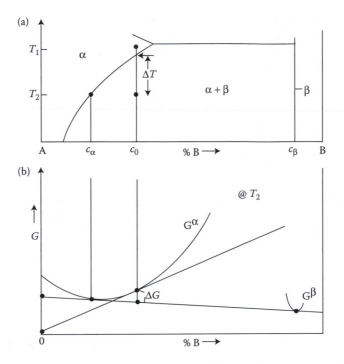

FIGURE 10.14 Illustration showing the decrease of free energy accompanying the precipitation of β. Phase diagram (a) and free energy diagram (b).

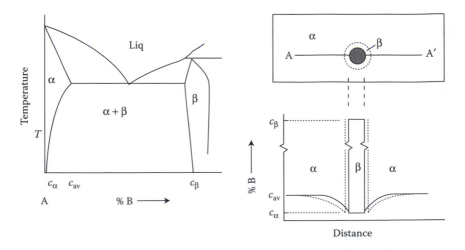

FIGURE 10.15 Phase diagram of A–B system (left) showing the alloy composition, c_{av}, and the compositions of α and β at the precipitation temperature, T. At the right is the composition profile through a β particle and the neighboring α. Particle growth requires diffusion of B through α. As the particle grows, the concentration gradient decreases.

10.9 PRECIPITATE GROWTH

Growth of β particles requires diffusion of B atoms to the particle. The concentration of B in the α-phase in equilibrium with β, c_α, is much lower than the average composition of α, c_o. Therefore, there is a concentration gradient, dc/dx, in the α as shown in Figure 10.16a. Fick's first law gives the diffusional flux as

$$J = -D\frac{dc}{dx}. \tag{10.2}$$

A mass balance requires that this flux equals $(c_\beta - c_o)dx/dt$, so

$$(c_\beta - c_o)\frac{dx}{dt} = -D\frac{dc}{dx}. \tag{10.3}$$

Figure 10.16b suggests a linear approximation to the concentration gradient,

$$\frac{-dc}{dx} = \frac{(c_o - c_\alpha)}{L}, \quad \text{but} \quad \left(\frac{1}{2}\right)L(c_o - c_\alpha) = x(c_\beta - c_o), \tag{10.4}$$

so $-dc/dx = [(1/2x)(c_o - c\alpha)^2/(c\beta - c_o)]$. Substituting,

$$(c_\beta - c_o)\frac{dx}{dt} = \left(\frac{1}{2x}\right)\frac{D(c_o - c_\alpha)^2}{(c_\beta - c_o)} \quad \text{or} \quad 2x\,dx = \left[\frac{(c_o - c_\alpha)^2}{(c_\beta - c_o)^2}\right]dt.$$

Integration gives

$$x = \left[\frac{c_o - c_\alpha}{c_\beta - c_\alpha}\right]\sqrt{(Dt)}. \tag{10.5}$$

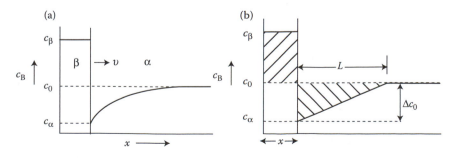

FIGURE 10.16 (a) The concentration gradient near a growing precipitate. (b) A linear approximation to the gradient.

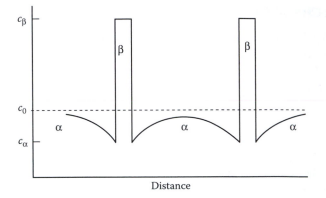

FIGURE 10.17   Overlapping of the diffusional fields of two particles slows the growth.

The growth rate is

$$G = \frac{dx}{dt} = \left(\frac{1}{2}\right)\left[\frac{c_0 - c_\alpha}{c_\beta - c_\alpha}\right]\sqrt{(Dt)}.$$ (10.6)

Thus the growth rate decreases with time and the particle radius is $\sqrt{(Dt)}$. The growth rate slows even more when diffusion fields overlap as shown in Figure 10.17.

10.10   PRECIPITATION KINETICS

As one might expect, precipitation can be regarded as a rate process and therefore the time, t, to peak hardness follows an Arrhenius equation:

$$t = A\exp\left(\frac{+Q}{RT}\right),$$ (10.7)

where A is a constant, Q is the activation energy, R is the gas constant, and T is the absolute temperature. The activation energy can be found by comparing the times to reach peak hardness at two different temperatures:

$$\left(\frac{t_2}{t_1}\right) = \exp\left[\left(\frac{Q}{R}\right)\left(\frac{1}{T_2} - \frac{1}{T_1}\right)\right], \quad \text{so}$$

$$Q = R\frac{\ln(t_2/t_1)}{(1/T_2 - 1/T_1)}.$$ (10.8)

10.11   STRAIN ENERGY

The strain energy associated with the formation of a particle depends on the fractional volume change, Δ, and the shape. Most particles can be characterized as spheroids having diameters a and c. Nabarro gives the elastic strain energy per volume as

$$\Delta G_S = \left(\frac{2}{3}\right)\mu\Delta^2 f\left(\frac{c}{a}\right),\tag{10.9}$$

where $f(c/a)$ is a maximum for spheres and minimum for thin discs. Figure 10.18 is a plot of Equation 10.9.

10.12 HARDENING MECHANISM

There are two causes of precipitation hardening. One, proposed by Orowan, involves the shear stress that is required to pass a dislocation between two particles, as sketched in Figure 10.19. The force on the dislocation segment that tends to force it between two particles is equal to $\tau b d$, where τ is the shear stress, d is the distance between the two particles, and $\mathbf{b}$ is the Burger's vector. Taking the line tension of the dislocation to be $G\mathbf{b}^2$, the restraining force is $2G\mathbf{b}^2 \sin\theta$, which has a maximum value of $2G\mathbf{b}^2$ when $\theta = 90°$. Equating, $\tau b d = 2G\mathbf{b}^2$, or the shear stress necessary to continue slip is

$$\tau = \frac{2G\mathbf{b}}{d}.\tag{10.10}$$

Thus the smaller the distance, d, between particles, the stronger and harder the material is. This explains why the hardness increases with time, as more precipitates particles form. When precipitation is nearly complete, however, the number of particles will start to decrease. The reason for this decrease is the decrease in total surface as the larger particles grow and the smaller ones shrink and disappear.

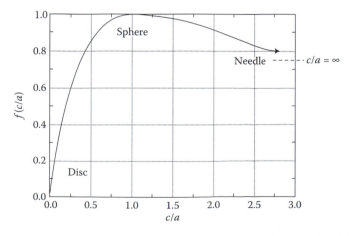

FIGURE 10.18 Relative strain energy associated with a spheroidal particle depends on its c/a ratio.

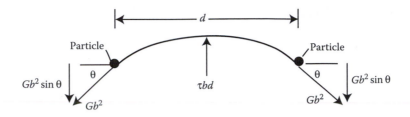

FIGURE 10.19   Force balance on a dislocation being forced between two particles.

The other possibility is that the particles are sheared. In this case the shear stress required for deformation is the volume average of the shear strengths of the matrix and the particles.

$$\tau_{ave} = f\text{ppt}\tau_{ppt} = (1 - f\text{ppt})\tau_{matrix} \tag{10.11}$$

10.13   OSTWALD RIPENING

As a microstructure of many fine particles is held for a long time at temperatures high enough for diffusion to occur, the smaller particles will dissolve and the larger ones will grow. This coarsening of the microstructure is called "Ostwald ripening." The driving force is the decrease of surface area and therefore surface energy. For the smaller particles to dissolve and the larger ones to grow, there must be diffusion of the solute. To understand how this occurs, consider the free energy versus composition curves for the α- and β-phases in the A–B system (Figure 10.20). Small β particles have a higher energy than large ones because of more surface area for the same volume. For small β particles, the tangent connecting the curves of α and β is tangent to the α curve at higher B concentrations. This means that there must be a diffusion of the solute and also that the solubility of B in small particles is higher than in large particles. This greater solubility creates a concentration gradient, dc/dx, between small and large particles, which causes B atoms to diffuse from the small particles to the large ones.

10.14   PRACTICAL AGE-HARDENING TREATMENTS

A practical heat treatment to harden an alloy consists of two heating stages, as shown in Figure 10.21. First the alloy should be solution treated at a temperature just below the eutectic temperature for long enough to allow solution of the second phase. Then it should be quenched to room temperature. Finally, it should be heated to a lower temperature to allow precipitation.

The reason why the solution treatment should be kept below the eutectic temperature is that if the alloy is heated above this temperature, local melting will occur where the average composition is locally equal to the eutectic composition. Where melting occurs, the S → L volume expansion on melting will expand the solid lattice around the molten region. On subsequent cooling, the L → S contraction will leave a void, lowering the mechanical properties. This is called "burning" and an example is shown in Figure 10.22.

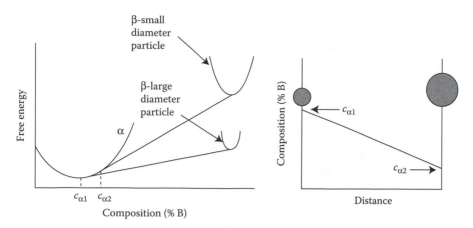

FIGURE 10.20 For small particles, the higher free energy curve (upper left) results in the solubility of B in the α-phase (phase diagram lower left). This in turn produces a concentration gradient in between small and large β particles. This concentration gradient (right) causes the diffusion needed to dissolve small particles and make large ones grow.

If the cooling from the solution treatment is not fast enough, coarse precipitates will form on the grain boundaries, causing the regions near the grain boundaries to be depleted in the solute. The result will be a precipitation-free zone (PFZ) near the grain boundaries, as shown in Figure 10.23. A slower quench increases the width of the PFZ, as indicated in Figure 10.24.

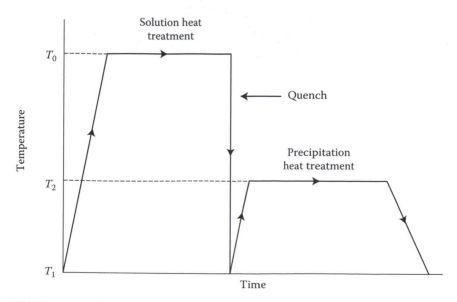

FIGURE 10.21 Schematic diagram of a practical heat treatment for precipitation hardening. The time between quenching from the solution treatment and reheating for precipitation is not important.

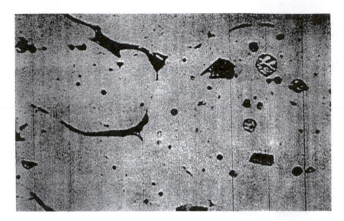

FIGURE 10.22   Microstructure of aluminum alloy 2024 showing evidence of "burning" along grain boundaries. (From *Metals Handbook*, 8th ed., v. 2 ASM, 1964. With permission.)

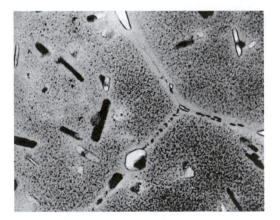

FIGURE 10.23   A PFZ along grain boundaries, resulting from depletion of solute by the formation of a few large precipitate particles during a slow quench. (From D. A. Porter and K. E. Easterling, *Phase Transformations in Metals and Alloys*, 2nd ed., Chapman and Hall, 1992. With permission.)

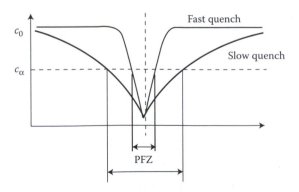

FIGURE 10.24   The width of the PFZ is narrower when the quench is rapid.

In selecting an aging treatment, it should be kept in mind that generally the lower the precipitation temperature, the higher are the hardness and yield strength. Figure 10.25 shows the aging response of aluminum alloy 2024. However, aging at lower temperatures requires longer times and usually there are economic limits on the length of the aging treatment. Times may vary but are usually between 30 min and 2 h. The lower time varies with part size and furnace load and is governed by the requirement to reach temperature in the center of all parts. The longer limit is governed by economy. The cooling rate from the aging treatment is unimportant.

Figure 10.25 also shows that at very long times, the alloy overages, becoming softer than in the as-quenched condition. The reason for this is that in the as-quenched condition, there is a considerable amount of solution hardening. As precipitation occurs, copper is removed from solid solution so the amount of solution hardening decreases. Initially, this loss of solution hardening is more than compensated for by the precipitation hardening. However, as the precipitate particles grow and become fewer in number the precipitation hardening decreases.

Alloys containing more second phases than can be dissolved during the solution treatment will still respond to precipitation hardening. The solute that does dissolve can be precipitated. In fact, many of the commercial alloys designed to be used as castings do contain excessive alloy. This is because the segregation that occurs during solidification creates some regions leaner in solute than others.

It is possible to combine the strengthening caused by precipitation and cold working. The saturated solid solutions formed by quenching from the solution treatment can be cold worked. On subsequent aging, the effects of cold work are not entirely lost. As a result, higher hardnesses can be achieved than by either cold working or precipitation alone.

For a given volume fraction, V_f, the finer the particles are, the smaller is the distance, L, between them. A rough estimate of the effect of particle size can be made by assuming that the particles intersecting the slip plane are arranged in a square

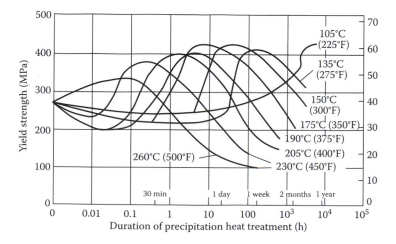

FIGURE 10.25 Age-hardening kinetics of aluminum alloy 2014. (From *Aluminum and Its Alloys*, ASM Int., 1993. With permission.)

pattern and that r is the radius of the particle as it is cut on the slip plane. Then the area fraction is equal to the volume fraction, $V_f = \pi r^2/d^2$. Solving for $d = r(\pi/V_f)^{1/2}$, and substituting into Equation 10.10,

$$\tau = \frac{2Gb(V_f/\pi)^{1/2}}{r}. \tag{10.12}$$

Since the increase of tensile yield strength caused by the precipitate particles, $\Delta\sigma$, is proportional to the shear stress, τ,

$$\Delta\sigma = \frac{\alpha Gb V_f^{1/2}}{r}. \tag{10.13}$$

10.15   ALUMINUM–COPPER ALLOYS

Aluminum alloys containing 4–4.5% copper are among the most important precipitation hardening alloys. Figure 10.26 is the aluminum-rich end of the aluminum–copper phase diagram. Although $\theta = Al_2Cu$ is the equilibrium precipitate, there are a number

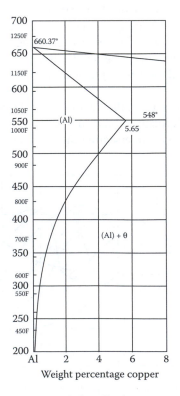

FIGURE 10.26   Aluminum-rich end of the aluminum–copper phase diagram showing the solubility of θ (Al$_2$Cu) in aluminum. (From *Metals Handbook*, 8th ed., v. 8 ASM, 1973. With permission.)

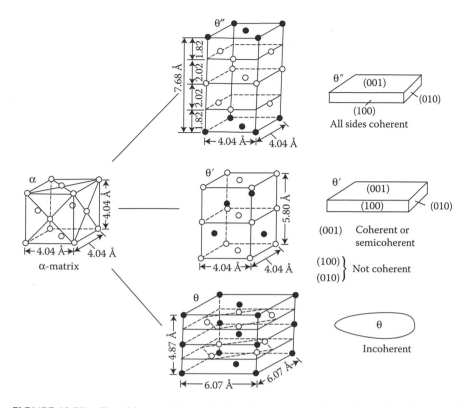

FIGURE 10.27 Transition precipitates in aluminum–copper alloys. (From D. A. Porter and K. E. Easterling, *Phase Transformations in Metals and Alloys*, 2nd ed., Chapman and Hall, 1992. With permission.)

of metastable transition precipitates that form faster than θ. Figure 10.27 shows the structures of these precipitates and Figure 10.28 is the phase diagram showing the metastable equilibrium. Figure 10.29 is a schematic representation of the variation of free energy with composition of these metastable phases.

Figure 10.30 shows how the hardness varies during precipitation at two different temperatures. In both cases, the highest hardness corresponds to the precipitation of θ′ and θ″, which are coherent with the aluminum-rich lattice. Note that the peak hardness is greater for the lower aging temperature, but a longer time is required to reach peak hardness. The peak hardnesses are also increased with the amount of copper in the alloy.

10.16 HABIT

Precipitates are usually in the form of platelets or rods. This minimizes strain energy. Furthermore, certain crystallographic planes in the precipitate are parallel to certain crystallographic planes in the matrix. This phenomenon is called *habit* and the matrix planes on which the precipitation occurs are called *habit planes*. It occurs to minimize the interface energies. Figure 10.31 shows θ′ precipitates in an Al-4% Cu alloy on the {100} planes. Table 10.2 gives some examples in other systems.

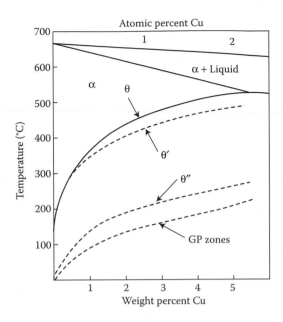

FIGURE 10.28 Metastable solubility of the transition phases in the aluminum–copper system superimposed on the equilibrium diagram.

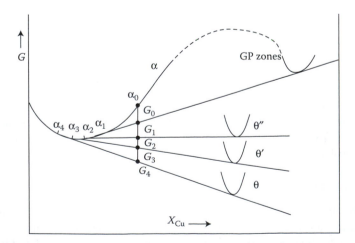

FIGURE 10.29 Schematic diagram of the free energy in the Al–Cu system. Note that the solubility is smaller for the more stable precipitates.

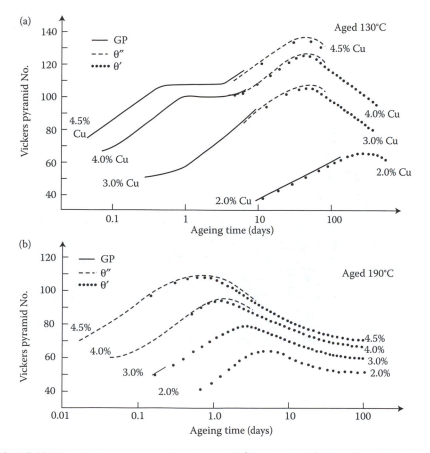

FIGURE 10.30 Hardness versus aging time at 130°C (a) and 190°C (b). (From D. A. Porter and K. E. Easterling, *Phase Transformations in Metals and Alloys*, 2nd ed., Chapman and Hall, 1992. With permission.)

10.17 MISCELLANY

At room temperature tin has a body centered tetragonal structure. This is called beta or "white tin." Below 13°C, tin's equilibrium structure is diamond cubic, and this form is called "gray tin." The volume expansion on the transformation $\beta \rightarrow \alpha\alpha$ transformation is so large (27%) and the α-phase so brittle that the transformation results disintegration into a powder. Hence, the name "gray." Although the transformation would ruin any part made from or joined by tin, fear of such catastrophe is unwarranted because the transformation is extremely sluggish and common impurities inhibit it. Tin cans left in the Antarctic by Scott's expedition show no signs of transforming.

There is an apocryphal story that attributes Napoleon's defeat at Moscow to this transformation. The Russian winter was so cold that the tin buttons on the French

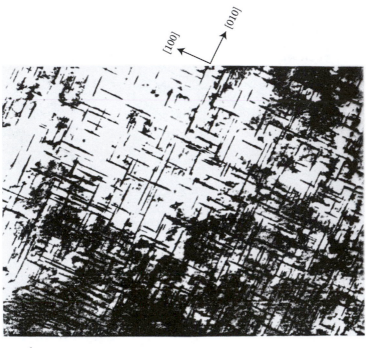

$[100]$ $[010]$

|⊢——— 1 μ ———⊣|

FIGURE 10.31 SEM micrograph showing θ′ precipitates in aluminum. Note that they are parallel to the (100) and (010) planes. The [001] direction is normal to the paper so (001) planes are not visible. (From W. F. Hosford and S. P. Agrawal, *Met. Trans*, v. 64A, 1975. With permission.)

TABLE 10.2
Precipitation Habits

Base Metal	Solute	Precipitate	Habit
Al	Cu	θ″, θ′	Platelets//{100}
Al	Ag	γ′	Platelets//{111}
Al	Mg, Si	Mg_2Si	Rods//<111>
Cu	Be	γ-CuBe	Platelets//{100}

uniforms transformed to gray tin and disintegrated, The French could not fight while holding up their pants.

PROBLEMS

1. An alloy of A–5% B has a microstructure consisting of an A-rich solid solution and particles of A_2B that are 0.5 μm diameter. Assume that the composition of A_2B is 40% B and that of the A-rich solid solution is 0.01% B. Also assume that A and A_2B have the same density.

a. How many A_2B particles are there per volume?
b. Assuming a simple distribution of these in space, find the distance between particles.
c. Assuming the Orowan equation, find the shear stress required to move dislocations through the material. Take the value of the shear modulus as 25 GPa and the atomic diameter as 0.28 nm.

2. The ratio of the activation energy for heterogeneous nucleation of a new phase on a grain boundary, ΔG^*_{hetero} to that for homogeneous nucleation, G^*_{hetero} is given by

$$S(\theta) = \frac{\Delta G^*_{hetero}}{\Delta G^*_{homo}} = \left(\frac{1}{2}\right)(2 + \cos\theta)(1 - \cos\theta)^2,$$

where θ is the wetting angle between the new phase, β, and the old one, α.
a. Calculate $S(\theta)$, if $\gamma_{\alpha\beta} = \gamma_{\alpha gb}$.
b. According to the attached plot (Figure 10.32) what would be the value of $S(\theta)$ for nucleation at the grain corners, if $\gamma_{\alpha\beta} = 1.5\gamma_{\alpha gb}$.

3. Data on a phase transformation is given in Figure 10.33.
a. Determine the exponent in the Avrami equation. At what time would you expect the fraction transformed to be 0.001?
b. At what time would you expect the fraction transformed to be 0.999?

4. To get combined benefits of precipitation and work hardening, the material is cold worked in the as-quenched condition before aging.
a. Why is the cold work not done before the solution treatment?
b. Why is the cold work not done after the precipitation treatment?

5. Sketch free energy versus composition diagrams for a material that undergoes a martensitic transformation.
a. At a temperature between the M_s and M_d,
b. At a temperature between the A_s and A_d.

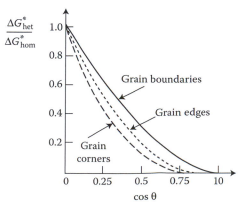

FIGURE 10.32 The ratio of ΔG^* for nucleation at boundaries, edges, and corners of grains to ΔG^* for homogeneous nucleation. (Data from J. W. Cahn, *Acta. Met.*, v 4, 1956.)

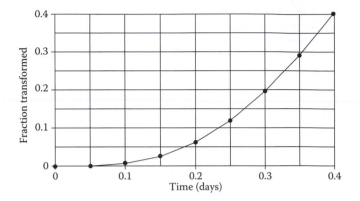

FIGURE 10.33   Plot for a phase transformation of the fraction transformed as a function of time.

6. Describe the conditions under which a reaction $\alpha \rightarrow \alpha_1 + \alpha_a$ will take place by a spinodal decomposition.

REFERENCES

Aluminum and Its Alloys, ASM Int., 1993.
J. W. Cahn, *Acta. Met.*, v 4, 1956.
W. F. Hosford and S. P. Agrawal, *Met. Trans*, v. 64A, 1975.
L. Kaufman and M. Cohen, *Trans. AIME*, 206, 1393, 1956.
Metals Handbook, 8th ed., v. 2 ASM, 1964.
Metals Handbook, 8th ed., v. 8 ASM, 1973.
D. A. Porter and K. E. Easterling, *Phase Transformations in Metals and Alloys*, 2nd ed., Chapman and Hall, 1992.

11 Crystallographic Textures

A random orientation of grains is very uncommon to have in polycrystals. Usually some orientations are more common than others. The term *crystallographic texture* refers to preferred orientations. Dependence of properties of crystals on direction causes textured polycrystals to be *anisotropic*. This chapter deals with the description of crystallographic textures, their origin, and effect on properties.

11.1 STEREOGRAPHIC PROJECTION

The angular relations between directions and planes in a crystal can be represented by stereographic projection. This projection system can be visualized by imagining a tiny (infinitesimal) crystal at the center of a sphere. All planes and directions of interest are extended until they intersect the surface of the sphere. Directions intersect the sphere as points and planes as great circles as shown in Figure 11.1. The problem of plotting these on a flat surface is exactly the same as the mapmaker's problem of plotting the spherical surface of the earth as a map. For crystals, it is necessary to plot only half of the spherical surface because the opposite hemisphere is identical. To visualize the stereographic projection, consider a glass sphere (Figure 11.2) resting on a flat plane (plane of projection) with a tangent at C and a light source (point of projection) at S on the sphere's surface diametrically opposite C. Planes and directions are projected onto the plane of projection as points and great circles. Directions are mapped as points and planes as great circles. By custom, only the features on the hemisphere nearest the plane of projection are mapped. The great circle that bounds the hemisphere (the equator) is called the reference circle.

Angular measurements may be made using a Wulff net (Figure 11.3), which is a projection of a reference sphere with lines of latitude (parallels) and lines of longitudes (meridians). The point of projection is on the equator so the north and south poles lie on the reference circle. Meridians and the equator are great circles. Angles between two directions can be made if both are on the same great circle. It is convenient to plot the directions on a piece of transparent paper laid over a Wulff net with a pin through the center of the paper and the center of the net. This allows the paper to be rotated relative to the net about the center. To find the angle between two directions, the paper should be rotated until both directions are on the same meridian. The angle between them is the difference in latitude (Figure 11.4). The plane common to both directions is the meridian that contains them. The normal to that plane lies on the equator 90° from where the plane intersects the equator.

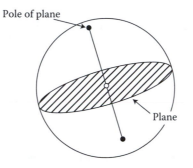

FIGURE 11.1   Spherical projection of a plane and its pole.

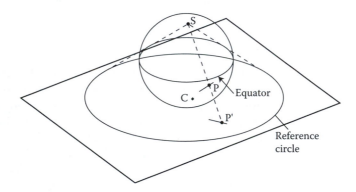

FIGURE 11.2   Stereographic projection of the lower hemisphere. (From W. F. Hosford, *The Mechanics of Crystals and Textured Polycrystals*, Oxford University Press, Oxford, 1993. With permission.)

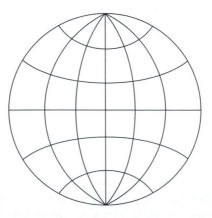

FIGURE 11.3   Wulff net (30°). (From W. F. Hosford, *The Mechanics of Crystals and Textured Polycrystals*, Oxford University Press, Oxford, 1993. With permission.)

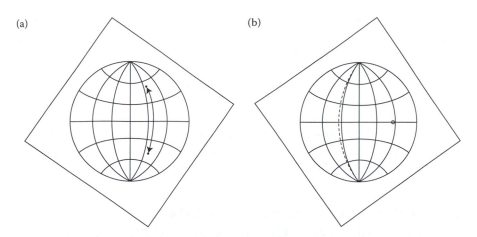

FIGURE 11.4 The angle between two directions is measured along a great circle (a). The pole of a plane is constructed by rotating the net, so the plane corresponds to a meridian. The pole is located 90° along the equator from the intersection of the plane with the equator (b). (From W. F. Hosford, *The Mechanics of Crystals and Textured Polycrystals*, Oxford University Press, Oxford, 1993. With permission.)

11.2 CONSTRUCTION OF A STANDARD CUBIC PROJECTION

Consider a stereographic plot for a cubic crystal with the [001] direction at the top (north pole) and with [100] in the center (on the equator) as shown in Figure 11.5. The other end of [001] is [00$\bar{1}$] and it is plotted as the south pole. The (100) plane is the reference circle at the periphery of the hemisphere being plotted, the (001) plane is the equator, and the (010) plane is plotted as a vertical line through [100] and [001]. The normals to (010) are [010] and [0$\bar{1}$0] which lie on the equator and the reference circle. Note that squares are used to indicate the $\langle 100 \rangle$ directions because both have fourfold rotational symmetry.

The [0$\bar{1}$1] and [011] directions (and their opposite ends, [01$\bar{1}$] and [0$\bar{1}$$\bar{1}$]) are indicated by ellipses because they have twofold symmetry (Figure 11.5b). They are on the reference circle, 45° from [010] and ± [001]. The corresponding (011) and (011) planes

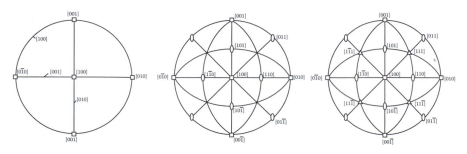

FIGURE 11.5 Construction of a standard cubic projection.

cut diagonally though the center. Note that [0$\bar{1}$1], [100], and [01$\bar{1}$] all lie on the (011) plane because they are 90° from [011]. The (110), (1$\bar{1}$0), (101), and (10$\bar{1}$) planes and their normals are similarly constructed and symbolized.

There are four points at which three great circles representing the {110} intersect. These points must be the directions that lie in all three of those planes, namely the ⟨111⟩ directions. Triangles are used as symbols for the ⟨111⟩ directions because they have threefold symmetry (Figure 11.5).

This construction divides orientation space into spherical triangles, each of which have ⟨100⟩, ⟨110⟩, and ⟨111⟩ corners. All of the triangles are crystallographically equivalent.

11.3 LOCATING THE *HKL* POLE IN THE STANDARD STEREOGRAPHIC PROJECTION OF A CUBIC CRYSTAL

Consider the standard projection with [100] at the center and [001] at the north pole (Figure 11.6). The value of *h* must be positive for all poles represented within the projected hemisphere because these poles must lie <90° from the center = [100] and therefore the dot product *hkl* with [100] is positive. Poles lying on the outer circumference are 90° from [100], so for these *h* = 0. Similarly, the projected hemisphere can be divided into four quadrants. In the first quadrant, *k* > 0 and *l* > 0 because all poles in this quadrant are <90° from both [010] and [001] (Figure 11.7). In the second quadrant, *k* < 0 because poles in this region are >90° from [010]. Both *k* and *l* are negative in the third quadrant because poles in this region are >90° from both [010] and [001]. Finally, *l* > 0 in the fourth quadrant because poles in this region are >90° from [001].

Similarly, each quadrant can be divided into half according to whether *h* > *k*, whether *k* > *l*, and whether *h* > *l* as shown below for the first quadrant.

These three bisections of the first quadrant split it into six triangles, as illustrated in Figure 11.8 with the appropriate ⟨123⟩ pole in each triangle.

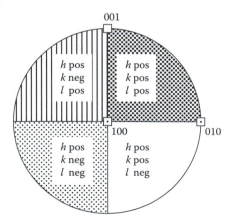

FIGURE 11.6   Signs of *h*, *l*, and *k* in the four quadrants of the standard projection.

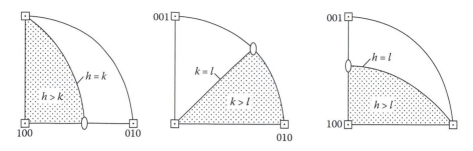

FIGURE 11.7 Relative values of *h*, *l*, and *k* in different regions of the first quadrant.

EXAMPLE PROBLEM 11.1

Locate [2$\bar{1}$1] on the standard projection.

SOLUTION

See Figure 11.9. The dot product of [2$\bar{1}$1] · [100] is positive, so [2$\bar{1}$1] must lie within 90° of [100] and therefore on the hemisphere represented (Figure 11.9a). Since [2$\bar{1}$1] · [010] is negative, [2$\bar{1}$1] must lie >90° from [010] so it is in the left-hand side of the plot (Figure 11.9b). Since [2$\bar{1}$1] · [001] is positive, [2$\bar{1}$1] must lie <90° from [001] and therefore in the top half of the plot. Thus [2$\bar{1}$1] is in the upper left quadrant (Figure 11.9c). The dot products [2$\bar{1}$1] · [0$\bar{1}$0] and [2$\bar{1}$1] · [001] are equal, so [2$\bar{1}$1] is equidistant from [0$\bar{1}$0] and [001] on the line connecting [100] and [0$\bar{1}$1] (Figure 11.9d). Finally, the dot product of [2$\bar{1}$1] with [1 0 0] is larger than the dot products of [2$\bar{1}$1] with either [0$\bar{1}$0] or [001]. This indicates that the angle between [2$\bar{1}$1] and [100] is less than the angles between [2$\bar{1}$1] and either [0$\bar{1}$0] or [001] (Figure 11.9e).

11.4   ORIENTATION REPRESENTATION

The orientation relation between a single crystal and an external coordinate system may be represented in two ways. One is to plot the orientation of the crystal on a

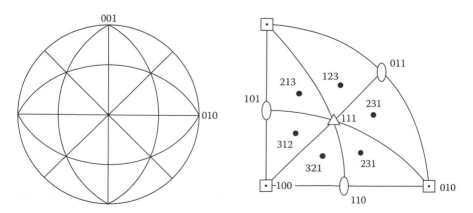

FIGURE 11.8 Locations of the ⟨123⟩ poles in the first quadrant.

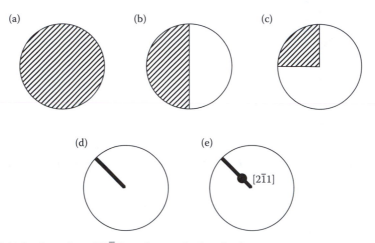

FIGURE 11.9 Location of [2$\bar{1}$1] on the standard projection.

projection where the external axes form the external coordinate system. For example, the cubic axes of a crystal may be plotted on a projection where the external x, and z are at the center and north pole as shown in Figure 11.10.

Alternatively, the external axes may be plotted on a standard projection as shown in Figure 11.11.

11.5 POLE FIGURES

Using the first way, a single orientation can be fully represented by plotting only one set of pole (e.g., only {100} or {111} poles). This type of representation can also be used to represent orientations of a large number of grains. If {100} poles are plotted,

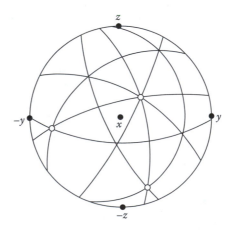

FIGURE 11.10 Crystal axes plotted relative to the x, y, and z axes. (From W. F. Hosford, *The Mechanics of Crystals and Textured Polycrystals*, Oxford University Press, Oxford, 1993. With permission.)

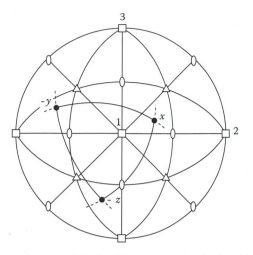

FIGURE 11.11 External axes plotted relative on a standard cubic projection. (From W. F. Hosford, *The Mechanics of Crystals and Textured Polycrystals*, Oxford University Press, Oxford, 1993. With permission.)

it is called a {100} pole figure. Figure 11.12a shows the hypothetical {100} pole distribution for a rolled sheet. The natural reference directions are the rolling direction, the transverse direction, and the sheet normal direction. Figure 11.12b with contour lines of density is the {100} pole figure. The corresponding {111} pole figure for the same texture is shown in Figure 11.12c. These can be represented by an ideal $\langle 100 \rangle \{1\bar{1}0\}$ orientation as shown in Figure 11.12d.

Usually, a number of ideal orientations are required to describe a texture. Most textures have orthotropic symmetry so many equivalent orientations are present. For example, an ideal orientation $\{h\,k\,l\}\langle u\,v\,w\rangle$ includes $(h\,\bar{k}\,l)[u\,\bar{v}\,w]$, $(h\,\bar{k}\,\bar{l})[u\,v\,\bar{w}]$, and $(h\,\bar{k}\,\bar{l})[u\,\bar{v}\,\bar{w}]$. Often textures have so many components that quantitative identification of them from pole figures is extremely difficult.

11.6 INVERSE POLE FIGURES

Because of crystal symmetry, a single reference direction may be plotted in any equivalent orientation triangle. Where additional reference directions are important, each must be plotted in its own orientation triangle. Figure 11.13 shows the inverse pole figures corresponding to the pole figures in Figure 11.12.

11.7 ORIENTATION DISTRIBUTION FUNCTIONS

Unless a texture is very simple, it is difficult to determine quantitatively from pole figures or inverse pole figures the amounts of each orientation. Only a three-dimensional plot can fully describe a complex texture. Such plots are called *orientation distribution function plots*. Three angles are required to fully describe an orientation. Figure 11.14 shows one system of plotting.

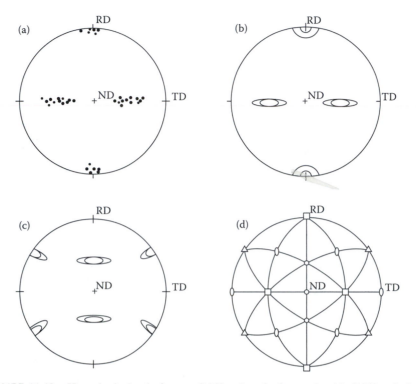

FIGURE 11.12 Hypothetical pole figures. {100} poles of a few grains (a), {100} pole figure for a textured polycrysrtal (b), {111} pole figure for the same texture (c), and ideal orientation for that texture (d). (From W. F. Hosford, *The Mechanics of Crystals and Textured Polycrystals*, Oxford University Press, Oxford, 1993. With permission.)

11.8   ORIGINS OF TEXTURE

When a liquid metal solidifies, certain orientations grow faster than others. This leads to a strong preferred orientation in columnar grains. In cubic metals, both bcc and fcc, the columnar grains have ⟨100⟩ axes. In hexagonal metals, the fastest

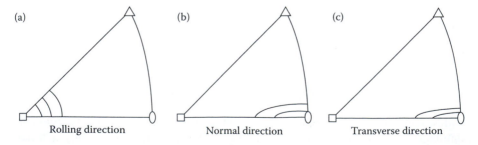

FIGURE 11.13 Inverse pole figures corresponding to the pole figures in Figure 11.12. (From W. F. Hosford, *The Mechanics of Crystals and Textured Polycrystals*, Oxford University Press, Oxford, 1993. With permission.)

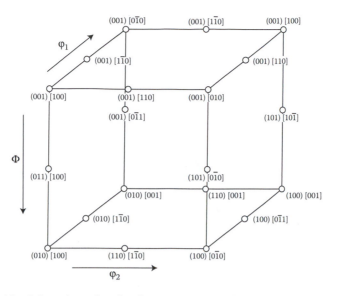

FIGURE 11.14 Orientation cube, showing a number of orientations using Bunge's notation system. (From W. F. Hosford, *The Mechanics of Crystals and Textured Polycrystals*, Oxford University Press, Oxford, 1993. With permission.)

growth direction is normal to the *c*-axis. There are also strong textures in electro-deposited metal.

The textures of wrought metals result from lattice rotations caused by slip or twinning. The original texture in a cast ingot is completely changed many times during the sequence of deformation and annealing steps.

Wire drawing, rod rolling, and extrusion of fcc metals tend to produce textures with complex ⟨100⟩ – ⟨111⟩ fiber textures. Figure 11.15 shows the inverse pole figure of an aluminum rod extruded to a 92% reduction. The relative amount of each component depends on the particular metal, its alloying additions, and the temperature of deformation. Wire drawing, rod rolling, and extrusion of bcc metals tend to produce textures with ⟨110⟩ fiber textures.

Heavy compression produces textures with ⟨110⟩ parallel to the axis of compression. Rolled textures are more complex while heavy compression produces textures with both ⟨100⟩ and ⟨111⟩ parallel to the axis of compression.

Rolling textures are more complex. Rolled sheets of fcc metals tend to develop either of two characteristic textures: copper or brass, which are shown in Figures 11.16 and 11.17. High stacking fault energy metals such as copper and aluminum tend to develop copper-like textures and low stacking fault energy metals such as brass, silver, and austenitic stainless steels tend to form brass-like textures. Low-temperature deformation favors the brass-like texture in metals with intermediate stacking fault energy.

Figure 11.18 shows the rolling texture of molybdenum. This rolling texture is typical of all bcc metals including iron, tantalum, vanadium, tungsten, and β-titanium.

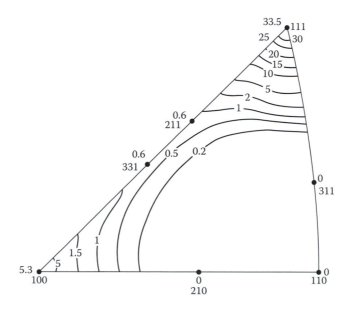

FIGURE 11.15 Inverse pole figure of an aluminum rod extruded to a 92% reduction. (Adapted from C. J. McHargue, L.K. Jetter, and J.C. Ogle, Trans, *Met. Soc. AIME*, 215, 831–837, 1959. With permission.)

11.9 RECRYSTALLIZATION TEXTURES

Usually, fiber textures after recrystalliztion are similar to those after cold working. Recrystallization changes the textures of rolled products. The changes depend on the amount of prior deformation, the rate of heating, and the temperature reached. Still other changes may occur during grain growth. Although development of recrystallization textures is not fully understood, several generalizations may be made. Both oriented-nucleation and oriented-growth theories have been proposed. Certain grain boundaries are more mobile than others. In particular, boundaries in fcc metals that involve a 25–40° rotation about a common $\langle 111 \rangle$ direction and those that correspond to a 15–60° rotation about a common $\langle 110 \rangle$ direction are highly mobile.

Recrytallized sheels of bcc metals (e.g., low-carbon steel) typically have a strong texture with {111} parallel to the rolling plane with some preference for $\langle 112 \rangle$ to be aligned with the rolling direction. Recrystallized sheets of hcp metals retain a strong basal texture.

11.10 EFFECT OF TEXTURE ON MECHANICAL PROPERTIES

How textures may affect the anisotropy of sheet metal is illustrated by the example of rolled hcp metals. These tend to develop textures with the basal (0001) plane parallel to the plane of the sheet. Easy slip occurs in hcp metals only by slip in the close-packed $\langle 11\bar{2}0 \rangle$ directions. For an ideal basal texture (Figure 11.19), slip will not cause any thinning parallel to the c-axis. In a tension test in any direction of the

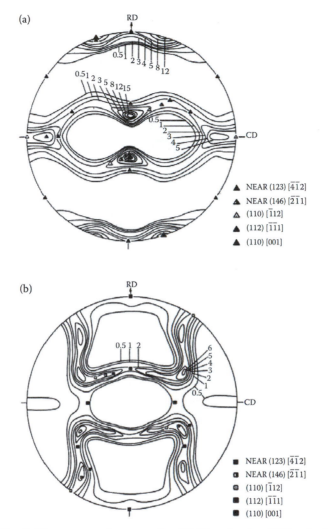

FIGURE 11.16 Rolling texture of copper after 96.5% reduction. {111} pole figure (a) and {1 0 0} pole figure (b). This is typical of high stacking fault energy metals such as copper and aluminum. (From Hsun Hu and S. R. Goldman, *Trans Met. Soc. AIME*, 227, 627–639, 1963. With permission.)

sheet, the elongation will be accompanied by a width contraction without any thinning of the sheet. The ratio

$$R = \frac{\varepsilon_w}{\varepsilon_t} \tag{11.1}$$

of the strains, ε_w, in the width and thickness directions, ε_t, will be very high. With some spread from this ideal texture, the R-values of α-titanium sheets are often about 5. High R-values are of considerable benefit in most sheet forming operations. The high R-values (typically 1.8) of low-carbon steel sheet lead to better formability than the R-values of aluminum sheet typically about 0.8.

(a)

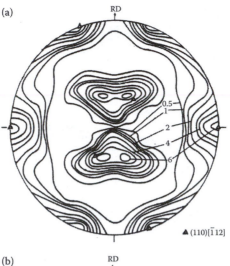

(b)

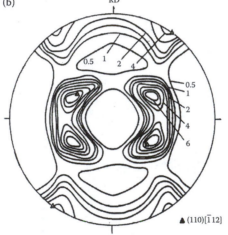

FIGURE 11.17   Rolling texture of silver after 91% reduction. {111} pole figure (a) and {100} pole figure (b). This texture is typical of low stacking fault energy metals such as silver, brass, and austenitic stainless steels and aluminum. (From Hsun Hu and R. S. Cline, *J. Appl. Phys*, 32, 760–763, 1961. With permission.)

11.11   EFFECT OF TEXTURE ON THE MICROSTRUCTURE OF WIRES OF BCC METALS

Peck and Thomas (1961) observed that the microstructures of drawn wires of tungsten metals contain grains that are elongated along the wire axis but seem to curl about one another as shown in Figure 11.22. A similar microstructure was observed in draw iron wires (Figures 11.20 and 11.21).

These microstructures suggest that the grain in the drawn wires have the shape of long ribbons, folded about the wire axis as illustrated schematically in Figure 11.22.

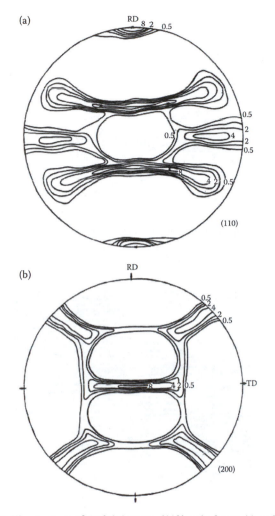

FIGURE 11.18 Rolling texture of molybdenum. {110} pole figure (a) and {200} pole figure (b). (From A. Segmüller and G. Wassermann, *Freiberger Forschungshefte*, 38, 38–44, 1964. With permission.)

The reason for this microstructure can be understood in terms of the ⟨110⟩ crystallographic texture in wires of bcc metals (Figure 11.23). With a ⟨110⟩ direction parallel to the wire axis, two of the ⟨111⟩ slip directions allow thinning parallel to the ⟨001⟩ lateral direction but not to the ⟨1$\bar{1}$0⟩ lateral direction. The other two ⟨111⟩ slip directions are perpendicular to the wire axis and should not operate.

This explains the appearance of a few wide grains in the longitudinal section like grain A in Figure 11.20. A grain appears to be wide when the plane of the section is parallel to [0$\bar{1}$1].

Swaged wires are produced by rotating dies reducing the wire diameter by hammering as illustrated in Figure 11.24. This produces the microstructure shown in

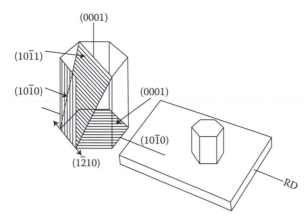

FIGURE 11.19 Basal texture common to sheets of hcp metals. (From W. F. Hosford and W. A. Backofen in *Fundamentals of Deformation Processing*, pp. 259–298, Syracuse University Press, 1964. With permission.)

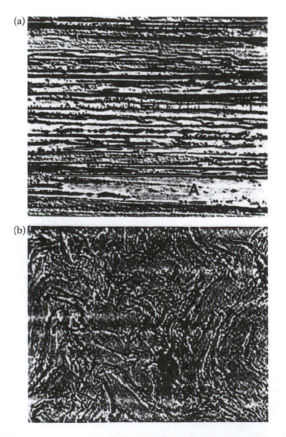

FIGURE 11.20 Microstructures of heavily drawn tungsten wires. (a) Longitudinal section with the wire axis horizontal and (b) curly grains viewed parallel to the wire axis. (From J. F. Peck and D. A. Thomas, *Trans. TMS-AIME*, 221, 1240–1246, 1961. With permission.)

FIGURE 11.21 Curly grain structure of an iron wire drawn to a strain of 2.7. (From J. F. Peck and D. A. Thomas, *Trans. TMS-AIME*, 221, 1240–1246, 1961. With permission.)

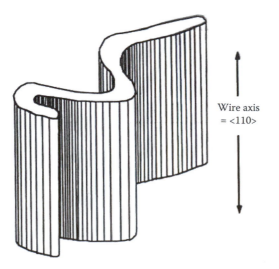

Wire axis
= <110>

FIGURE 11.22 Schematic drawing of the shape of grains in drawn wires of bcc metals. (From W. F. Hosford, *The Mechanics of Crystals and Textured Polycrystals*, Oxford University Press, Oxford, 1993. With permission.)

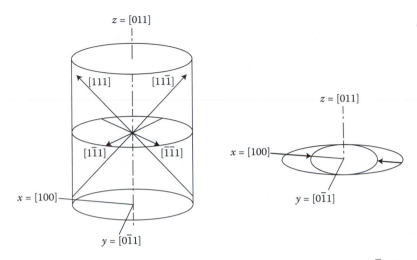

FIGURE 11.23 With a [011] direction parallel to the wire axis, the [111] and [11$\bar{1}$] slip directions cause thinning parallel to the lateral [100] direction without any thinning parallel to [0$\bar{1}$1]. The other two ⟨111⟩ slip directions are unstressed. (From W. F. Hosford, *The Mechanics of Crystals and Textured Polycrystals*, Oxford University Press, Oxford, 1993. With permission.)

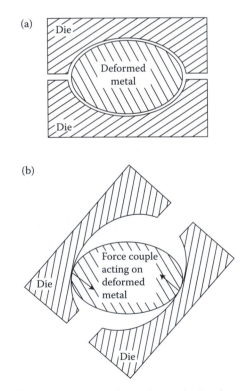

FIGURE 11.24 Swaging of a wire causes it to twist producing the microstructure in Figure 11.25. (From J. F. Peck and D. A. Thomas, *Trans. TMS-AIME*, 221, 1240–1246, 1961. With permission.)

FIGURE 11.25 Cross-sectional microstrucrure of an iron wire swaged to an 87% reduction of area. Note that the thin [001] directions are aligned radially. (From J. F. Peck and D. A. Thomas, *Trans. TMS-AIME*, 221, 1240–1246, 1961. With permission.)

Figure 11.25. All of the grains are oriented so that their ⟨100⟩ directions are oriented in the radial direction, producing a ⟨011⟩⟨001⟩ cylindrical texture.

The subgrain structure in drawn iron wires is similar to the grain structure as indicated in Figure 11.26.

Langford et al. observed that the strain hardening caused by drawing was linear with strain as shown in Figure 11.27 (Leslie, p. 119). Probably this is because the critical dimension of the grains decreases linearly with elongation instead of with the square root of elongation.

The microstructure of pearlite after wire drawing is similar to that of pure bcc metals (Figure 11.28). During wire drawing the ferrite phase develops a ⟨110⟩ wire texture.

Hexagonal close-packed magnesium develops a similar microstructure during wire drawing as a result of a wire textures in which only two slip directions contribute to elongation (Wonsiewitcz, private communication). There is a similar relation between texture and microstructure in heavily compressed fcc metals (Hosford, 1964). With the ⟨110⟩ compression texture, slip should occur on only two of the four {111} planes with the result that lateral spreading occurs only in the ⟨001⟩ direction and grains must curl about each other.

11.12 MISCELLANY

E. Schmidt did much of the pioneering work developing the use of pole figures. He published a text with W. Boas, *Kristallplasticität* (Springer, 1935). Inverse pole figures were first used to describe textures by C. J. McHargue, et al. in 1959.

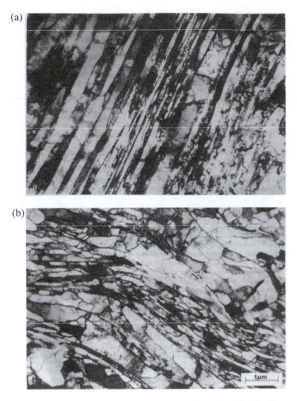

FIGURE 11.26 Subgrains in iron wire drawn to a strain of 3.8. Longitudinal section (a) and cross section (b). (From G. Langford and M. Cohen, *Trans. ASM*, 62, 629, 1969. With permission.)

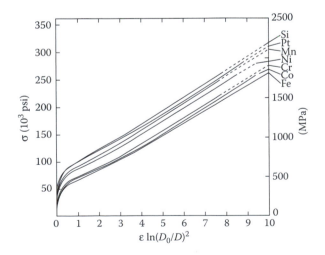

FIGURE 11.27 Increase of tensile strength of drawn wires with drawing strain for iron containing 3 atom% of several elements. Note that the tensile stress increases linearly with strain. (From G. Langford and M. Cohen, *Trans. ASM*, 62, 629, 1969. With permission.)

FIGURE 11.28   SEM microstructure of pearlite drawn to a strain of 3.2. (From W. C. Leslie, *The Physical Metallurgy of Steels*, McGraw Hill, 1981 and attributed to G. Langford. With permission.)

The method of completely describing textures by an orientation distribution function was independently developed by R. J. Roe in 1966 and H.-J. Bunge in 1968. They are very similar but use different notation systems. Bunge's system is described in *Mathematische Methoden der Texturanalyse*, published by Akademie-Verlag, 1969 and translated into English as *Mathematical Methods of Texture Analysis*, Butterworth, 1982. Roe's analysis was published with W. R. Krigbaum in the *Journal of Applied Physics,* 35, 2215, 1964.

PROBLEMS

1. Locate [312] on a stereographic projection with [001] at the north pole and [100] at the center.
2. How many different <210> directions are there? How many stereographic triangles does each <210> direction shares?
3. Sketch the {111}-pole figure for an ideal ⟨100⟩ − {011} texture.
4. Sketch the inverse pole figures corresponding to a ⟨211⟩{111} texture.
5. Consider that a bcc crystal is forced to deform with axial symmetry while being extended in the [110] direction so that $\varepsilon_{[100]} = \varepsilon_{[0\bar{1}1]} = -(1/2)\varepsilon_{[110]}$. There is slip in the [1$\bar{1}$1] and [$\bar{1}\,\bar{1}$1] directions to cause contraction in the lateral [0$\bar{1}$1] direction and undo some of the lateral contraction in the [100] direction. Deduce the ratio of $(\gamma_{[1\bar{1}1]} + \gamma_{[\bar{1}\bar{1}1]})/(\gamma_{[111]} + \gamma_{[11\bar{1}]})$.
6. Write equations describing the stress–strain curves in Figures 11.28.

REFERENCES

C. S. Barrett and T. B. Masalski, *Structure of Metals*, McGraw-Hill, 1966.
H.-J. Bunge, *Z. Metallk.*, 56, 872–874, 1965.

W. F. Hosford Jr. *Trans. TMS-AIME*, 230, 12–15, 1964.

W. F. Hosford, *The Mechanics of Crystals and Textured Polycrystals*, Oxford University Press, Oxford, 1993.

W. F. Hosford and W. A. Backofen in *Fundamentals of Deformation Processing*, pp. 259–298, Syracuse University Press, 1964.

Hsun Hu and R. S. Cline, *J. Appl. Phys*, 32, 760–763, 1961.

Hsun Hu and S. R. Goldman, *Trans Met. Soc. AIME*, 227, 627–639, 1963.

F. J. Humphries and M. Hatherly, *Recrystallization and Related Annealing Phenomenon*, 2nd ed., Pergamon Press, 2004.

U. F. Kocks, C. Tomé, and H.-R. Wenk, Eds., *Texture and Anisotropy*, Cambridge University Press, Cambridge, 1998.

G. Langford and M. Cohen, *Trans. ASM*, 62, 629, 1969.

G. Langford, *Met. Trans.* 8A, 1977.

W. C. Leslie, *The Physical Metallurgy of Steels*, McGraw Hill, 1981.

C. J. McHargue, L. K. Jetter, and J. C. Ogle, *Trans, Met. Soc. AIME*, 215, 831–837, 1959.

J. F. Peck and D. A. Thomas, *Trans. TMS-AIME*, 221, 1240–1246, 1961.

R.-J. Roe, *J. Appl. Phys*, 36, 2024–2031, 1965.

E. Schmidt and W. Boas, *Kristallplasticät*, Springer-Verlag, Berlin, 1935.

A. Segmüller and G. Wassermann, *Freiberger Forschungshefte*, 38, 38–44, 1964.

B. C. Wonsiewitcz, private communication.

12 Aluminum and Its Alloys

Aluminum has an fcc crystal structure. Its melting point is 660°C. Its density, ρ, is 2.7, which is about 1/3 of the density of iron. Young's modulus is 70 GPa (10×10^6 psi), which is also about 1/3 that of iron. The unique properties of aluminum that account for most of its usage are:

1. A very good corrosion and oxidation resistance
2. Very good electrical thermal conductivities
3. Very low density
4. High reflectivity
5. High ductility and reasonably high strength.

12.1 USES

The uses of aluminum include foil, beverage cans, cooking and food processing, boats and canoes, aircraft and other transportation (sheet, engine blocks, and wheels), die castings, and thermit reactions. Its high reflectivity accounts for its use as foil for insulation and as reflective coatings on glass. It is used for power transmission line and some wiring because of its high electrical conductivity. On an equal weight of cross section and equal cost basis, it is a better conductor than copper. Its high thermal conductivity is advantageous in its applications for radiators, air-cooled engines, and cooking utensils. The low density is important for lawn furniture, hand-held tools, and in cars, trucks, and aircraft. Aluminum's good strength and ductility is important in all structural uses where wrought products are used. Its chemical reactivity is important principally in its use in photoflash bulbs and the thermit reaction ($Al + Fe_2O_3 \rightarrow Fe + Al_2O_3$). Its corrosion and oxidation resistance are important in packaging (foil and cans), architectural applications, and watercraft.

12.2 ALLOYS

The principal alloying elements are manganese, copper, silicon, magnesium, and zinc. The aluminum–copper phase diagram was presented in Figure 10.26, Figures 12.1 through 12.5 are the phase diagrams of these systems. Of these systems, the precipitation-hardening ones are copper, magnesium in the ratio of approximately Mg_2Si, and zinc.

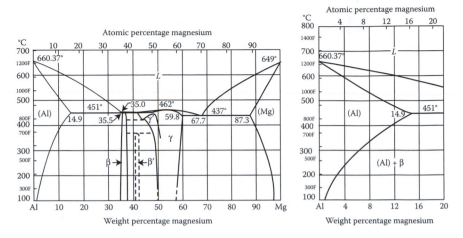

FIGURE 12.1 Aluminum–magnesium phase diagram. The aluminum-rich end is enlarged at the right. (From *Metals Handbook*, 8th ed., Vol. 8, p. 261, 1973. With permission.)

Alloys may be classified as to whether they are used in the wrought condition or as castings. They may also be classified as to whether they are precipitation-hardenable or not. Figure 12.5 summarizes these classifications.

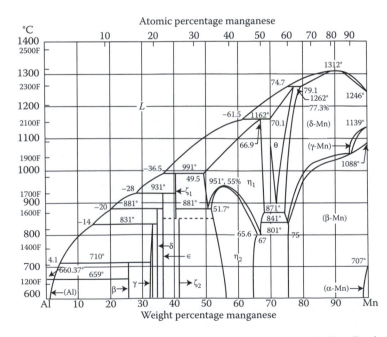

FIGURE 12.2 Aluminum–manganese phase diagram. (From *Metals Handbook*, 8th ed., Vol. 8, p. 262, 1973. With permission.)

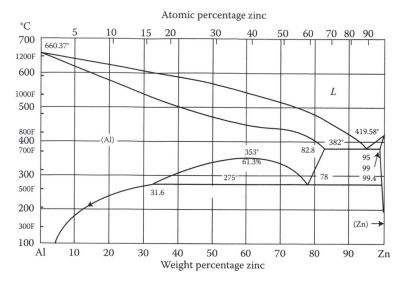

FIGURE 12.3   Aluminum–zinc phase diagram. (From *Metals Handbook*, 8th ed., Vol. 8, p. 265, 1973. With permission.)

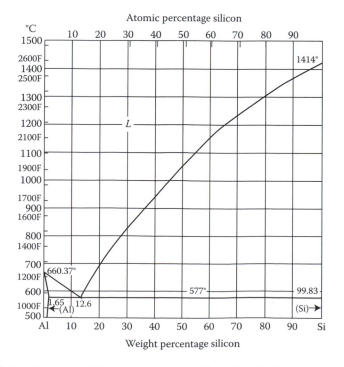

FIGURE 12.4   Aluminum–silicon phase diagram. (From *Metals Handbook*, 8th ed., Vol. 8, p. 264, 1973. With permission.)

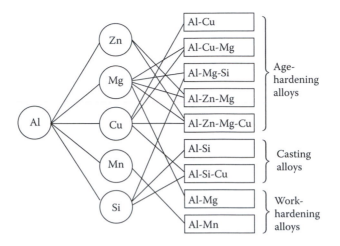

FIGURE 12.5   Classification of major alloying systems for aluminum-base alloys. (From J.E. Hatch, Ed., *Aluminum: Properties and Physical Metallurgy*, ASM, 1984. With permission.)

12.3   WROUGHT ALLOYS

Table 12.1 shows the alloy and temper designation systems used for wrought alloys.

12.3.1   TEMPER DESIGNATIONS

F   As fabricated
O   Annealed
H   Strain hardened
H-1x Strain hardened only
H-2x Strain hardened and partially annealed
H-3x Strain hardened and stabilized

The second digit (1 through 9) refers to the degree of hardening with 8 corresponding to the hardness achieved by a 75% reduction.
A third digit may be added to identify the surface condition.

W   Solution treated (used only with alloys that will age at room temperature)
T   Heat treated to produce a stable temper (other than O, F, H, and W)
T-3   Solution treated, strain hardened, and naturally aged
T-4   Solution treated and naturally aged
T-5   Artificially aged only (after fast cooling from hot working)
T-6   Solution treated and artificially naturally aged

The numbers 4xxx were originally set aside for alloys with silicon as the major alloy, but because of low ductility, there are no commercial alloys of this type.
Table 12.2 gives the maximum solid solubility of various elements.
Because of the limited solid solubility of Fe, Ni, Cr, and Mn, the microstructures of aluminum alloys contain many intermetallic phases. In particular, commercially pure aluminum (1xxx) contains stable $FeAl_3$, Fe_3SiAl_{12}, $Fe_2Si_2Al_9$, and metastable $FeAl_6$.

TABLE 12.1

Designation System for Wrought Aluminum Alloys

Number	Principal Alloy	Examples	Uses
EC	99.45 + % Al	1100 (0.12% Cu)	Electrical grade
1xxx	99.0 + % Al	1060 (≥99.6% Al)	Architectural, cookware
2xxx	Copper	2014 (4.5 Cu, 0.5 Mg, 0.8 Mn, 0.8 Si)	Structural forgings
		2024 (4.5 Cu, 1.5 Mg, 0.6 Mn)	Aircraft, hardware
3xxx	Manganese	3003 (1.2 Mn)	Food and chemical processing
		3004 (1.2 Mn, 1 Mg, 0.2 Fe)	Beverage cans, roofing
5xxx	Magnesium	5052 (2.5 Mg, 0.25 Cr)	Boats, trucks, buses
6xxx	Mg + Si	6061 (1 Mg, 0.5 Si, 0.3 Cu)	Trucks, furniture, boats
		6063 (0.7 Mg, 0.4 Si)	Extrusions, architectural, irrigation
7xxx	Zn	7075 (5.6 Zn, 1.6 Cu, 2.5 Mg)	Aircraft
8xxx	misc. (including Li)		

The phases $(Mn,Fe)Al_6$ and $(FeMn)_3SiAl_{12}$ are commonly found in the 3xxx series alloys and Mg_2Al_3, Mg_2Si if $Mg > 3.5\%$, and Mg_2Al_3 $Cr_cMg_3Al_{18}$ in the 5xxx series alloys.

The principal precipitate in the 6xxx alloys is Mg_2Si, which is responsible for precipitation hardening, but Fe_3SiAl_{12}, $Fe_2Si_2Al_9$, and $(Fe,Mn,Cr)_3SiAl_{12}$ are also found. In the 7xxx series, the equilibrium precipitates are $MgZn_2$ and $Mg_3Zn_3Al_2$, but $(Fe,Cr)_3SiAl_{12}$, Mg_2Si, and $Mg(Zn,Cu,Al)_2$ may also be present.

Figures 12.6 and 12.7 show the microstructures of 1100-O and 2024-O aluminum. Note the number of intermetallic phases.

Manganese is a strong solution strengthener as shown in Figure 12.8. The yield strength is nearly tripled by the addition of 1% Mn. Magnesium is also a solid solution strengthener (Figure 12.9). Aluminum beverage cans account for about 1/4 of the total aluminum production. They are made from a special alloy with a composition close to that of 3004 but with tighter composition control.

TABLE 12.2

Solid Solubility of Several Elements in Aluminum

Element	Maximum Solubility (%)
Cr	0.44
Cu	2.5
Fe	0.025
Mg	18.9
Mn	0.7
Ni	0.023
Si	1.59
Zn	66.5

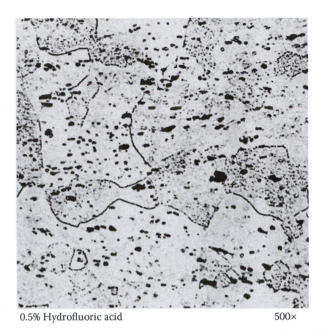

0.5% Hydrofluoric acid   500×

FIGURE 12.6   Microstructure of aluminum alloy 1100-O. The black particles are FeAl$_3$. (From *Metals Handbook*, 8th ed., Vol. 7, p. 242, ASM, 1973. With permission.)

Keller's reagent   100×

FIGURE 12.7   Microstructure of alloy 2024-T6. The particles are CuMgAl$_2$. (From *Metals Handbook*, 8th ed., Vol. 7, p. 246, ASM, 1973. With permission.)

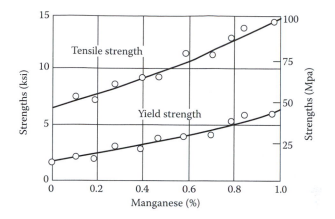

FIGURE 12.8 Solid solution strengthening effect of manganese in aluminum. (From J.E. Hatch, Ed., *Aluminum: Properties and Physical Metallurgy*, ASM, 1984. With permission.)

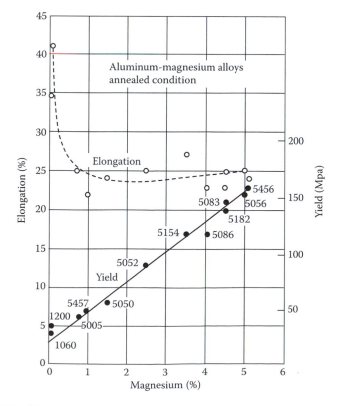

FIGURE 12.9 Magnesium has a strong solid solution strengthening effect in aluminum alloys. (From J.E. Hatch, Ed., *Aluminum: Properties and Physical Metallurgy*, ASM, 1984. With permission.)

12.4   CASTING ALLOYS

Table 12.3 lists a few common casting alloys. Most of the casting alloys contain major amounts of silicon. Alloys with silicon undergo less shrinkage on freezing than most alloys. With 18% silicon, there is no shrinkage. This is because the silicon appears as elemental silicon with a diamond cubic crystal structure, which is less densely packed than the liquid. The as-cast microstructure of alloy 356 is shown in Figures 12.10 and 12.11. Sodium additions cause a finer eutectic structure.

Aluminum-rich dendrites form the primary phase. Silicon particles in the eutectic are sharp needles. Figure 12.11 shows how 0.025% Na blunts and decreases the size of the silicon particles.

Figure 12.12 shows the microstructure of a hypereutectic alloy 392 with large primary silicon particles. Figure 12.13 shows the same alloy except with the addition of a small amount of phosphorus, which reduces the size of the primary silicon.

TABLE 12.3
Designations of Some Aluminum Casting Alloys

Number	Composition (%)	Uses
296	4.5 Cu	Aircraft fittings, pumps
356	7 Si, 3.5 Cu	Engine blocks, transmissions, wheels
380	8 Si, 3.5 Cu	Die castings
390	17 Si, 4.5 Cu,1 Fe, 0.5 Mg	Die castings

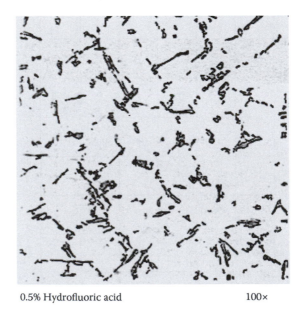

0.5% Hydrofluoric acid   100×

FIGURE 12.10   As-cast alloy 356. The eutectic with coarse silicon particles surrounds the primary aluminum-rich dendrites. (From *Metals Handbook*, 8th ed., v. 7 ASM, 1972. With permission.)

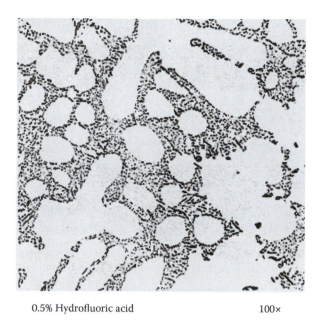

0.5% Hydrofluoric acid   100×

FIGURE 12.11 The same alloy as in Figure 12.10, with the addition of 0.25% Na. Note the much finer eutectic structure. (From *Metals Handbook*, 8th ed., v. 7 ASM, 1972. With permission.)

Three types of casting processes are used with aluminum alloys: Sand castings for large pieces and for pieces of limited production, permanent mold castings for large production particularly for small and medium size pieces, and die castings for small parts made in large production runs. Gas porosity can be a problem in aluminum casting. Molten aluminum picks up hydrogen from moisture. Release of this during freezing causes porosity.

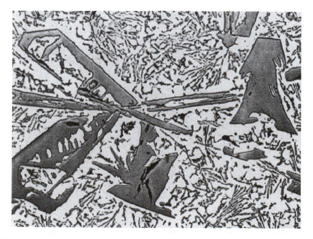

FIGURE 12.12 As-cast structure of alloy 392. Note the large primary silicon particles. (From *Metals Handbook*, 8th ed., v. 7 ASM, 1972. With permission.)

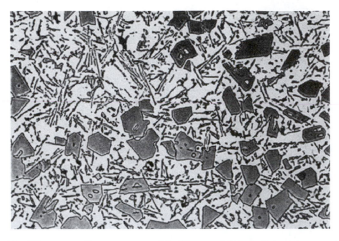

FIGURE 12.13 The same alloy except modified by an addition of phosphorus which reduces the size of the primary silicon particles. (From *Metals Handbook*, 8th ed., v. 7 ASM, 1972. With permission.)

12.5 POWDER PROCESSING

Parts made from sintered aluminum powder (SAP) have good creep resistance and are very resistant to recrystallization. The aluminum oxide, which coated the powder, prevents growth of recrystallizing grains and adds strength at high temperatures. Figure 12.14 compares the high-temperature yield strength of a P/M SAP alloy with other aluminum alloys.

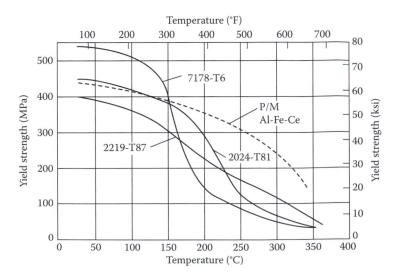

FIGURE 12.14 Yield strength of powder metallurgy aluminum–iron–cerium alloy and other aluminum alloys after a 100-h exposure to the indicated temperatures. (From J.E. Hatch, Ed., *Aluminum: Properties and Physical Metallurgy*, ASM, 1984. With permission.)

12.6 CORROSION RESISTANCE

Aluminum's corrosion and oxidation resistance are a result of a very adherent Al_2O_3 film and the ability of the film to heal. This requires oxidizing conditions. Aluminum has very poor corrosion resistance in bases or reducing acids. Aluminum is passive in neutral solution, but corrodes in both basic and acid conditions as illustrated in Figures 12.15 and 12.16.

Single-phase alloys tend to be more corrosion resistant than multiphase alloys. With multiple phases, corrosion cells can arise between the phases. Table 12.4 shows the electrode potential of various phases that occur in aluminum alloys. Figure 12.17 summarizes these potentials graphically.

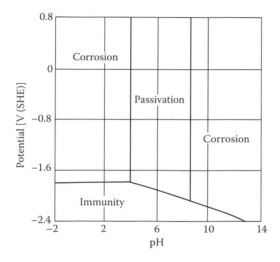

FIGURE 12.15 Pourbaix diagram showing the potential on the corrosion of pure aluminum.

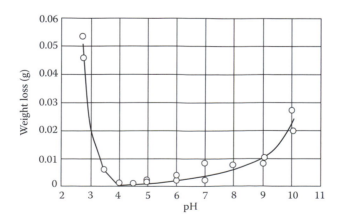

FIGURE 12.16 Weight loss of plates of 3004-H14 combinations of pH and electrode after 1-week exposures in solutions of different pHs. (From J.E. Hatch, Ed., *Aluminum: Properties and Physical Metallurgy*, ASM, 1984. With permission.)

TABLE 12.4

Electrode Potentials

Phase	Potential (v)
Mg_5Al_8	−1.24
4% Zn in solution	−1.05
1% Zn in solution	−0.96
7% Mg in solution	−0.89
5% Mg in solution	−0.88
3% Mg in solution	−0.87
$MnAl_8$	−0.85
Al (99.95 + %)	−0.85
1% Mg_2Si in solution	−0.83
1% Si in solution	−0.81
2% Cu in solution	−0.75
$CuAl_2$	−0.73
4% Cu in solution	−0.69
$FeAl_3$	−0.56
Ni_3Al	−0.52
Si	−0.26

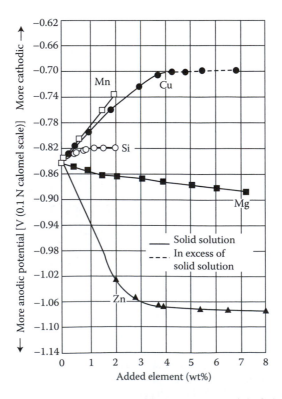

FIGURE 12.17 Effect of various solutes on the electrode potential of aluminum. Note that copper raises the electrode potential while zinc lowers it. Magnesium has little effect.

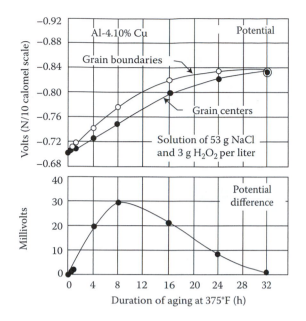

FIGURE 12.18 Effect of aging at 375°F on the electrode potential of an Al-4.1% Cu alloy. (From *Metals Handbook*, 8th ed., Vol. 1, p. 918, ASM, 1961. With permission.)

The 3xxx and 5xxx alloys have very good corrosion resistance because the presence of $MnAl_8$ in the 3xxx alloys and variations of the amount of Mg in solution in the 5xxx alloys have very little effect on the electrode potential. On the other hand, the corrosion resistance of the 2xxx and 7xxx alloys is very sensitive to variations of the amount of solute. Concentration gradients arising from improper heat treatment can lead to severe corrosion problems as indicated in Figures 12.18 and 12.19. In Copper-bearing alloys, precipitation of $CuAl_2$ at grain boundaries causes a depletion of copper near the grain boundaries, making those regions anodic to the grain centers. The result can be rapid grain boundary corrosion. In high Mg alloys, precipitation of continuous zones of Mg_5Al_8 (which is anodic) on grain boundaries and slip bands can also cause grain boundary corrosion (Figure 12.20).

Alloy 2014 is often galvanically protected by a pure aluminum cladding. Commercially pure aluminum (1xxx series), Mn, and Mg-base alloys have good corrosion resistance because they are single phases. The electrode potential does not change greatly with Mg, Mn, and Si. Of the precipitation-hardenable alloys, the 6xxx series has the best corrosion resistance.

12.7 ALUMINUM–LITHIUM ALLOYS

In recent years, there has been considerable research on aluminum–lithium alloys. Lithium is the only alloying element (other than beryllium) that both decreases the density of aluminum and increases its elastic modulus. One weight percent Li decreases the density by 3% and increases Young's modulus by 6% (Figures 12.21

(a)

−0.69   −0.69

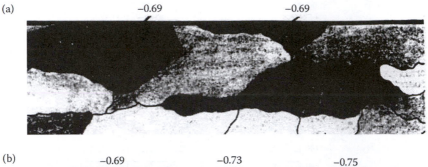

(b)

−0.69   −0.73   −0.75

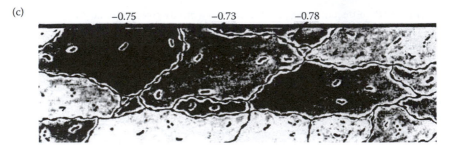

(c)

−0.75   −0.73   −0.78

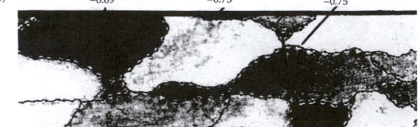

(d)

−0.84   −0.73   −0.84

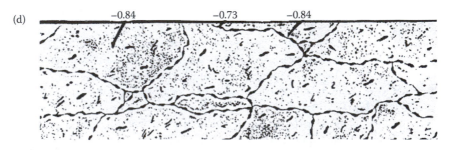

FIGURE 12.19 Schematic illustration of how heat treatment can set up corrosion cells in an aluminum alloy containing 4% Cu. (a) A uniform solid solution after quenching from a solution treatment. (b) The same alloy after heating to a temperature at which grain boundary precipitation occurs. (c) The same alloy heated long enough for continuous copper depleted zone at the grain boundaries. (d) The same alloy heated for a long enough time for grain centers and grain boundary regions to have the same composition. (From *Metals Handbook*, 8th ed., Vol. 1, p. 918, ASM, 1961. With permission.)

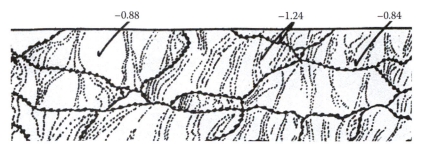

FIGURE 12.20 Precipitation of Mg$_5$Al$_8$ on grain boundaries and slip bands causes these regions to be anodic. (From *Metals Handbook*, 8th ed., Vol. 1, p. 918, ASM, 1961. With permission.)

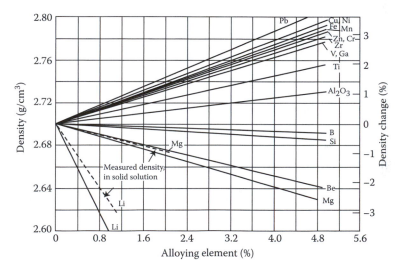

FIGURE 12.21 Effect of various alloying elements on the density of aluminum. Note that a small amount of lithium has a large effect. (From J.E. Hatch, Ed., *Aluminum: Properties and Physical Metallurgy*, ASM, 1984. With permission.)

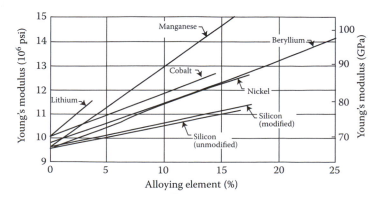

FIGURE 12.22 Effect of various alloying elements on the elastic modulus of aluminum. Note that a small amount of lithium has a large effect. (From J.E. Hatch, Ed., *Aluminum: Properties and Physical Metallurgy*, ASM, 1984. With permission.)

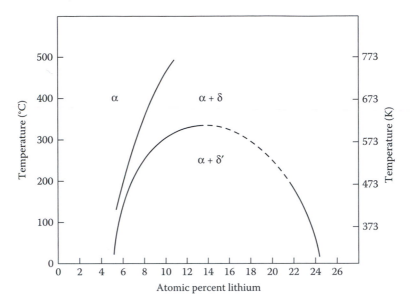

FIGURE 12.23 The aluminum-rich end of the aluminum–lithium phase diagram. (From T. H. Sanders Jr. and E. A. Starke, *Aluminum–Lithium Alloys*, p. 63, TMS, 1981. With permission.)

and 12.22). However Al–Li alloys have poor ductility and toughness. For this reason the alloy is produced by pressing and sintering of powder. The aluminum–lithium phase diagram is given in Figure 12.23. At low temperatures δ and δ' can precipitate in alloys containing over 5% lithium.

12.8 FOAMS

Aluminum can be produced in the form of foams. Metal foams are useful because they can reduce weight without loss of rigidity and because of their ability to absorb mechanical energy. Foams of aluminum alloys 6101 and A356 are produced in a density range of 3–12% of solid aluminum. They can be compressed to a density up to 60% solid aluminum. They have a high specific surface area and low density. Open-celled foams can be used to uniformly mix gasses. A controlled pressure drop and tortuous path through the foam provides the essential microturbulent blending for thorough mixing. Their stress–strain characteristics can be tailored to specific applications by varying the density and alloy of the foam. The high strength-to-weight ratio results in superior performance and greater shear strength than honeycomb of the same density.

Foams are finding application in the automobile industry as materials for energy absorption.

12.9 MISCELLANY

Paul L.T. Héroult in France and Charles M. Hall at Oberlin College in Ohio independently developed the same process for electrolytic reduction of aluminum in molten salts process in 1886. Aluminum was very expensive until after the development of the Hall–Heroult process. For example, when the material for the cap of the Washington monument was selected, aluminum was the choice because it was more expensive than platinum. With the introduction of the Hall–Heroult process, the price dropped dramatically from $8 per pound in 1888 to $0.65 per pound in 1894.

Hall's work led to the formation in 1886 of the Pittsburgh Reduction Company, which later became the Aluminum Company of America. The name *aluminum* in America rather than *aluminium*, which is used in the rest of the world, dates back to an early advertisement. It is not known whether the dropping of the conventional *i* in the name of a metal was intentional or a mistake.

In 1906, Alfred Wilm undertook a series of experiments in an attempt to produce a hardening effect in aluminum alloys akin to that in steel. He heated and quenched a series of alloys. One of the alloys was an alloy containing 4% copper and 0.5% Mg. On quenching, the initial hardnesses were not remarkable. However, after a weekend interruption, he discovered much to his surprise that the hardnesses were much higher than before the weekend. No observable change in the microstructure could be observed because the precipitate was much too fine. His results were published in 1911. Most of the world did not take much notice of his work. However, when the British were able to shoot down a German dirigible that had been bombing London during the First World War, they found its superstructure was made of Wilm's alloy.

In 1919, Wilm's results were explained in a paper by Merica, Waltenberg, and Scott. They showed that the solubility of copper in aluminum decreased at lower temperatures and proposed that precipitation from supersaturated solid solution was the basis of the hardening.

PROBLEMS

1. Your boss has asked you to specify the heat treatment (furnace temperature and times) to achieve a maximum hardness in parts made from aluminum alloy 2014. Production schedules limit the aging time to a maximum of 2 h. Temperature variations from place to place and time to time are ±10°C. Aluminum alloy 2014 contains 4.5% copper. Neglect the minor amounts of the other elements. The aluminum-rich end of the Al–Cu phase diagram is shown in Figure 10.26. The aging characteristics of 2014 are shown in Figure 10.30.

2. Consider the aging of aluminum alloy 2014 contains (0.5 wt% Cu. Its aging characteristics are indicated in Figure 12.24. Assume that in the solution treated condition (before aging) all of the copper is in solution. Also assume that none is in solution after aging at 650°F for 100 h and that the precipitate has a negligible effect on the strength in this overaged condition. Assume that the as-quenched yield strength is 20 ksi.

   a. Solid solution strengthening is proportional to the amount of solute, $\Delta YS = \alpha C$, where C is the atom fraction solute. Determine the constant α.

   b. Now consider the aging condition that gives maximum strengthening (20 h at 300°F). Assume that all of the copper is in the precipitates (none is in solution). Find the value of ΔYS attributable to the precipitates.

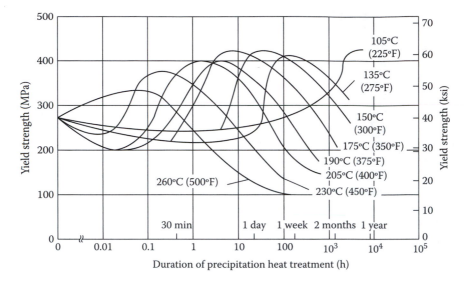

FIGURE 12.24 The dependence of yield strength of Al-2014 on temperature and time. (From J.E. Hatch, Ed., *Aluminum and Aluminum Alloys, ASM Speciality Handbook*, ASM International, 1993. With permission.)

    c. Assuming the Orowan mechanism of hardening, $\Delta YS \approx 4Gb/L$, estimate the particle spacing, L.

3. Read as carefully as possible the times to reach peak hardness in Figure 12.24. In reading a logarithmic scale it should be remembered that 2 is located about 0.3 times the distance between 1 and 10, and 5 is located at about 0.7 times the distance.

    a. Make a plot of time (log scale) versus $1/T$ and determine whether the data can be represented by an Arrhenius equation.

    b. If it can, determine the activation energy.

    c. Estimate the time to reach peak hardness at 212°F.

4. The Lorentz number, L is defined as $k/(\sigma T)$, where k is the thermal conductivity (W/mK), σ is the electrical conductivity (m^{-1}ohm^{-1}) and T is the temperature (K). For most metals, L is in the range of 2 to 2.5×10^{-8} v^{2}T^{-2}s^{-1}. Using the data in Table12.1, calculate L for Ag, Cu, Al, Fe, Pb and Mg and compare with this typical range.

5. Using the data in Table 12.1 and Figure 12.3 calculate the ratio of the electrical conductivity of a copper alloy containing 0.2 % P and 0.5% Si to that of pure copper.

6. There is a great deal of interest in substituting aluminum for steel in auto bodies to save weight. Use the data below to answer the following:

    a. When a flat sheet or plate is bent elastically, the stress, σ, at the outer surface is given by $\sigma = 6(M/w)/t^2$, where M is the bending moment, and w and t are the width and thickness of the sheet. If yielding is to be avoided, the stress, σ, must be less than the yield stress, Y. Calculate the weight saving when aluminum is substituted for steel, with a thickness increase so that it has the same resistance to denting.

b. The elastic deflection, δ, of a flat sheet under load is given by $\delta = C(F/w)L^3/(Et^3)$, where E is Young's modulus, F is the applied force, L is the unsupported span, and C is a constant that depends on how the load is distributed and how the sheet is supported. Calculate the weight saving when aluminum is substituted for steel, with a thickness increase so that it has the same stiffness.

c. For each of the above cases, calculate the ratio of the cost of the aluminum sheet to that of the steel sheet.

Data:

	Young's Modulus (GPa)	Yield Strength (MPa)	Density (Mg/m³)	Price ($/lb)
Steel	205	195	7.87	0.255
Aluminum	70	140	2.70	1.55

7. Aluminum alloy 2017 is used for rivets. These are kept refrigerated before use. Why?

8. Alclad 2024 is a high-strength plate or sheet of aluminum alloy used in corrosive environments. It consists of a core of 2024-T4 clad with pure aluminum. The pure aluminum corrodes very slowly, but it galvanically protects the 2024 core. There is no similar clad product for the 7xxx series of alloys. Why?

9. Pure aluminum shrinks 6% as it solidifies. For an Al-18% Si alloy there is no shrinkage. Estimate the shrinkage for a eutectic Al–Si alloy.

REFERENCES

Aluminum: Properties and Physical Metallurgy, ASM, 1984.
Aluminum and Aluminum Alloys; ASM Specialty Handbook, ASM International, 1993.
John E. Hatch Ed., *Aluminum: Properties and Physical Metallurgy*, ASM, 1983.
Metals Handbook, 8th ed., Vol. 7, ASM.
Metals Handbook, 8th ed., Vol. 8.
T. H. Sanders Jr. and E. A. Strake, *Aluminum–Lithium Alloys*, p. 63, TMS, 1981.

13 Copper and Nickel Alloys

Copper is among the few metals known to ancient civilizations. Copper, gold, silver, platinum, and meteoric iron are the only metals found in their native state. However, about 90% of copper comes from sulfide ores. Most copper ores contain 0.5% Cu or less, iron being the principal metallic impurity. The ores are ground to a fine powder and concentrated (usually by floatation) to 20–25% Cu. This concentrate is melted as a matte of mixed Cu and Fe sulfides containing up to 60% Cu. The matte is oxidized to remove the iron as an oxide and burn off the sulfur. The product is called blister copper and contains 98.5% Cu. Blister copper is fire refined to tough-pitch (99.5% Cu). It is then often electrolytically refined to 99.95 + % Cu. Au, Ag, and Pt metals are recovered from slime.

Copper has good strength and ductility as well as a reasonably good corrosion resistance. The unique features of copper are its color and its very high electrical and thermal conductivities.

Its excellent electrical conductivity accounts for most of its use, principally as wire. Corrosion resistance, thermal conductivity, formability, and unique color account for most of the rest of the consumption of copper.

13.1 ELECTRICAL PROPERTIES

The IACS standard of conductivity was set up with what was thought to be the conductivity of pure copper as 100%. Later it was found that copper had a higher conductivity than originally thought, but the standard was not changed. Pure copper has a conductivity of 104% according to the IACS standard.

Only silver has higher electrical and thermal conductivities than copper. Table 13.1 lists the conductivities of various metals. The resistivity increases by 43% between 0°C and 100°C.

All impurities raise the electrical resistivity of copper. At low concentrations, the increase is proportional to the concentration of the impurity. In general, the effect of solutes in raising the resistivity is greater for large differences between the atomic diameter of the solute and that of copper as shown in Figure 13.1. Also, solutes of high valence (P, Si, and As) have a greater effect on resistivity than solutes of low valence such as Ag, Au, and Cd.

13.2 COMMERCIAL GRADES OF COPPER

There are several commercial grades of copper. The principal use of oxygen-free copper (C10100) with a minimum of 99.99% Cu is for wire. Fire-refined tough-pitch

TABLE 13.1

Electrical and Thermal Conductivities of Several Metals

Metal	Electrical Resistivity (nΩm)	Electrical Conductivity (% IACS)	Thermal Conductivity (Wm⁻¹ K⁻¹)
Silver	14.7	108.4	428 (20°C)
Copper	16.73	103.06%	398 (27°C)
Gold	23.5	73.4	317.9 (0°C)
Aluminum	26.55	64.94	247 (25°C)
Beryllium	40	43	190
Magnesium	44.5	38	418 (20°C)
Zinc	59.16	28.27	113 (25°C)
Nickel	68.44	25.2	82.9 (100°C)
Iron	98	17.59%	80 (20°C)
Platinum	106	16.3	71.1 (0°C)

Source: From *Metals Handbook*, ASM, 9th ed., Vol. 2, 1979. With permission.

copper (C2500) is fire-refined copper containing between 0.02% and 0.05% oxygen in the form of Cu_2O and about 0.5% other elements. The copper–oxygen phase diagram is shown in Figure 13.2. There is a eutectic at 0.38% oxygen and 1066°C. Figure 13.3 is a typical microstructure of cast tough-pitch copper showing primary copper dendrites surrounded by a copper–copper oxide eutectic.

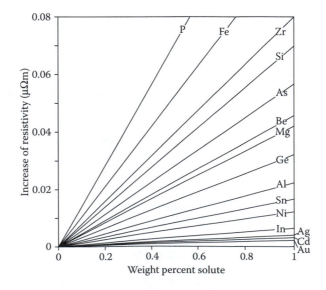

FIGURE 13.1 All impurities increase the resistivity of copper. The increase is proportional to the concentration. (Adapted from *Copper and Copper Alloys, ASM Specialty Handbook*, 2001.)

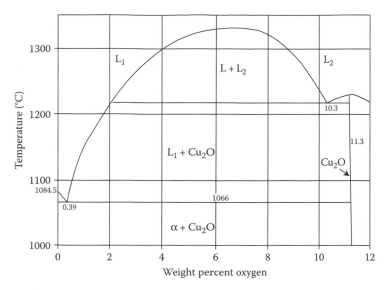

FIGURE 13.2 Copper–oxygen phase diagram. (From *ASM Metals Handbook*, 8th ed., Vol. 8, p. 296, 1973. With permission.)

Often tellurium (~0.5%) or sulfur (~0.5%) is added to copper to promote free machining. Copper is frequently deoxidized with phosphorus. Deoxidation leaves about 0.01% residual phosphorus in solid solution. This lowers the conductivity below that of oxygen-free copper. Lead is often added to copper and copper alloys to

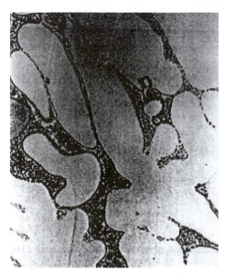

FIGURE 13.3 Microstructure of cast tough-pitch copper. Note the primary copper dendrites and the Cu–Cu_2O eutectic. (From R. E. Reed-Hill and R. Abbaschian, *Physical Metallurgy Principles*, 3rd ed., p. 690, PWS, 1994. With permission.)

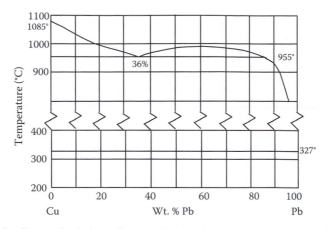

FIGURE 13.4   Copper–lead phase diagram. (Adapted from *ASM Metals Handbook*, 8th ed., Vol. 8, 1973.)

form free-machining characteristics. Lead is virtually insoluble in copper as shown by the copper–lead phase diagram (Figure 13.4). Lead particles appear as a separate phase in the grain boundaries (Figure 13.5).

Zirconium copper (C-15000) contains 0.13–0.20% Zr. It can be heat treated to yield strengths of 400 MPa while retaining a conductivity of 84% ICAS. Copper dispersion strengthened by 0.2–0.7% Al_2O_3 retains reasonable strength at temperatures up to 1000°C. It finds applications as electrodes for resistance welding.

Bismuth is a very detrimental impurity. It completely wets the grain boundaries and because it is brittle, its presence renders copper and copper-base alloys brittle. It must be kept below 0.003%.

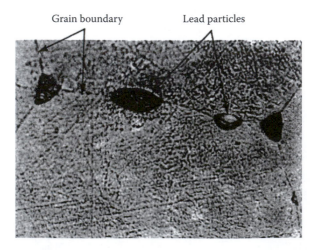

FIGURE 13.5   Lead appears as a separate phase in copper alloys. (From R. M. Brick, R. B. Gordon, and A. Philips, *Structure and Properties of Alloys*, McGraw-Hill, 1965. With permission.)

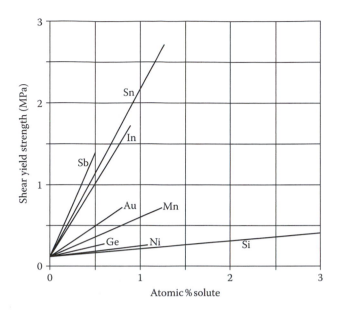

FIGURE 13.6 Effect of solute concentration on shear yield strength of dilute copper-base alloys. Note that the strength is proportional to concentration. (Adapted from J. O. Linde, B. O. Lindell, and C. H. Stade, *Arkiv för Fysik*, Vol. 2, p. 89, 1950.)

Copper is too soft for structural applications. Its strength is markedly increased by alloying. For dilute substitutional solutions (Figure 13.6), the yield strength increases proportional to the solute concentration. The rate of this increase is proportional to the 4/3 power of a misfit parameter defined as $\varepsilon = (da/a)/dc$, where da/a is the fractional change in lattice parameter with concentration, c, expressed as atomic fraction of solutes,

$$\frac{\Delta \tau}{\Delta c} = CG\varepsilon^{4/3}, \tag{13.1}$$

where G is the shear modulus of copper and C is a constant. The misfit, ε, is the fractional difference of atomic diameters of copper and solute. On an atomic basis, Sb, Sn, and In are the most potent hardeners.

Although different measurements of the stacking fault energy of pure copper have ranged from 40 to 169 mJ m^{-2}, the best estimate is probably 80 mJ m^{-2}. Zinc, tin, and aluminum form solid solutions that lower the stacking fault energy to less than 5 mJ m^{-2} at high concentrations as shown in Figure 13.7. The stacking fault energy depends on the ratio of valence electrons to atoms. The lower stacking fault energy leads to more strain hardening. The value of exponent n in the power-law equation,

$$\sigma = K\varepsilon^n, \tag{13.2}$$

increases from about 0.4 for pure copper to about 0.6 or 0.65 at 35% Zn.

The greater strain hardening raises the tensile strength as shown in Figure 13.8.

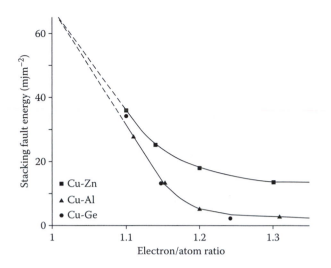

FIGURE 13.7 Stacking fault energy of copper alloys. (Data from J. A. Venables, in *Deformation Twinning*, R. Reed-Hill, J. Hirth, and H. Rogers, Eds, AIME, 1963.)

Finer grain size also increases strength. The grain-size dependence of the yield strength follows the Hall–Petch relation.

It is significant that copper is not embrittled at low temperatures and does not fracture by cleavage. Because of this property, copper is used in cryogenic equipment.

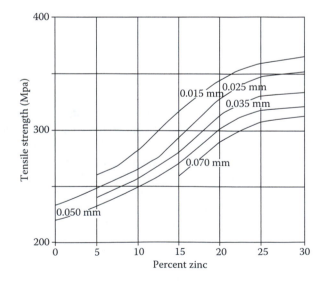

FIGURE 13.8 Effect of zinc content on increasing the tensile strength. Finer grain size has a similar effect. (From *Copper and Copper Alloys, ASM Specialty Handbook*, ASM International, Materials Park, OH, 2001. With permission.)

13.3 COMMERCIAL COPPER ALLOYS

Copper alloys are classified as being either wrought alloys or casting alloys, depending on their usage. The unified number system (UNS) designations for some of wrought alloys are listed in Table 13.2.

Sheets produced by cold rolling and rods or wire produced by drawing are sold with a temper designation indicating the amount of cold reduction. The system is outlined in Table 13.3.

13.4 BRASS

Brasses are alloys of copper and zinc. Figure 13.9 is the copper–zinc phase diagram. The solubility of zinc in the fcc lattice of copper is over 35%. At nearly equal atomic percentages of copper and zinc there is a bcc intermetallic phase, β, that has a wide range of solubility. At temperatures below about 460°C, β-brass undergoes an ordering reaction to form an ordered bcc phase β' with a B2 (CsCl) structure. Each copper atom is surrounded by eight zinc atoms and eight copper atoms. A eutectoid transformation $\beta' \rightarrow \alpha + \gamma$ at 250°C has been reported. However the transformation is rare if ever encountered.

Zinc provides solid solution hardening to copper. Both the yield strength and the tensile strength increase with zinc content as indicated in Figures 13.10 and 13.11.

As the lower stacking fault energy increases the exponent, n, in the power-law approximation of the stress–strain curve, the uniform elongation also increases.

Another effect of the lower stacking fault energy is that annealing twins are much more frequent. Figure 13.12 compares the annealed microstructure of copper with a brass containing 30% Zn.

TABLE 13.2
Commercial Wrought Copper Alloys

Designation	Composition	Common Name	Uses
C21000	5% Zn	Gilding metal	Coinage, jewelry
C22000	10% Zn	Commercial bronze	Architectural, jewelry
C23000	15% Zn	Red brass	Architectural, plumbing
C24000	20% Zn	Low brass	Ornamental, musical
C26000	30% Zn	Cartridge brass	Lamps, cartridge cases
C27000	35% Zn	Yellow brass	Architectural, lamps
C28000	40% Zn	Muntz metal	Architectural, heat exchangers
C36000	35.5% Zn, 2.5% Pb	Free cutting brass	Screw machine parts
C44300	28% Zn, 1% Sn	Admiralty brass	Heat exchangers
C51000	5% Sn, 0.2% P	Phosphor bronze	Hardware
C71500	30% Ni	Cupronickel	Heat exchangers
C75200	18% Ni, 17% Zn	Nickel silver	Hardware, jewelry
C172000	1.8–2.0% Be, 0.2–0.8% Ni + Co	Beryllium copper	Springs, instruments

TABLE 13.3

Temper Designations

Temper Designation		B&S Gage No's.	Rolled Sheet Thickness Reduction (%)	Strain	Drawn Wire Diameter Reduction (%)	Strain
H01	1/4 Hard	1	10.9	0.116	10.9	0.232
H02	1/2 Hard	2	20.7	0.232	20.7	0.463
H03	3/4 Hard	3	29.4	0.347	29.4	0.694
H04	Hard	4	37.1	0.463	37.1	0.926
H06	Extra hard	6	50.1	0.696	50.1	1.39
H08	Spring	8	60.5	0.925	60.5	1.86
H10	Extra spring	10	68.6	1.16	68.6	2.32

The microstructures of brass containing 40% zinc or more consists of two phases, α and β. The two-phase microstructure of Muntz metal (40% Zn) is shown in Figure 13.13. At 800°C the microstructure is entirely β, but as the alloy is cooled, α precipitates.

The effects of other alloying elements to brass on the solid solubility of zinc can be approximated by *zinc equivalents*. A concentration of 1% tin is equivalent to 2%

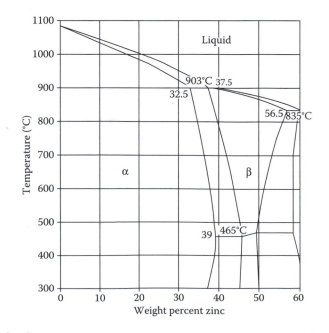

FIGURE 13.9 Copper–zinc phase diagram. (Adapted from *Copper and Copper Alloys, ASM Specialty Handbook*, Materials Park, OH, 2001.)

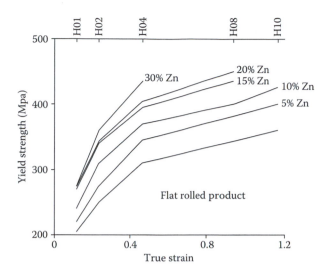

FIGURE 13.10   Yield strength of brasses increases with the amount of cold work and with the zinc content. The true strain $\varepsilon = \ln(t_o/t)$, where t is the thickness and t_o is the thickness before cold rolling. (Data from *ASM Metals Handbook*, 9th ed., Vol. 2, 1979.)

zinc, so tin has a zinc equivalent of 2. Table 13.4 lists the zinc equivalents of the common alloying elements.

The relative values of the zinc equivalents of the elements can be understood in terms of the number of valence electrons per atom. Silicon and aluminum have high valences (4 for silicon and 3 for aluminum) and low atomic masses (28 for silicon and

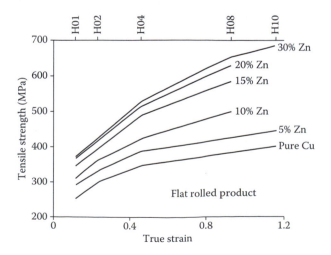

FIGURE 13.11   Tensile strength of brasses increases with the amount of cold work and with the zinc content. The true strain $\varepsilon = \ln(t_o/t)$, where t is the thickness and t_o is the thickness before cold rolling. (Data from *ASM Metals Handbook*, 9th ed., Vol. 2, 1979.)

(a)   (b)

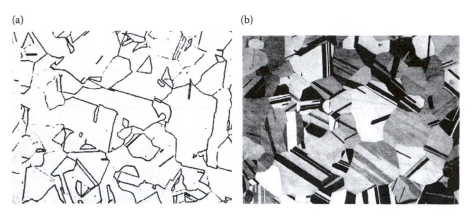

FIGURE 13.12 Microstructures of a pure copper (a) and 70/30 brass (b) (200×). (From W. F. Hosford, *The Mechanics of Crystals and Textured Polycrystals*, Oxford Sci. Pub. 1993. With permission and courtesy of A. Graf.)

27 for aluminum), so 1 wt% causes a large increase of the electron-to-atom ratio. Magnesium, like zinc, has a valence of 2, but the atomic mass of magnesium is less than half that of zinc (24 for magnesium versus 65 for zinc), so 1 wt% magnesium has over twice the effect of 1 wt% zinc. Tin has a valence of 4, but its atomic mass (119) is roughly twice that of zinc.

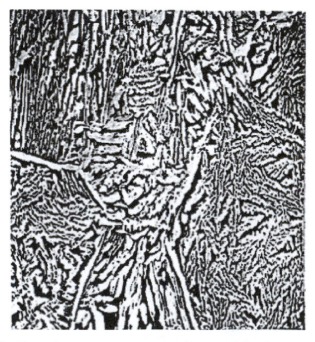

FIGURE 13.13 Photomicrograph of muntz metal (40% Zn) after air cooling. The dark phase is b and the light phase is a, which has precipitated during cooling. (From R. M. Brick, R. B. Gordon, and A. Phillips, *Structure and Properties of Alloys*, p. 135, McGraw-Hill, 1965. With permission.)

TABLE 13.4

Zinc Equivalents of Several Alloying Elements in Brass

Element	Zn Equivalent	Element	Zn Equivalent
Si	10	Al	6
Mg	2	Sn	2
Fe	0.9	Mn	0.5
Co	−0.8	Ni	−1.3

13.5 OTHER COPPER-BASED ALLOYS

True bronzes are copper–tin alloys. Tin is much less soluble in copper than zinc as shown in the copper–tin phase diagram (Figure 13.14). The ε-phase is very sluggish to form and is rarely encountered. Tin is a potent solid solution strengthener. However, alloys with more than 8% tin are too brittle to be formed mechanically. Because bronzes have a much greater difference of the liquidus and solidus temperatures than brass, they are much easier to cast. Casting alloys often contain up to about 20% Sn. These are used extensively for pipe fittings and bells.

Alloys of copper with aluminum and silicon are called aluminum bronze and silicon bronze, even though they contain no tin. Silicon bronzes are more easily cast than brass and have better resistance to oxidation and acid corrosion.

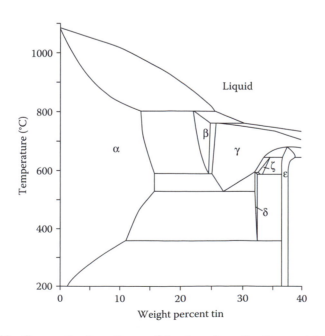

FIGURE 13.14 Copper–tin phase diagram β is a bcc phase. In alloys containing less than about 20% Sn, formation of the ε-phase is so sluggish that it is seldom encountered.

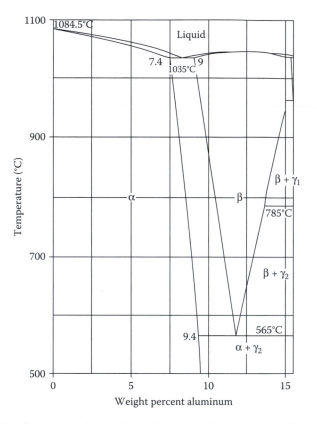

FIGURE 13.15 Copper–aluminum phase diagram. b is a bcc phase. Note the eutectoid transformation at 565°C.

Manganese bronze is a brass with only 1% Sn and 0.1% Mn.

Aluminum bronzes containing up to 7% Al behave much like brass. The phase diagram (Figure 13.15) shows that a bcc β-phase occurs at about 12% Al.

There is a eutectoid transformation of $\beta \rightarrow \alpha + \gamma_2$ at 565°C and 11.8% Al. Figure 13.16 shows the pearlite-like microstructure of alternating platelets of α and γ_2 in an alloy of Cu–11.8% Al after slowly cooling from 800°C. With rapid cooling the eutectoid transformation is suppressed and the alloy transforms by martensitic shear to a new hexagonal phase, β′. Figure 13.17 shows the martensitic β′ needles in the same alloy after rapid cooling. The martensitic reaction is almost instantaneous. It starts when the temperature of the β reaches the M_s temperature and is virtually complete when the temperature is at the M_f. The M_s and M_f temperatures depend on the composition as shown in Figure 13.18. The martensite is not particularly hard.

The oxidation resistance of aluminum bronzes is somewhat superior to brass.

Copper and nickel are completely miscible in both the liquid and solid state as shown in the copper–nickel phase diagram (Figure 13.19). The low-temperature miscibility gap is rarely encountered.

Nickel imparts improved corrosion and oxidation resistance to copper. Cupronickels containing 20–30% nickel are widely used for heat exchangers.

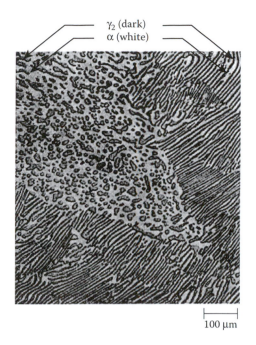

FIGURE 13.16 Pearlite-like structure in Cu–11.8% Al after slow cooling from 800°C. (From *Copper and Copper Alloys*, *ASM Specialty Handbook*, p. 48, ASM International, Materials Park, OH, 2001. With permission.)

FIGURE 13.17 Martensitic structure of Cu–11.8% Al after rapid cooling from 800°C. (From *Copper and Copper Alloys*, *ASM Specialty Handbook*, p. 48, ASM International, Materials Park, OH, 2001. With permission.)

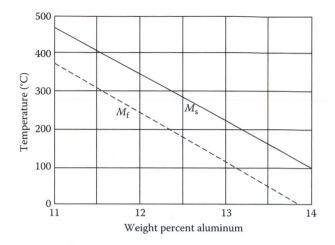

FIGURE 13.18 M_s and M_f temperatures of aluminum bronzes. (Adapted from R. A. Flinn, *Copper, Brass, and Bronze Castings*, p. 20, The Non-Ferrous Founders Society, Cleveland, OH, 1961.)

Alloying copper with 20% nickel causes a loss of the copper's familiar yellow-reddish color. Coinage accounts for about 1% of the consumption of copper. The five-cent pieces ("nickels") in the United States and Canada are alloys of 75% copper and 25% nickel. The 10-cent ("dime") and 25-cent ("quarter") pieces are sandwiches of pure copper core clad with 75% copper–25% nickel alloy as shown in Figure 13.20.

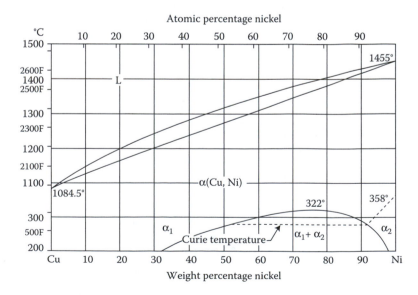

FIGURE 13.19   Copper–nickel phase diagram shows complete miscibility. (Data from *ASM Handbook*, 8th ed., Vol. 8, p. 294, 1973.)

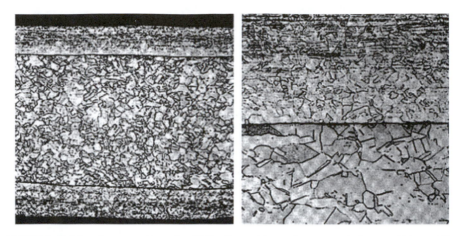

FIGURE 13.20 Cross section of a U.S. dime. (From *Copper and Copper Alloys*, *ASM Specialty Handbook*, ASM International, Materials Park, OH, 2001. With permission.)

The Susan B. Anthony U.S. dollar is a sandwich of pure copper clad with 77Cu–12-Zn–7Mn–4Ni. European coins are also either copper-base alloys or combinations of two copper-base alloys.

Constantan (45% Ni) has a nearly zero temperature coefficient of resistivity, so it is useful in instruments.

Nickel–silvers are ternary alloys containing 12–18% nickel and 17–27% zinc. They have good strength and good corrosion resistance. They are used for springs and corrosion-resistant hardware. Costume jewelry is often made from nickel–silver because of its pleasant silvery color.

Copper alloys containing 1.8–2.0% beryllium and 0.2–0.8% Ni + Co are precipitation hardenable. The alloy C172000 is solution treated at about 800°C and rapidly cooled. The precipitation is carried out between 300°C and 400°C. Figure 13.21 shows the copper-rich end of the Cu–Be phase diagram. Precipitation hardening produces very high strengths. Even higher strengths can be achieved by cold working after the solution treatment and before the precipitation as shown in Table 13.5. The principal drawback is that the toxicity of beryllium is an extreme danger in preparing this alloy.

There are two groups of shape memory alloys: One group consists of Cu–Zn–Al alloys with 10–30% Zn and 5–10% Al. Cu–Al–Ni alloys with 11–14.5%Al and 3–5%Ni form the other group. The recoverable strain for both groups is 4%. The martensitic transformation temperature is less than 120°C for the Cu–Zn–Al alloys and less than 200°C for the Cu–Al–Ni alloys. The exact temperatures depend on composition. The temperature hysteresis is in the range of 15–25°C.

13.6 CASTING OF COPPER ALLOYS

In general, alloys with a greater separation of liquidus and solidus are regarded as easier to cast. In a fixed thermal gradient, the length of dendrite, L, arms is

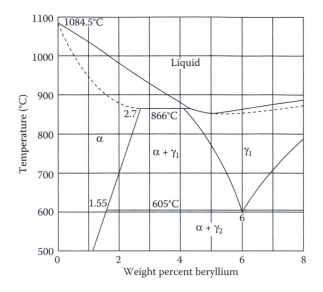

FIGURE 13.21   Copper–beryllium phase diagram. (Adapted from *ASM Handbook*, 8th ed., Vol. 8, 1973.)

proportional to this temperature difference, ΔT. The liquid-to-solid shrinkage can occur interdendritically. This simplifies risering. Lower liquidus temperatures are also beneficial.

The solubility of hydrogen, like other gasses, is much greater in liquid copper than solid copper. Figure 13.22 shows the solubility of hydrogen in copper and

TABLE 13.5
Strength of Beryllium–Copper Sheet (C17200)

Condition[a]	Yield Strength (MPa)	Tensile Strength (MPa)
Annealed	250	465
1/4H	485	570
1/2H	555	605
H	690	730
Annealed & HT	1060	1260
1/4H & HT	1125	1290
1/2H & HT	1180	1325
H & HT	1195	1360

Source: Adapted from *Copper and Copper Alloys*, *ASM Specialty Handbook*, 2001.

[a] 1/4H, 1/2H, and H indicate increasing strengths by work hardening, HT indicates age-hardening treatment. (Solution treatments were carried out prior to work hardening.)

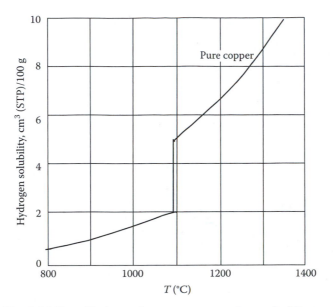

FIGURE 13.22 Solubility of hydrogen in copper, measured as cm³ of H₂ at standard temperature and pressure per 100 g copper.

copper–aluminum alloys. Alloying with aluminum and some other elements decreases the solubility.

As the metal solidifies, dissolved gas is released. For hydrogen,

$$H_2O + 2Cu \rightarrow 2\underline{H} + Cu_2O. \tag{13.3}$$

This is likely to cause gas porosity. The source of dissolved hydrogen may be from wet scrap, or from furnace gases.

13.7 CORROSION OF COPPER ALLOYS

Many of the applications of copper and copper-base alloys depend on their generally good corrosion resistance. A few specific forms of corrosion should be noted. One is pitting corrosion. The attack is very localized. Pitting attacks cause much more damage than the same amount of metal loss uniformly spread over the corroded surface.

A second type of corrosion is dezincification of brasses containing more than 15% Zn. Both copper and zinc atoms go into solution but copper reprecipitates as a porous plug. An example of dezincification is shown in Figure 13.23.

Brass containing 15% or more zinc is susceptible to stress corrosion or season cracking. It occurs only when the brass is under tensile stresses and only under specific environments. For brass, ammonia is the most common agent. Figure 13.24 shows examples of stress-corrosion cracking of brass. The propensity to stress

FIGURE 13.23 A plug of dezincified metal on a brass pipe. (From *Copper and Copper Alloys*, *ASM Specialty Handbook*, p. 48, ASM International, Materials Park, OH, 2001. With permission.)

corrosion increases with zinc content and with temperature. Removal of residual stresses by stress-relief anneals is sufficient for many applications.

Hydrogen is not a problem for most copper alloys. However, if tough-pitch copper containing Cu_2O is exposed to hydrogen at high temperatures, the following reaction occurs:

$$2\underline{H} + Cu_2O \rightarrow H_2O + 2Cu. \tag{13.4}$$

FIGURE 13.24 Examples of stress-corrosion cracking in brass in a condenser tube of drawn C12200 copper. (From *Copper and Copper Alloys*, *ASM Specialty Handbook*, p. 347, ASM International, Materials Park, OH, 2001. With permission.)

Such exposure could, for example, occur during torch welding. The formation of H_2O in the form of steam causes embrittlement.

13.8 NICKEL-BASED ALLOYS

Nickel has an fcc crystal structure. Its melting point is 145°C and its density is 8.9.

Its main uses are in the chemical industry for its corrosion resistance, for heat-resisting alloys including the super alloys and for magnetic alloys.

Relatively pure Ni, with or without small amounts of Mn, Al, and Si, has very good corrosion resistance in air and seawater. The oxidation resistance is lowered by sulfur. *Monel* contains 67% Ni and 33% Cu, which is approximately the Ni/Cu ratio in Sudbury Ontario deposits. It finds widespread use because of its corrosion resistance.

The solubility of chromium is over 30% as shown in Figure 13.25. Chromium additions provide oxidation resistance at elevated temperatures. Nickel–chromium alloys are widely used for electrical heating elements. These contain typically 20% Cr and 1.2–2% Si with or without smaller amounts of Fe and Al. These are used for furnace windings up to 1200°C as well as for toaster. Figure 13.26 shows the temperature dependence of resistivity for a number of Ni–Cr alloys. *Nichrome* is a trade name.

Ni-base superalloys are alloys used for service at temperatures in the range of 1500–2000°F (800–1100°C) in jet engines and power-generating turbines. Figure 12.27 compares these alloys with other superalloys. Among the alloys are Waspalloy, Udimet 500 and 700, and René 41. These are all nickel-base alloys with 13–20% Cr, 10–18% Co, and 2–5% Al and/or Ti. Their microstructures consist of γ', $Ni_3(Al, Ti)$,

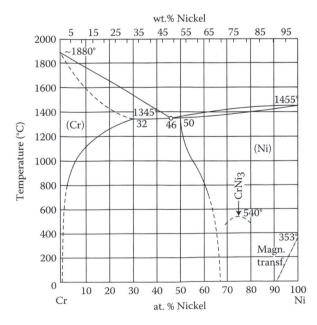

FIGURE 13.25 Nickel–chromium phase diagram. Chromium has a great solubility in the fcc lattice of nickel. (From *Copper and Copper Alloys, ASM Specialty Handbook*, p. 347, ASM International, Materials Park, OH, 2001. With permission.)

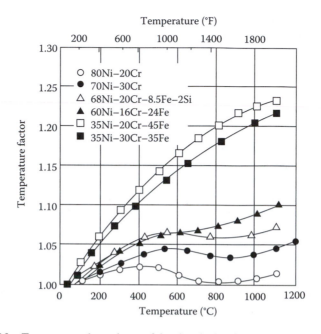

FIGURE 13.26 Temperature dependence of the electrical resistance for several Ni–Cr alloys. The temperature factor is the ratio of the resistance at temperature to that at room temperature. (From *ASM Specialty Handbook, Nickel and Cobalt Alloys*, ASMI, 2000. With permission.)

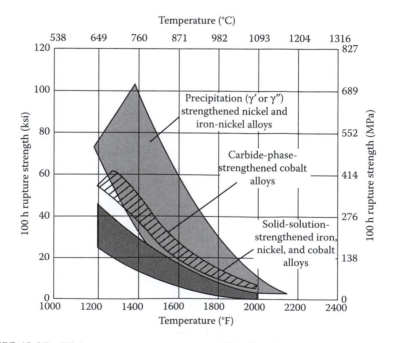

FIGURE 13.27 High-temperature stress-rupture life of various superalloys. Note that the Ni-base alloys strengthened by γ′ precipitates offer superior properties. (From *ASM Specialty Handbook, Nickel and Cobalt Alloys*, ASMI, 2000. With permission.)

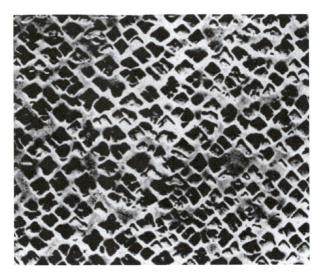

FIGURE 13.28   Microstructure of a cast superalloy with γ′ precipitates in a γ matrix. (From *ASM Specialty Handbook, Nickel and Cobalt Alloys,* ASMI, 2000. With permission.)

and precipitates in a γ matrix (Figure 13.28). Both γ′ and γ matrices are fcc, with slightly different lattice parameters.

Nickel is one of the three common ferromagnetic metals (Fe and Co are the other two). It is a principal component of several soft magnetic alloys including *permalloy 45* and *hypernik,* which contain 45–50% Ni with iron as most of the balance. It is also used in permanent magnetic alloys. One of these is alnico, which is an iron-base alloy with typically 20% Ni, 8–12% Al, and 5–24% Co. Chapter 24 treats ferromagnetic alloys.

13.9   MISCELLANY

Cyprus was a major source of copper in the ancient world. The name "Cyprus" comes from the Latin "cuprus." The oldest worked copper artifact is a pendant dating about 9500 BC. The Egyptians and the Sumerians reduced ores with charcoal about 4000 BC. About 3500 BC, copper was alloyed with tin to produce bronze and this harder metal was so universal in early history that one period is known as the Bronze Age. The first bronzes were really copper containing up to 8% arsenic. Because of this, many of the early smelters suffered from arsenic poisoning. True tin bronzes came later.

Columbus's ships, Nina, Pinta, and Santa Maria, had copper skins below the water line to increase the hull life by protecting against barnacles. Paul Revere was a colonial coppersmith and he supplied the brass fittings and the ship's bronze bell for the battleship US Constitution. When it was built in 1797, the US Constitution was sheathed with copper sheet imported from England to prevent biofouling. The federal government loaned Revere $10,000 to build the first American mill to roll copper. When the US Constitution was refitted in 1802, the sheathing was replaced with copper from his mill. The Statue of Liberty contains 179,000 lbs of copper.

The name, *nickel*, comes from the German, *kupfernickel*, meaning false copper. Nickel coins were minted as early as 235 BC in China. A. Cronstedt purified Ni in 1751. Nickel is a constituent of meteorites. The Sudbury district of Ontario supplies 30% of world's production. Nickel is produced also in Russia, New Caledonia, Australia, Indonesia, and Cuba.

PROBLEMS

1. Figure 13.3 is a photomicrograph of electrolytic tough-pitch copper. Showing a Cu–Cu$_2$O eutectic.
   a. Find the area percent eutectic by either a line or a point count method. The density of the eutectic is almost the same as that of copper, so this is approximately the wt% eutectic.
   b. Use the phase diagram (Figure 13.2) together with the lever law to determine the wt% oxygen in the material.
2. Use the copper–lead phase diagram (Figure 13.4) to explain the following:
   a. Copper–lead alloys are not used at temperatures in excess of 350°C.
   b. Copper–lead alloys contain less than 30% Pb.
3. Figure 13.29 is a photomicrograph of a manganese bronze (Cu alloy containing 39.5% Zn, 1% Fe, 1% Al, and 0.5% Sn) as cast. Determine which phase is the darker one.

FIGURE 13.29 Cast manganese bronze containing 39.5% Zn, 1% Fe, 1% Al, and 0.5% Sn. (From *Metals Handbook*, 8th ed., Vol. 7, p. 296, 1973. With permission.)

REFERENCES

ASM Metals Handbook, 8th ed., Vol. 7, p. 296.

ASM Metals Handbook, 8th ed., Vol. 8, p. 296.

R. M. Brick, R. B. Gordon, and A. Philips, *Structure and Properties of Alloys*, McGraw-Hill, 1965.

Copper and Copper Alloys, ASM Specialty Handbook, ASM International, 2001.

R. A. Flinn, *Fundamentals of Metal Casting*, p. 210, Addison-Wesley, 1963.

W. F. Hosford, *The Mechanics of Crystals and Textured Polycrystals*, Oxford Sci. Pub. 1993.

J. O. Linde, B. O. Lindell, and C. H. Stade, *Arkiv för Fysik*, Vol. 2, p. 89, 1950.

Metals Handbook, ASM, 9th ed., Vol. 2, 1979.

Nickel and Cobalt and Their Alloys, ASM Specialty Handbook, ASM International, 2000.

R. E. Reed-Hill and R. Abbaschian, *Physical Metallurgy Principles*, 3rd ed., p. 690, PWS, 1994.

Understanding Copper Alloys, J. H. Mendenhall, Ed., Olin Brass, 1977.

J. A. Venables, in *Deformation Twinning*, R. Reed-Hill, J. Hirth, and H. Rogers, Eds, AIME, 1963.

14 Hexagonal Close-Packed Metals

14.1 GENERAL

The hexagonal close-packed metals of commercial importance are Be, Mg, Zn, Ti, and Zr. A summary of the important properties is given in Table 14.1. The main uses of beryllium are related to its combination of a high elastic modulus and very low density. Magnesium's uses are mainly a result of its low density. Zinc is widely used in the die casting industry. Titanium is used in the chemical industry because of its corrosion resistance and in the aerospace industry because of its combination of its low density and high strength at elevated temperatures. Zirconium is used in the nuclear industry for its low neutron cross section.

14.2 ZINC

The main uses of zinc are (1) as an alloying element in aluminum and copper alloys, (2) for galvanizing steel, and (3) for die castings. The principal die casting alloy is *Zamak* 4Al, 0.2 Mg. It is used widely for parts of locks and other small parts, handles, carburetor, and fuel pump housings. Die casting accounts for approximately 1/3 of the annual consumption of zinc. Of the three metals used for most die castings, zinc, magnesium, and aluminum, zinc has the lowest melting point. Wear of the dies is least, so zinc die castings usually have the best surface finish. However, because of its relatively high density, zinc is being replaced by lighter metals and plastics in applications where weight is of concern.

About 40% of all the zinc consumed in the United States is used to protect steel from corrosion. *Galvanizing* is the plating of steel with zinc either electrolytically or by hot dipping. Even if there are holes in the plating, the zinc coatings protect the steel because zinc is anodic to iron. The protection is effective because zinc itself corrodes very slowly in neutral solutions as indicated by Figure 14.1.

EXAMPLE PROBLEM 14.1

Consider two neighboring grains, A and B, in zinc as shown in Figure 14.2. Assume that the dimensions in the *x* direction remain equal in both grains as the temperature

TABLE 14.1
HCP Metals

	MP	ρ (g/cm³)	E (10^6 psi)	Price ($1/lb)
Be	1277	1.85	40	375 (powder)
				240.00 (bulk 98.5%)Be
Mg	650	1.74	6.5	0.98
Zn	419	6.5	x	0.53
Ti	1668	4.5	17	13.75 (ingot)
Zr	1852	6.5	13.7	10.00 (sponge)

is changed and the grains have the same cross-sectional area. Calculate σ_x for a 1°C temperature change.

The coefficients of thermal expansion of zinc are $\alpha_A = 61.5 \times 10^{-6} °C^{-1}$ parallel to the c-axis and $\alpha_B = 15 \times 10^{-6} °C^{-1}$ perpendicular to the c-axis. The values of Young's moduli parallel and perpendicular to c are $E_A = 35$ GPa and $E_B = 120$ GPa.

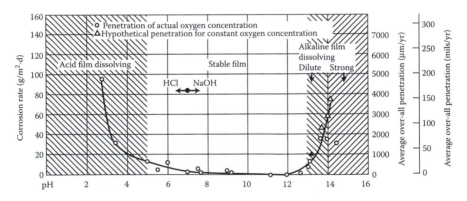

FIGURE 14.1 The effect of pH on the corrosion rate of zinc. (From *Metals Handbook*, 9th ed., Vol. 2, ASM, 1979. With permission.)

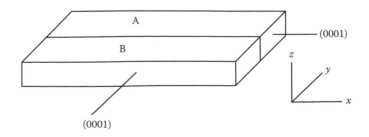

FIGURE 14.2 Bicrystal of zinc.

SOLUTION

Substituting $\sigma_x = \sigma_A = -\sigma_B$ into $\varepsilon_x = \sigma_A/E_A + \alpha_A \Delta T = \sigma_B/E_B + \alpha_B \Delta T$

$$\frac{\sigma_A}{E_A} + \frac{\alpha_B}{E_B} = (\sigma_B - \sigma_A)\Delta T, \qquad \sigma_x = \frac{(\sigma_B - \sigma_A)\Delta T}{1/E_A + 1/E_B T}.$$

Substituting $\Delta T = 1°C$ and values for α and E, $\sigma_x = -1.95$ MPa.

This suggests that with appreciable temperature changes, the stresses in some grains will be high enough to cause yielding. Therefore, it is impossible to stress relieve zinc, because cooling from such an anneal will create high stresses in some grains. Tables of elastic moduli rarely, if ever, list the modulus of zinc. The reason is that stress–strain curves for zinc do not have an initial linear region. Some grains yield as soon as any external stress is applied.

An alloy containing about 30% Al undergoes a eutectoid transformation on cooling. The fine stable eutectoid structure permits superplastic forming.

Both zinc and cadmium have high vapor pressures at their melting temperatures. Good ventilation is required when welding galvanized steel. Exposure to the zinc vapor causes the *zinc shakes*. The major use of metallic cadmium is plating of bolts for corrosion resistance.

14.3 MAGNESIUM

The density of magnesium is about 2/3 that of aluminum and a quarter of that of steel. Its low density accounts for most of the use of magnesium-base alloys. Its use in automobiles, kitchen appliances, and garden tools is increasing. Injection molding of semisolids (*thixomolding*) is finding application for laptop computers, cameras, and cell phones. Magnesium is commonly alloyed with zinc and aluminum. Both have appreciable solubility above 300°C. Decreasing solubility at lower temperatures permits precipitation hardening. The magnesium–aluminum phase diagram is shown in Figure 14.3 and the magnesium–zinc phase diagram in Figure 14.4. Other alloying elements include thorium, manganese, zirconium, and rare earths.

The alloy designation system uses two letters to designate the two most abundant alloying elements followed by two numbers, which indicate the amount of the two alloying elements to the nearest percent. Table 14.2 indicates the letters used for each element.

For example, AZ63 contains 6% Al, 3% Zn and AM100 contains 9.5% Al, 0.1% Mn. The temper designations are same as for aluminum alloys, for example, T6 is solution treated and artificially aged and H24 is strain hardened and partially annealed.

Zinc and aluminum gives solid solution hardening. Although much less soluble, the rare earths are potent solid solution strengtheners as indicated in Figure 14.5. Manganese increases the corrosion resistance in seawater. Rare earths are alloyed as *mischmetal* (a natural mixture of rare earths containing 50% Ce) or as *didymium* (a mixture of 85% neodymium and 15% praseodynium). Rare earths and cerium form precipitation-hardening systems. Mg_3RE and Mg_3Ce form planar precipitates on the prism planes. As indicated in Figure 14.6, these block slip on the basal plane. At 1%,

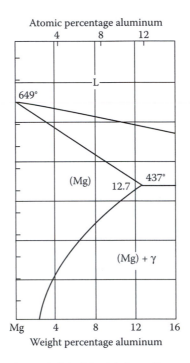

FIGURE 14.3 Magnesium-rich end of the Mg–Al phase diagram. (From *Metals Handbook*, 8th ed., Vol. 8, 1973. With permission.)

rare earth additions increase strength and reduce the tendency to crack during welding. Silicon increases fluidity. Thorium and yttrium improve creep resistance. Tin increases the ductility. Zinc improves strength and with Zr. Rare earth and thorium additions form precipitation-hardenable alloys. Zirconium is added for grain refinement during casting. Lithium reduces the density. With over 11%, Mg–Li alloys are bcc. There has been great interest in Mg–Li alloys but they have found little or no application.

Seawater is the principal source of magnesium, though some is produced from dolomite. The single largest use of magnesium is as alloy in aluminum alloys. However, the use of magnesium-base alloys is increasing. At one time Volkswagon used more than half of the world's production of magnesium for engine blocks, transmission housings, and other parts. Recently, the automobile industry has begun to use more magnesium components in its effort to reduce weight. Most of its use has been in die casting. A recent development of a process of casting of metals in a two-phase liquid–solid condition is called *rheocasting* or *thixomolding*. AZ91 cast by this process has found growing use for laptop computers.

The compositions of common die-casting alloys are given in Table 14.3.

For wrought products (sheet and bar) the tensile yield strength is usually higher than the compressive yield strength. This is because of the textures formed during rolling and extrusion in which the basal (0001) plane is aligned with the sheet or

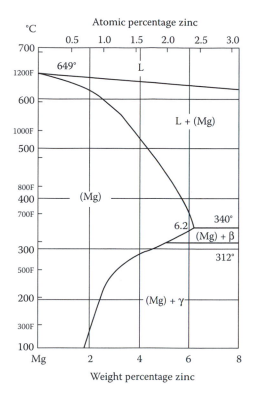

FIGURE 14.4 Magnesium-rich end of the Mg–Zn phase diagram. Note the appreciable solid solubility of Zn in Mg. (From *Metals Handbook*, 8th ed., Vol. 8, 1973. With permission.)

TABLE 14.2
Magnesium Alloy Designation System

A = **A**luminum

E = rare **E**arth elements (misch metal)

H = t**H**orium

K = zir**K**onium

M = **M**anganese

Z = **Z**inc

C = **C**opper

F = iron (**F**errous)

N = **N**ickel

Q = silver (**Q**uick silver)

R = ch**R**omium

T = **T**in

Y = **Y**ttrium

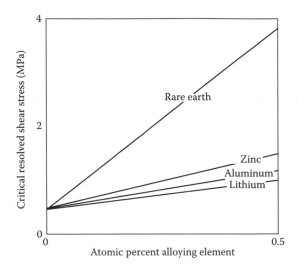

FIGURE 14.5 Effect of several substitutional alloying elements on the critical resolved shear stress for basal slip. Note the large effect of rare earths.

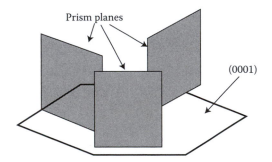

FIGURE 14.6 Mg_3RE and Mg_3Ce precipitate as plates on the prism planes making basal slip more difficult.

TABLE 14.3
Compositions of Common Die-Casting Magnesium Alloys

Alloy (%)	Al	Zn	Mn	Si	Rare Earths
AE42	4.2		0.2		2.5
AM20	2.1		0.4		
AM50A	4.9		0.4		
AM60	6.0		0.4		
AS21	2.2		0.2	1.0	
AS41	4.25		0.2	1.0	
AZ91	9.0	0.7	0.15		

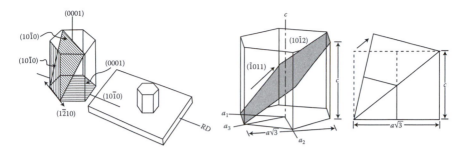

FIGURE 14.7 After rolling, the (0001) basal plane is closely aligned with the plane of the sheet. Compression in the sheet can cause $10\bar{1}2\ \langle10\bar{1}1\rangle$ twinning but tension cannot, so yielding must occur by slip on unfavorably oriented (0001) planes. (Adapted from W. F. Hosford and W. A. Backofen, in *Fundamentals of Deformation Processing: Proceedings of the 9th Sagamore AMRA Conference*, Syracuse University Press, 1964.)

parallel to the rod axis. With this orientation, compression applied in the plane of the sheet (or parallel to the rod axis) will cause $\{10\bar{1}2\}\ \langle10\bar{1}1\rangle$ twinning, whereas tension will not (Figure 14.7), so yielding must occur by slip on the unfavorably oriented (0001) plane. The four-digit index system is explained in Appendix A2. The yield locus of a textured magnesium sheet is shown in Figure 14.8.

Magnesium alloys corrode readily in acid solutions but become passive in basic solutions as indicated in Figure 14.9. Because of this, it is used for hand tools for working concrete, an application that would cause rapid corrosion of aluminum alloys.

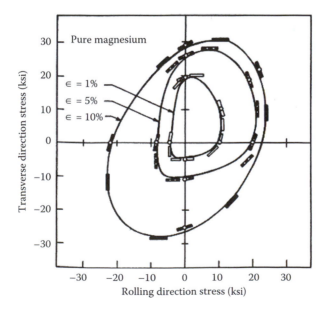

FIGURE 14.8 Yield loci for a sheet of pure magnesium. The strength in compression in the plane of the sheet is low because of $\{10\bar{1}2\}\ \langle10\bar{1}1\rangle$ of twinning in compression. (From E. W. Kelley and W. F. Hosford, *Trans TMS-AIME*, 242, 5–13, 1968. With permission.)

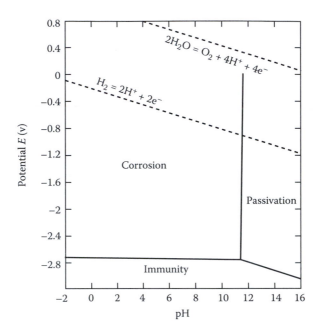

FIGURE 14.9 Poubaix diagram for magnesium. Note that it is passive in basic solutions. (From *Magnesium and Magnesium Alloys, ASM Specialty Handbook*, ASMI, 1999. With permission.)

14.4 TITANIUM

At room temperature, titanium has an hcp crystal structure (α-phase). Above 882°C it is bcc (β-phase) up to its melting point at 1670°C. Its density, 4.5, is intermediate between aluminum and steel.

EXAMPLE PROBLEM 14.2

α-titanium at 20°C has an hcp crystal structure with $a = 0.2950$ nm and $c = 0.4683$ nm at 20°C. At 900°C, it transforms to β-titanium with a bcc crystal structure with $a = 0.332$. The average coefficient of thermal expansion of α-titanium is 8.4×10^{-6} K^{-1}. Calculate the % volume change as β-titanium transforms to α-titanium on cooling. Indicate whether this is a contraction or expansion.

SOLUTION

The volume of the α-hexagonal cell is $(3/2)\sqrt{3}a^2c$.

At 20°C, $V_\alpha 20 = ((3/2)\sqrt{3})(0.2950)^2(0.4683) = 0.10588$ nm^3.

The volume at 900°C is $V_{\alpha900} = V_{\alpha20}[(1 + 8.4 \times 10^{-6}\text{K}^{-1})(880 \text{ K})]^3 = 0.10588$ nm^3. There are six atoms in the hexagonal cell, so the volume per atom is 0.01804 nm^3.

The volume of the β-bcc cell at 900°C is $0.332^3 = 0.03659$ nm^3. There are two atoms per cell, so the volume per atom is 0.018295 nm^3. The volume change as $\beta \rightarrow \alpha$ is $(0.01804 - 0.018295)/(0.018295) = -0.014$ or -1.4%. This is a contraction.

Titanium is very reactive. It is produced by reduction of titanium oxide by sodium. The resultant product is sponge that must be melted to produce useful products.

Because titanium readily dissolves both oxygen and nitrogen, it must be processed in vacuum or an inert atmosphere. This includes melting as well as hot working. Because of this requirement, the cost of finished titanium parts is very high relative to the cost of sponge.

Alloying elements can be classified as to whether they stabilize the α-phase or the β-phase (e.g., whether they raise or lower the temperature at which, α → β). Interstitial oxygen and nitrogen have much higher solubilities in hcp α than bcc β, and are, therefore, strong α stabilizers. Carbon, which also dissolves interstitially, is a weaker α stabilizer. The only other important α stabilizer is aluminum. Figures 14.10 through 14.13 are the phase diagrams for these α stabilizers. Tin and Zr are almost neutral with regard to stabilizing either phase. Figure 14.14 is the Ti–Sn phase diagram. Aluminum and tin are strong solid solution strengtheners.

β stabilizers include Cr, Mo, Nb, Ta, V, and W, all of which have bcc crystal structures. Figures 14.15 through 14.17 show the titanium-rich ends of the Ti–Cr, Ti–Mo, and Ti–V phase diagrams.

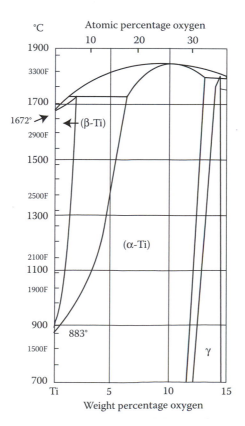

FIGURE 14.10 Ti–O phase diagram. (From T. B. Massalski, *Binary Alloy Phase Diagrams*, ASM, 1986. With permission.)

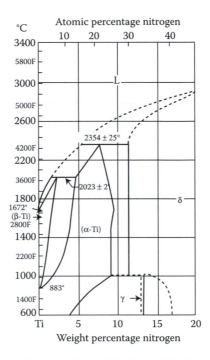

FIGURE 14.11   Ti–N phase diagram. (From T. B. Massalski, *Binary Alloy Phase Diagrams*, ASM, 1986. With permission.)

Titanium alloys are classified as to whether the room temperature structure is normally α, β, or α + β. This is illustrated by a schematic phase diagram (Figure 14.18). Some common titanium alloys are listed in Table 14.4.

Interstitials and Sn and Al have a potent solid-solution strengthening effect on α. However, interstitials are regarded as undesirable because of their embrittling effect. Solid solution strengthening of the β-phase by transition elements is much less.

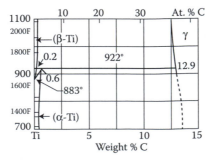

FIGURE 14.12   Ti–C phase diagram. (From T. B. Massalski, *Binary Alloy Phase Diagrams*, ASM, 1986. With permission.)

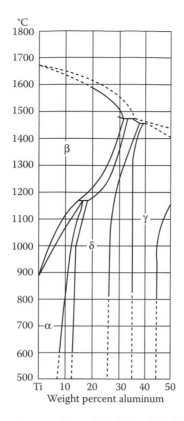

FIGURE 14.13   Ti–Al phase diagram. (From T. B. Massalski, *Binary Alloy Phase Diagrams*, ASM, 1986. With permission.)

α-titanium alloys like other hcp metals develop on rolling, a strong crystallographic texture in which the basal (0001) plane tends to be aligned with the rolling plane (Figure 14.7). With this orientation, the <11$\bar{2}$0> are in the rolling plane and therefore slip does not contribute to thinning of the sheet. This texture causes greater strengths in biaxial tension (which requires thinning) than uniaxial tension as shown in Figure 14.19.

The α – β alloys may be heat treated to obtain high strengths. Some α – β alloys may be quenched from an all β temperature to retain a supersaturated β. Later they can be aged in the α + β region to precipitate α. If there is an insufficient amount of β stabilizer, quenching may result in a martensitic transformation. Tempering of the martensite results in precipitation of β. Figures 14.20 and 14.21 show the microstructures of some α – β alloys.

The density of titanium (4.5 Mg/m³) is intermediate between the density of aluminum (2.7 Mg/m³) and that of steel (7.9 Mg/m³). Titanium is used in aircraft and sporting goods because of its high strength-to-weight ratio. The other major use of titanium is in the chemical industry and desalination plants where its use depends on its resistance to corrosion. Figure 14.22 shows that titanium is passive over a wide pH range for low potentials.

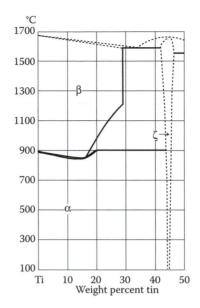

FIGURE 14.14   Ti–Sn phase diagram. (From T. B. Massalski, *Binary Alloy Phase Diagrams*, ASM, 1986. With permission.)

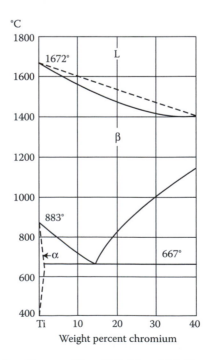

FIGURE 14.15   Ti–Cr phase diagram. (From T. B. Massalski, *Binary Alloy Phase Diagrams*, ASM, 1986. With permission.)

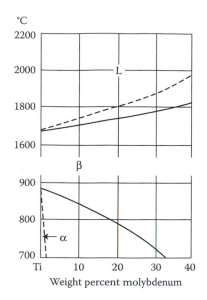

FIGURE 14.16 Ti–Mo phase diagram. (From T. B. Massalski, *Binary Alloy Phase Diagrams*, ASM, 1986. With permission.)

At high temperatures, the oxide, TiO_2, will dissolve in titanium. If exposed to oxygen at high temperatures, titanium will absorb oxygen forming a hard brittle surface.

An interesting processing technique for titanium is the diffusion bonding/superplastic forming (DB/SPF) developed for aircraft skin. Figure 14.23 shows a panel of the B-1 bomber produced by this process. Three sheets of very fine grain titanium

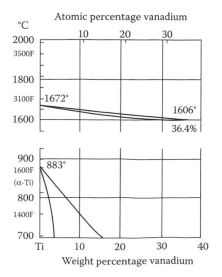

FIGURE 14.17 Ti–V phase diagram. (From T. B. Massalski, *Binary Alloy Phase Diagrams*, ASM, 1986. With permission.)

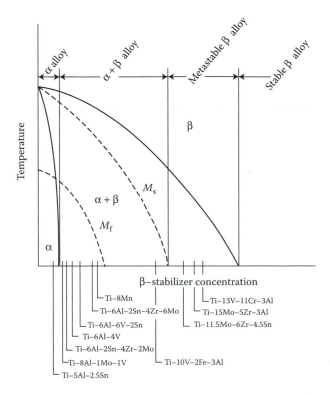

FIGURE 14.18 Some common alloy compositions relative to a schematic phase diagram. (From M. Donachie, *Titanium: A Technical Guide*, 2nd ed., ASM, 2000. With permission.)

TABLE 14.4

Titanium Alloys

Grade	N (max)	C (max)	O (max)	Al	Sn	Zr	Mo	Other
α								
ASTM 1	0.03	0.10	0.18	—	—	—	—	—
Ti code 12	0.03	0.01	0.25	—	—	—	−0.3	0.8Ni
Ti–5Al–2.5Sn	0.05	0.08	0.20	5	2.5			
Ti–8Al–1Mo–1V				8	—	—	1	1V
Ti–6Al–2Nb–1Ta–0.8Mo				6			0.8	2Nb, 1Ta
α – β								
Ti–6Al–4V				6	—	—	—	4V
Ti–6Al–2Sn–2Zr–2Mo–2Cr			6	2	2	2		2Cr, 0.25Si
Ti–10V–2Fe–3Al				3	—	—	—	10V, 2Fe
β								
Ti–13V–11Cr–3Al								
Ti–3Al–8V–6Cr–4Mo–4Zr				3	—	4	8	8V, 6Cr
Ti–11.5Mo–6Zr–4.5Sn				—	4.5	6		11.5

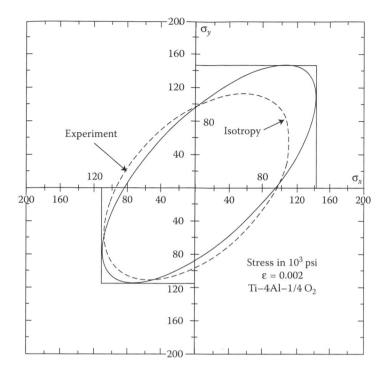

FIGURE 14.19 Yield locus for a sheet of Ti–4Al–1/4O. Note that in biaxial tension, which requires thinning, the yield strength is lower than that in uniaxial tension. (From D. Lee and W. A. Backofen, *Trans TMS-AIME*, 242, 1968. With permission.)

FIGURE 14.20 Ti–5Al–2.5Sn that has been air cooled from above the $\alpha - \beta$ transus. Prior β grains are outlined by α. Acicular α is in the interior. (From *Metals Handbook*, 8th ed., Vol. 7, p. 323, 1972. With permission.)

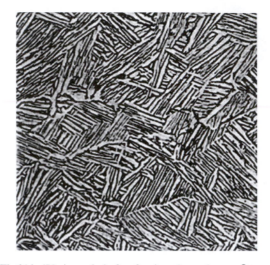

FIGURE 14.21 Ti–6Al–4V air cooled after forging above the $\alpha - \beta$ transus. Note the plate-like α and the intergranular β. (From *Metals Handbook*, 8th ed., Vol. 7, p. 323, 1972. With permission.)

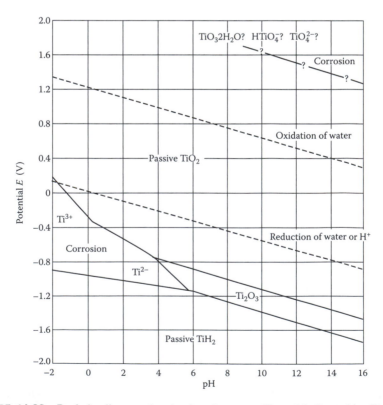

FIGURE 14.22 Poubaix diagram for titanium in water. (From M. Donachie, *Titanium: A Technical Guide*, 2nd ed., ASM, 2000. With permission.)

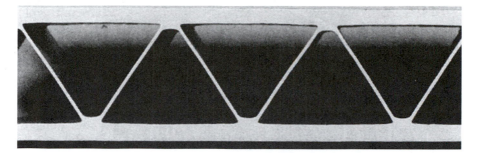

FIGURE 14.23 Cross section of a panel produced by diffusion bonding. (From W. F. Hosford and R. M. Caddell, *Metal Forming, Mechanics and Metallurgy,* 2nd ed., Prentice Hall, 1993. With permission.)

are stacked, one on another. An inert oxide is silk-screened onto two of the sheets to prevent bonding. The stack is heated under external pressure. Because titanium dissolves its own oxide, the sheets bond where they are in contact. Then the unbonded channels are pressurized, stretching the interior sheet and forcing the outer two sheets against a mold. The very great elongation of the interior sheet is possible because the very fine grain size promotes superplasticity.

14.5 ZIRCONIUM

Zirconium like titanium undergoes an α(hcp) $\rightarrow$ β(bcc) as it is heated. For zirconium the transformation temperature is 862°C. The principal alloy of zirconium is *zircaloy-2*, which contains 1.5% Sn. It is used to clad nuclear fuel elements in reactors. It is used for this purpose because of its low neutron cross section, for example, it absorbs few neutrons. However, it is embrittled by hydrogen, which reacts with zirconium to form hydrides. The hydrides are in the form of plates that precipitate on (0001) planes. Special effort is made to align the plates with the walls of the tube, where they will have the least effect as shown in Figure 14.24.

14.6 BERYLLIUM

Beryllium has a very low density (1.85) and a very high elastic modulus (275 GPa). It is extremely brittle and must be processed by compacting and sintering powder. Because of its lightness and stiffness it has found use in gyroscopes and other instruments. It is also used for x-ray windows. It is extremely toxic. Exposure to fumes or dust of Be or BeO can cause berylliosis that affects the lungs. Its high cost and toxicity limit its use.

14.7 MISCELLANY

In 1808, Sir Humphry Davy discovered magnesium in its oxide, although it is not certain that he isolated the metal. Pure magnesium was isolated substantially by A. A. B. Bussy in 1828 by chemical reduction of the chloride. Magnesium was first

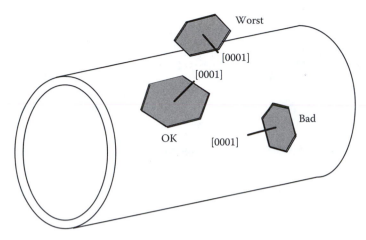

FIGURE 14.24 Possible orientations of hydride platelets in zircaloy tube walls. Cracking under internal pressure is most likely if the *c*-axis is in the hoop direction and least likely when the (0001) planes are aligned with the tube walls.

isolated electrolytically by Michael Faraday in 1833. The technology of magnesium extraction and alloying was developed in Germany during the 1930s. Because magnesium is extracted from seawater, the Germans knew that blockades could not cut off magnesium sources as it could cut off aluminum sources. The Germans called magnesium, *elektron*, possibly because of the need for reduction by electrolysis. The Volkswagon Beetle was designed to make maximum use of magnesium. They had magnesium blocks and transmissions as well as many smaller parts. After the Second World War, when the Volkswagon Beetle sales soared, Volkswagon consumed over half of the world's magnesium production.

Titanium is the fourth most abundant structural metal after Fe, Al, and Mg, making up 0.6% of the earth's crust. Although it was first discovered about 1790, it was not until about 1950 that it was produced commercially. It is produced as a sponge by reducing TiO_2 with sodium. This must then be melted in vacuum.

PROBLEMS

1. α-titanium at 20°C has an hcp crystal structure with $a = 0.2950$ nm and $c = 0.4683$ nm. At 900°C, it transforms to β-titanium with a bcc crystal structure with $a = 0.332$. The coefficients of thermal expansion between 20°C and 900°C is 13×10^{-6} K^{-1} parallel to the *c*-axis and 11.2×10^{-6} K^{-1} parallel to the *a*-axis. Calculate the percent volume change as β-titanium transforms to α-titanium on cooling. Indicate whether this is a contraction or expansion.

2. A part is now being die cast from aluminum alloy 356. The boss has suggested die casting the same part from magnesium alloy AZ91. Assume that the same dies will be used, so the shape and size of the part will not change. What would be the percent increase or decrease of the material cost? Assume that prices of aluminum and magnesium are $0.85 and $1.45 lb^{-1}, respectively.

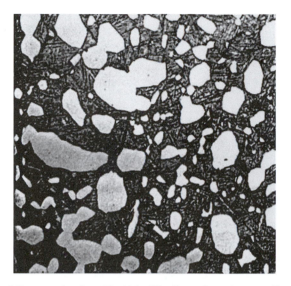

FIGURE 14.25 Micrograph of a Ti–6Al–4V alloy after slow cooling. (From *Metals Handbook*, 8th ed., Vol. 7, p. 323, 1972. With permission.)

3. A micrograph of an alloy Ti–6Al–4V is shown in Figure 14.25. Identify the light and dark phases.

4. For extruded bars of magnesium alloy AZ61A, the *Metals Handbook* (ASM Vol. 1 8th ed., 1961.) lists the tensile yield strength as 35,000 psi and the compressive yield strength as 19,000 psi. Assuming that the difference is caused by the directionality of twinning, deduce how the *c*-axis is oriented relative to the rod axis.

5. Calculate the *c/a* ratio for hexagonal close packing of spheres.

REFERENCES

E. W. Collins, *The Physical Metallurgy of Titanium Alloys*, ASM, 1984.

M. Donachie, *Titanium: A Technical Guide*, 2nd ed., ASM, 2000.

W. F. Hosford and R. M. Caddell, *Metal Forming, Mechanics and Metallurgy*, 2nd ed., Prentice Hall, 1993.

E. W. Kelley and W. F. Hosford, Plane-strain compression of magnesium and magnesium alloy crystals, *Trans TMS-AIME*, 242, 5–13, 1968.

D. Lee and W. A. Backofen, *Trans TMS-AIME*, 242, 1968.

Magnesium and Magnesium Alloys, ASM Specialty Handbook, ASM, 1999.

T. B. Massalski, *Binary Alloy Phase Diagrams*, ASM, 1986.

Metals Handbook, 8th ed., Vol. 1, 1961.

Metals Handbook, 8th ed., Vol. 7, p. 323, 1972.

Metals Handbook, 8th ed., Vol. 8, 1973.

Metals Handbook, 9th ed., Vol. 2, ASM, 1979.

15 Other Nonferrous Metals

15.1 LEAD

Lead has an fcc crystal structure. Its low melting point (327°C) makes it easy to cast. It has a high density (11.2) and an excellent corrosion resistance in basic solutions. Lead is very soft. It is frequently alloyed with Sb, Sn, Ca, or As to harden it. Its largest usage is in storage batteries.

Another use is soft *solder*, an alloy with tin. Lead and 62% tin form a eutectic that is often used for soldering wires. The lead–tin phase diagram is shown in Figure 7.4. Lead *babbitts* (15Sb/10Sn) are used as bearings. *Terne* is steel coated with lead for roofing.

15.2 TIN

Tin has a body-centered tetragonal crystal structure at room temperature. Below 13.2°C, the equilibrium structure is diamond cubic (gray tin) but the transformation is so sluggish that it can normally be ignored. When tin does undergo this transformation, the volume expansion is so great that the brittle diamond cubic form turns to powder (hence the name *gray tin*). Tin, like lead, has a low melting point (232°C). Other than solder, the notable tin alloys include pewter (7Sb, 2Cu) and tin babbitts (8Sb, 4–8Cu). It is illegal to put lead into *pewter*. Coating of steel for tin cans was once a major use of tin in the past. Today, however, because tin costs over $3.00 lb^{-1}, tin cans contain almost no tin. *Float glass* window panes is made by floating molten glass on molten tin.

15.3 GOLD

Gold has an fcc crystal structure. Its melting point (1063°C) is close to that of copper. Its price, about $400 (troy oz)$^{-1}$,* is equivalent to $4800 lb^{-1}. It is very soft. The tensile strength of annealed pure gold is about half of that of annealed pure copper. It is almost always alloyed to increase its hardness.

The principal uses of gold are for coinage, dentistry, jewelry, and semiconductor leads. For jewelry, it is commonly alloyed with Cu, Zn, Ni, and Ag. The purity of gold is described in terms of karats. One karat = 1/24 part gold by weight, so 18 karat gold contains 75% gold by weight. By alloying gold with nickel, the gold color is lost with relatively small amounts of nickel.

---

* A troy ounce is 31.103 g versus an avoirdupois ounce, which equals 28.349 g.

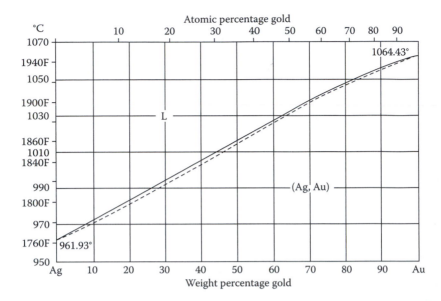

FIGURE 15.1 Gold–silver phase diagram. (From *Metals Handbook*, 8th ed., Vol. 8, ASM, 1973. With permission.)

White gold is such an alloy. Gold and silver are mutually soluble in all proportions. Figure 15.1 is the gold–silver phase diagram. The use of gold in coins started as early as 1899 BC. In the Unites States, gold coins contained 10% copper.

Gold leaf is made from annealed sheet by hammering. Small pieces are placed between sheets of paper or membranes to form stacks, which are hammered to a thickness of about 0.1 μm. Gold leaf is applied to other metals, leather, or other materials by simply rubbing.

15.4 SILVER

Silver has an fcc crystal structure. Its price, roughly $5 (troy oz)$^{-1}$, is equivalent to about $62 lb^{-1}. It has a higher electrical conductivity than any other metal. It is widely used for jewelry. Sterling silver contains 92.5% Ag and 7.5% Cu. An alloy of 90% Ag–10% Cu (*Coin silver*) has been used for coinage. Figure 15.2 is the silver–copper phase diagram.

In the past, a major use has been for dental amalgams (33Ag, 52Hg, 12.5Sn, 2Cu, and 0.5Zn). The dentist mixed the mercury with a powder containing other elements and worked the mixture into the cavity before the mercury completely reacted with the powder. It is also used in photographic emulsions, but this is declining with the advent of digital cameras.

Silver is also used for electrical contacts. An interesting contact alloy contains 2.5% Mg in solid solution. If the contact is heated in air, oxygen will dissolve and diffuse into the silver and react with the dissolved magnesium to precipitate MgO. This internal oxidation reaction hardens the contact and leaves nearly pure silver with a high conductivity.

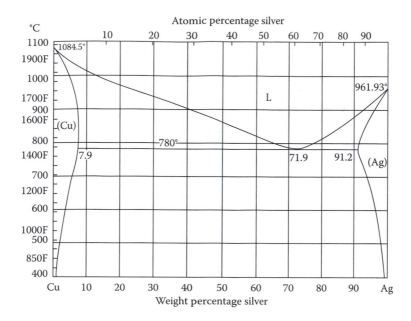

FIGURE 15.2 Silver–copper phase diagram. (From *Metals Handbook*, 8th ed., Vol. 8, ASM, 1973. With permission.)

A final important use is as the base for silver solders. These are alloys with copper and zinc that are used for higher-strength joints than can be achieved with lead–tin solders.

15.5 COBALT

The crystal structure of cobalt is fcc(α) at room temperature. Above 417°C it is hcp (b), its melting point is 1495°C. The stacking fault energy is very low (25 mJ m^{-2}), which is consistent with it having an hcp structure at high temperatures. The principal uses are as magnets (both permanent and soft), superalloys, hard-facing and wear-resistant alloys, and in carbide tools. Cobalt is one of the three common ferromagnetic metals. Many of the best permanent magnetic alloys contain cobalt. The magnetic application of cobalt is discussed in Chapter 24 and the use of cobalt in carbide tools is discussed in Chapter 23. Cobalt also finds use as a catalyst.

There are cobalt-base superalloys, but these are more expensive than nickel-base superalloys. The principal advantage that cobalt offers is superior resistance to attack by sulfur. Most of the cobalt-base superalloys contain 40–65% Co, 20–25% Cr, 0–20% Ni, 1–2% Fe, 0–10% W, and 0.4–0.5% C.

15.6 PLATINUM METALS

The platinum metals include ruthenium, rhodium, palladium, osmium, iridium, and platinum. Figure 15.3 shows their relative positions in the periodic table and lists their melting points and densities. Pd, Pt, Rh, and Ir are fcc. Ru and Os are hcp.

Ru mp = 2500 hcp, ρ = 12.2	Rh mp = 1966 fcc, ρ = 12.4	Pd mp = 1552 fcc, ρ = 12.0
Os mp = 2700 hcp, ρ = 22.6	Ir mp = 2454 fcc, ρ = 22.5	Pt mp = 1769 fcc, ρ = 21.5

FIGURE 15.3   Platinum metals.

They are all produced principally as by-products in refining of copper, gold, and silver. Jewelry consumes about half of the platinum production (Figure 15.4). The other uses of the platinum metals are based on their chemical inertness and high melting points. About 1/4 of the platinum and nearly all of the rhodium is used as a catalyst, in automobile exhaust systems, and in the chemical industry. It is also used as crucibles, particularly for glass and molten oxides. Platinum coupled with a platinum–rhodium alloy is used for furnace windings and high-temperature thermocouples. Alloys of platinum with Rh, and alloys of Hf with Al and Ru and an alloy of Ir with Nb have excellent mechanical properties at high temperatures. Were it not for their very high cost they would find applications now filled by the superalloys. Palladium is similar to platinum and it is used for similar applications. It is cheaper to use than platinum because its density is only 56% of platinum's. It is often alloyed with platinum. They form a continuous series of solid solutions.

Palladium can absorb up to 900 times its volume of H_2. It is alloyed with gold to make white gold. Sixty percent of its use in 1999 was as a catalyst for catalytic converters in autos.

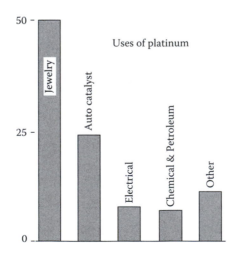

FIGURE 15.4   Uses of platinum in 2000. (Adapted from A. P. Patrick, *JOM*, October 2000.)

Rhodium is used as an alloying element to harden platinum and palladium for use in sparkplugs, furnace windings, thermocouples, and jewelry. Rhodium platings are very hard.

Other uses of the platinum metals include electrical contacts, pen nibs, and in glass making. The first metal lamp filaments were made from osmium because of its high melting point.

15.7 GROUP VA AND VIA METALS

Figure 15.5 shows a portion of the periodic table containing the group VA and VIA metals. All have bcc crystal structures. All except chromium have very poor oxidation resistance. The oxides of tungsten and molybdenum are volatile. Group VA metals have very good ductility, but Group VIA metals are very brittle. Molybdenum and tungsten suffer grain boundary embrittlement. Figure 15.6 shows an intergranular fracture in tungsten.

Niobium, tantalum, molybdenum, and tungsten along with rhenium are regarded as *refractory* metals because of their very high melting points. "Refractory" means resistant to high temperatures. Because of their very high melting points and their reactivity, there are no crucible or mold materials that can be used for melting and casting the refractory metals. They may be arc-melted in a water-cooled mold arranged in such a way that the liquid metal contacts only the previously solidified metal. Powder metallurgy is an alternative process used for tungsten and molybdenum. The elements V, Mo, W, Ta, and Nb are used as alloying elements in steel. Steel makers still use the name *columbium* and the chemical symbol *Cb* for niobium. Unlike the other refractory metals, both niobium and tantalum are very ductile. Tantalum however has a tendency to cold weld (seize) to tooling.

Group	VA	VIA	VIIA
	V bcc 1900°C $\rho = 6.1$	Cr bcc 1875°C $\rho = 7.19$	
	Nb bcc 2468°C $\rho = 8.57$	Mo bcc 2610°C $\rho = 10.2$	
	Ta bcc 2996°C $\rho = 16.6$	W bcc 3410°C $\rho = 18.2$	Re hcp 3180°C $\rho = 21.0$

FIGURE 15.5 Portion of the periodic table containing metals of groups VA and VIA.

FIGURE 15.6 Intergranular fracture in tungsten that had been recrystallized at 2600°C. (From *Metals Handbook*, 8th ed., Vol. 9, ASM, 1974. With permission.)

15.8 TUNGSTEN AND MOLYBDENUM

Because of the brittleness of the grain boundaries in tungsten, tungsten wires for incandescent lamps are made by pressing and sintering tungsten powder. The powder compacts are first swaged at a high temperature. As the grain boundaries become aligned with the rod or wire axis, the temperatures can be gradually reduced. The final reductions can be made by wire drawing. A typical microstructure is shown in Figure 15.7. Heating will cause recrystallization. For pure tungsten, the recrystallized microstructure has grain boundaries normal to the wire axis (Figure 15.8) and hence is very brittle. To control grain growth, an inert dopant is normally added to the tungsten powder before compaction. The dopant, KAS, is a powder composed of K_2O, Al_2O_3, SiO_2, and ThO_2. Particles of the dopant tend to be aligned with the wire axis during the wire manufacture, so recrystallized grains are also aligned with the wire axis as shown in Figure 15.9. Without grain boundaries normal to the wire axis, the strength and ductility are greatly improved. Figure 15.10 shows the improvement of rupture life.

FIGURE 15.7 Microstructure of a 0.007-in diameter tungsten wire as drawn (250×). (Courtesy of General Electric Company.)

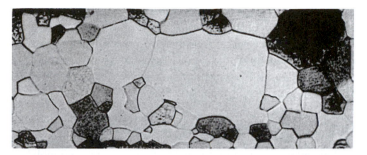

FIGURE 15.8   Microstructure of a 0.007-in diameter undoped tungsten wire after recrystallization (250×). (Courtesy of General Electric Company.)

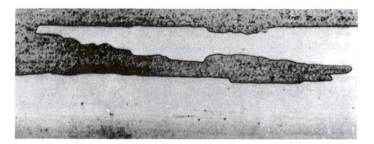

FIGURE 15.9   Microstructure of a 0.007-in diameter doped tungsten wire after recrystallization (250×). (Courtesy of General Electric Company.) Note the alignment of the grain boundaries with the wire axis.

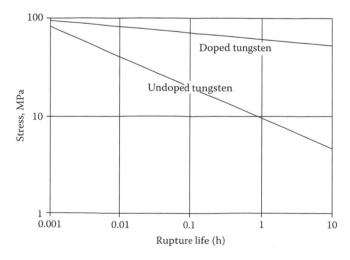

FIGURE 15.10   Effect of doping on the rupture life at 3000 K.

Because its oxide is volatile, tungsten can be used at high temperatures only in an inert atmosphere.

Molybdenum is very much like tungsten. Its grain boundaries are brittle, so it must be processed by arc melting or by powder metallurgy. Its oxide is volatile so its high temperature use is limited to inert atmospheres. Metallic molybdenum is used for support wires in lamps and for heat shields.

15.9 TANTALUM, NIOBIUM, AND RHENIUM

Tantalum and niobium are much more ductile than tungsten and molybdenum. The high density of tantalum makes it of interest for armor-piercing projectiles. Very thin tantalum foil is used for capacitors. One of the uses of niobium is in niobium–tin superconductors. Rhenium is used as a catalyst, for electrical contact, and as an alloying element in tungsten and molybdenum alloys.

15.10 ALLOYS WITH SPECIAL PHYSICAL PROPERTIES

There are some alloys that are used because of special physical properties. Among these are *invar*, which has a nearly zero coefficient of thermal expansion at room temperature. The composition is iron with 36% nickel. This corresponds to a minimum of expansion coefficient versus composition (Figure 15.11). This alloy is very

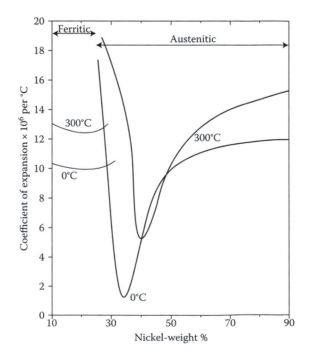

FIGURE 15.11 Effect of nickel content on the coefficient of thermal expansion of iron–nickel alloys. There is a minimum at 36% Ni. (From E. R. Petty, *Physical Metallurgy of Engineering Materials*, p. 68, American Elsevier, 1968. With permission.)

TABLE 15.1
Metals and Alloys with Low Melting Points

Composition (wt%)	MP (°C)
69.7Hg, 16In, 14.3Th	− 63.3
91.5Hg, 8.5Th	− 60
Hg	− 38
76Ga, 24In	0.6
62.5Ga, 21.5In, 16Sn	10.7
76Ga, 24In	16
Ga	30
44.7Bi, 22.6Pb, 19.1In, 8.3Sn, 5.3Cd	47
50Bi, 22.7Pb, 13.3Sn, 10Cd[a]	70
49Bi, 26.7Th, 23.4Sn, 18.2Cd	95
In	156.6
Bi	271.4

[a] Woods metal.

useful in precise instruments where dimensional change would cause errors. It is interesting that the standard meter stick in Paris is made from invar.

The Young's modulus of *elinvar*: (Fe-36Ni-12Cr) is temperature independent ($dE/dT = 0$). This is useful for springs in precision instruments. The electrical resistance of *constantan* (copper–45% nickel) is nearly temperature independent at room temperature ($d\rho/dT = 0$). It is used for precision resistors. Constantan also finds use in thermocouples.

There are a series of alloys with very low melting points. One of the more common is Wood's metal, a quaternary eutectic (50Bi, 26.7Pb, 13.3Sn, and 10Cd) that melts at 70°C. With additions of indium, even lower melting points can be obtained. Several other eutectic alloys with melting points below 100°C are listed in Table 15.1. These find uses in chucking parts with complicated shapes for machining and as fusible elements in water sprinklers. Bismuth expands on freezing. With high bismuth content, the dimensional change on melting is low.

Gallium will melt in ones hand as shown in Figure 15.12. The difference between its melting and boiling points (30–2237°C) is greater than any other metal. The principal use of indium is in low-melting alloys. Bismuth expands on freezing, so many of the bismuth alloys have very low or zero-dimensional changes on freezing.

15.11   MISCELLANEOUS METALS

Uranium has a very high density. Because of this, depleted uranium was used for armor-piercing missiles. However, this use has ceased because of uranium's toxicity.

FIGURE 15.12   Gallium's melting point (30°C) is less than body temperature, so it will melt in one's hand. (From B. W. Gonswer, *J. Mater.*, 1, ASTM, 1966. With permission.)

Antimony is used as an alloying element in lead to harden.

Because of its high melting point (2222°C), hafnium is used as an alloying element in some high-temperature alloys.

Lanthanides (rare earths) including Ce are used in steels to control sulfides. Other uses of rare earths include catalysts for automobile exhaust systems and petroleum (31%) and in permanent magnets (31%). Mischmetal is the result of reducing a naturally mixed ore of rare earth elements. It is used as the "flint" in cigarette lighters and an alloying ingredient in magnesium alloys.

15.12   MISCELLANY

The English word *tungsten* comes from the Swedish words *tungsten* meaning heavy stone. The Swedish, like the rest of the world, use the name wolfram for tungsten.

Our word cobalt comes from the German *Kobold* who is a mischievous goblin related to *Nickel*.

Brandt discovered cobalt in 1735. *Kobald* is German for evil spirit or goblin. A cobalt-base alloy, vitallium was developed in the 1930s as an alloy for body implantation and dentistry—but after the development of jet engines, it became the first superalloy. It is now known as Stellite 31.

Gold—the term *karat*, which is a measure of the purity of gold, should not be confused with the term *carat*, which is used for the weight of a gem. (1 carat = 200 mg.)

The name "Troy ounce" comes from the city of Troyes, France, where it was a standard unit of weight.

The first knowledge that Europeans had of platinum was as nuggets in rivers of Columbia, where it was often associated with gold. The Spaniards considered it an undesirable contaminant of gold and were known to have thrown it away. The very name derives from the Spanish word *platina* meaning small *plata* (silver) implying that it was less valuable than silver.

As early as 3000 BC, Sumerians were making intricate gold jewelry. In China, gold became legal tender in 1091 BC.

Steel makers still use the name *columbium* and the chemical symbol *Cb* for niobium.

In Greek mythology, Niobe was the daughter of Tantalus.

The symbol, Sn, for tin comes from Latin *stannum*. Tin was known to the ancients and is mentioned in the Old Testament. Early metal workers found it too soft for most purposes but mixed with copper it gives the alloy bronze, of Bronze Age fame. Today most of the tin comes from Malaya, Bolivia, Nigeria, Zaire, and Thailand. The Cornish mines were almost completely wiped out by the price collapse of 1985.

PROBLEMS

1. How many grams of gold would be required to make 1 m² of gold leaf, 0.1 µm thick?
2. Explain why the ductility of tungsten is increased by cold working and drastically decreased by recrystallization.
3. What is the value of the metal in 10 g of silver?
4. Using the cost data in Chapter 1, calculate the cost of the metals in eutectic lead–tin solder and compare it with the cost of 50–50 lead–tin solder.

REFERENCES

B. W. Gonswer, *J. Mater.*, 1, ASTM, 1966.
Metals Handbook, 8th ed., Vol. 8, ASM, 1973.
Metals Handbook, 8th ed., Vol. 9, ASM, 1974.
A. P. Patrick, *JOM*, October 2000.
E. R. Petty, *Physical Metallurgy of Engineering Materials*, p. 68, American Elsevier, 1968.

16 Steels

16.1 MICROSTRUCTURES OF CARBON STEELS

Steels are iron-base alloys. Most common are carbon steels, which may contain up to 1.5% carbon. Cast irons typically contain between 2% and 4% carbon. Figure 16.1 is the iron–iron carbide phase diagram. Below 912°C, pure iron has a bcc crystal structure and is called *ferrite* and is designated by the symbol α. Between 912°C and 1400°C, the crystal structure is fcc. This phase, called *austenite*, is designated by the symbol γ. Between 1400°C and the melting point, iron is again bcc. This phase is called δ-ferrite, but it is really no different than α-ferrite. The maximum solubility of carbon in α (bcc) iron is 0.02% C and in γ (fcc) iron is about 2%. Iron carbide, Fe_3C, is called *cementite* and has a composition of 6.67% C. The structure developed by the eutectoid reaction, $\gamma \rightarrow \alpha + Fe_3C$, consists of alternating platelets of ferrite and carbide (Figure 16.2). This structure is called *pearlite*.

Steels containing <0.77% carbon are called *hypoeutectoid* and those with >0.77% C are called *hypereutectoid*. The microstructures of medium carbon steels (0.2–0.7% C) depend on how rapidly they are cooled from the austenitic temperature. If the cooling is very slow (furnace cooling), the proeutectoid ferrite will form in the austenite grain boundaries, surrounding regions of austenite that subsequently transforms to pearlite as shown in Figure 16.3. With somewhat more rapid (air cooling), there is not enough time to allow all of the carbon to diffuse away from the austenite grain boundaries into the center of the austenite grains. The proeutectoid ferrite will form as plates (that look like needles in a microstructure) that penetrate into the center of the old austenite grains. This decreases the distance the carbon has to diffuse. The resulting microstructure is often called a Widmanstätten structure (Figure 16.4.).

For a hypoeutectoid steel, the term *full anneal* is used to describe heating 25–30°C into the austenite field followed by a furnace cool. This produces very coarse pearlite. Heating 50–60°C into the austenite range the austenite field followed by air cooling is called *normalizing*. It produces a much finer pearlite. *Spheroidizing* is a heat treatment used on medium carbon steels to improve cold formability. This can be accomplished by heating just below the eutectoid temperature for 1 or 2 h. At this temperature surface tension causes the carbides to spheroidize. Some companies use a more complicated process involving first heating above the eutectoid temperature for 2 h and then slowly cooling below it and holding 8 h or more. The reason for this process is unclear.

257

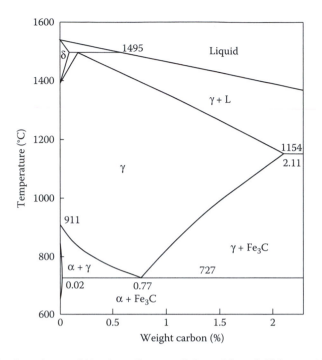

FIGURE 16.1 Iron–iron carbide phase diagram. (Adapted from J. Chipman, *Met. Trans.*, 3, 55, 1972.)

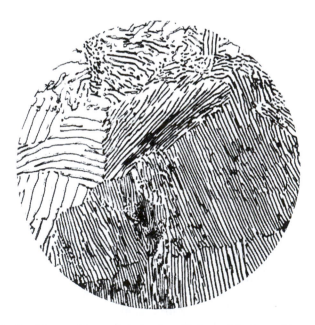

FIGURE 16.2 Pearlite consisting of alternating platelets of ferrite and iron carbide. (From *Making, Shaping and Treating of Steel*, 9th ed., U.S. Steel Corporation, 1971. With permission.)

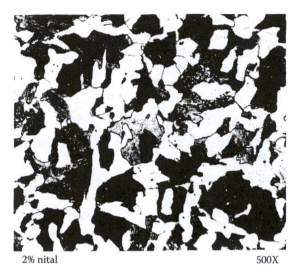

2% nital   500X

FIGURE 16.3 Hypoeutectoid 1045 steel normalized by air cooling. Note that the proeutectoid ferrite is formed in the prior austenite grain boundaries. (From *Metals Handbook*, 8th ed., Vol. 7, ASM. With permission.)

All steels contain manganese and silicon in addition to carbon. Silicon is present as an impurity. Manganese is added to react with sulfur, which is always present as an impurity, to form MnS. If manganese were not present, the sulfur would form FeS. Iron sulfide wets the austenite grain boundaries and is molten at the hot working temperatures. This results in the steel being hot short (having no ductility at high temperatures).

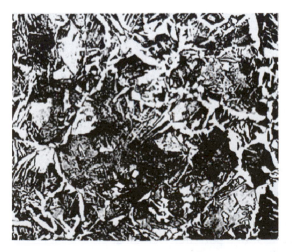

FIGURE 16.4 Widmanstätten structure formed in a hypoeutectoid steel during a rapid air cool. (From *Metals Handbook*, 8th ed., Vol. 7, p. 27, ASM. With permission.)

FIGURE 16.5 Banded microstructure of a medium carbon steel. (From *Metals Handbook*, 8th ed., Vol. 7, p. 27, ASM. With permission.)

When a steel is first cast, manganese and other alloying elements will tend to lie in the interdendritic regions. Carbon is more strongly attracted to manganese than iron. If the steel has been hot rolled into a plate or tube, the initial regions of inter-dendritic segregation will be aligned with the rolling direction. The proeutectoid ferrite will form in the carbon-lean regions and the pearlite in the carbon-rich regions. The resulting microstructure will have a very directional microstructure, called a banded microstructure (Figure 16.5).

The directionality of the microstructure would suggest that the ductility transverse to the rolling direction would be worse than that parallel to the rolling direction. However, the principal reason is not the banding, but the elongation of MnS inclusions that occurs during the hot rolling. Figure 16.6 is of the same steel as in Figure 15.5, but at a much higher magnification. The elongated MnS inclusions are apparent.

Additional manganese is often added to promote lower ductile–brittle transition temperatures. Silicon is present as an impurity in amounts up to 0.25%. Other common impurities include S and P in amounts up to 0.05%.

Figure 16.7 shows the microstructure of a slowly cooled hypereutectoid steel.

16.2   KINETICS OF PEARLITE FORMATION

Figure 16.8 shows the eutectoid reaction in Fe–C alloys. On cooling through the eutectoid temperature $\gamma \rightarrow \alpha +$ carbide forming pearlite. At the austenite–pearlite interface carbon must diffuse from in front of the ferrite to in front of the carbide as shown in Figure 16.9. In a pure iron–carbon alloy, it is this diffusion that controls how rapidly pearlite can form. The growth rate of pearlite depends on the temperature at which the austenite transforms. As the temperature is lowered, the diffusivity

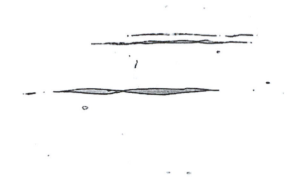

FIGURE 16.6 Magnified view of the structure in Figure 16.5 as polished. Note the elongated manganese sulfide stringers. (From *Metals Handbook*, 8th ed., Vol. 7, p. 27, ASM. With permission.)

of carbon decreases. Somewhat compensating for this is the fact that the spacing of the pearlite decreases at lower transformation temperatures.

Figure 16.10 shows that the pearlite spacing, λ, is inversely proportional to ΔT, the difference between the actual and equilibrium transformation temperatures.

$$\lambda = K/\Delta T. \qquad (16.1)$$

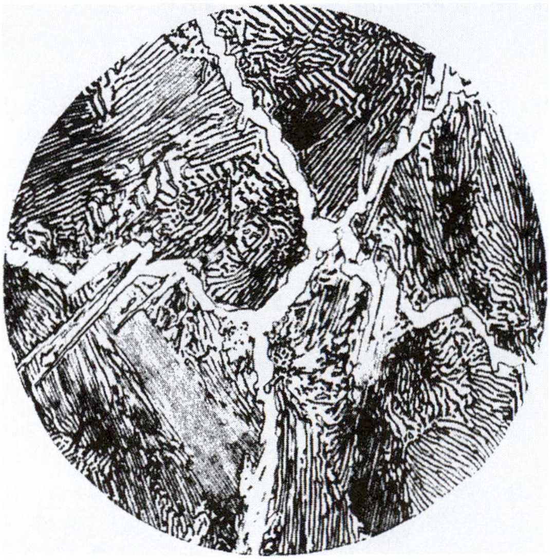

FIGURE 16.7 Hypereutectoid steel. Note the proeutectoid cementite formed in the prior austenite grain boundaries. (From *Making, Shaping and Treating of Steel*, 9th ed., U.S. Steel Corporation, 1971. With permission.)

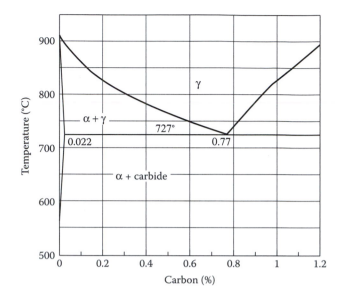

FIGURE 16.8 Eutectoid transformation in the iron–carbon system.

The lower the transformation temperature, the finer is the pearlite. Fine pearlite is harder than coarse pearlite, so the lower the transformation temperature, the harder is the transformation product.

EXAMPLE PROBLEM 16.1

Calculate the relative widths of the ferrite and carbide platelets using the lever law.

SOLUTION

The pearlite contains 0.77% C, ferrite 0.02%, and carbide 6.67%. The weight fraction ferrite $f_\alpha = (6.67 - 0.77)/(6.67 - 0.02) = 0.889$, so the fraction carbide = 0.111. The densities of ferrite and carbide are almost equal so volume fractions are nearly identical to the weight fractions. The ratio of the thicknesses of the ferrite and carbide lamellae is therefore 0.889/0.111 = 7.99.

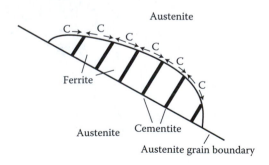

FIGURE 16.9 The growth of pearlite into austenite requires carbon to diffuse away from the growing ferrite toward the growing carbide.

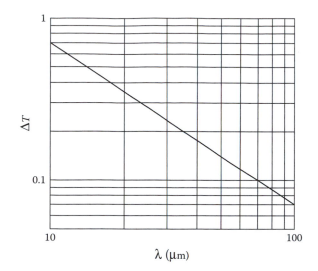

FIGURE 16.10 The lamellae spacing in pearlite is finer the lower the transformation temperature. (Data from D. D. Pearson and J. D. Verhoeven, *Met. Trans. A*, 15A, 1037, 1984.)

When hypoeutectoid steels undergo the transformation at a temperature below the equilibrium transformation temperature, carbide cannot precipitate unless the austenite is saturated with respect to carbide. This is illustrated in Figure 16.11. Consider a steel containing 0.40% C undergoing transformation at 650°C. An extrapolation of the line representing the solubility of carbon in austenite (the Hultgren extrapolation) indicates that carbon will not precipitate from the austenite. Initially only ferrite can

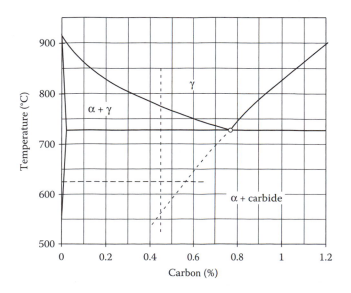

FIGURE 16.11 The Hultgren extrapolation of the solubility of austenite for carbon. At 625°C, carbide cannot form until the carbon content of the austenite reaches 0.57% C.

form. However, the precipitation of ferrite leaves the remaining austenite enriched in carbon. Carbide will form when the carbon content reaches 0.62%, so the pearlite that forms will contain 0.62% carbon. The ratio of the thickness of the ferrite and carbide platelets in the pearlite will be greater than the value of 7.99 calculated for transformation at 727°C.

There are several common heat treatments that produce microstructures that are not particularly hard. In normalizing, a hypereutectoid steel is air cooled after being heated 50–60°C into the fully austenitic range. This results in fine pearlite. Full annealing is furnace cooling from 25°C in the austenitic range and results in coarse pearlite.

16.3 ALLOYING ELEMENTS IN STEEL

The effect of alloying elements in steel on the temperature and composition of the eutectoid is shown in Figure 16.12. All alloying elements in steel lower the carbon content at the eutectoid. Alloying elements can be classified as austenite stabilizers and ferrite stabilizers, depending on whether they tend to dissolve more readily in austenite or ferrite. Nickel and manganese are the two most important austenite stabilizers. They lower the eutectoid temperature bcc elements (molybdenum, vanadium, and tungsten) as well as titanium and silicon are ferrite stabilizers. They raise the eutectoid temperature. Chromium up to 7% tends to open the γ-loop, but greater amounts close it. Above 13%, it makes steels ferritic at all temperatures. Figures 16.13 through 16.17 are the phase diagrams of Fe, Cr, Mo, Mn, Mo, and W.

EXAMPLE PROBLEM 16.2

Calculate the eutectoid temperature and composition for a 4140 steel, containing 0.40% C, 0.83% Mn, 0.30% Si, 1.0% Ni, and 0.20% Mo.

SOLUTION

Estimating the slopes, $\Delta T/\%$, in Figure 15.12a at low percentages, $\Delta T/\%$ Mn = (650 – 723)/5 = –15, $\Delta T/\%$ Si(850 – 723)/5 = +25, $\Delta T/\%$ Ni = (655 – 723)/5 = –14, and $\Delta T/\%$ Mo = (1000 – 7230)/3 = 92. ΔT = –15(0.83) + 25(0.30) – 14(1.0) + 62(0.20) = –6. The eutectoid temperature is 717°C.

Estimating the slopes, $\Delta\%C/\%$, in Figure 15.12b at low alloy concentrations, $\Delta\%C/\%$ Mn = (0.50 – 0.77)/5 = –0.054, $\Delta\%C/\%$ Si = (0.50 – 0.77)/2.5 = –0.108, $\Delta\%C/\%$ Ni = (0.50 – 0.77)/9 = –0.03, and $\Delta\%C/\%$ Mo = (0.50 – 0.77)/0.8 = –0.337.

$\Delta\%C$ = –0.054(0.83) – 0.108(0.30) – 0.03(1) – 0.337(0.2) = –0.175. The eutectoid composition is 0.77 – 0.175 = 0.60% C.

16.4 ISOTHERMAL TRANSFORMATION DIAGRAMS

Figure 16.18 is the isothermal transformation diagram for a 1045 steel. It describes the transformation of austenite when austenite is suddenly cooled to a temperature below that at which it is the equilibrium phase. It should be noted that for temperatures above about 600°C, ferrite forms before any pearlite can form. The rate of transformation depends on the rate of nucleation of pearlite colonies and the rate of their growth. In general, the nucleation rate increases with greater cooling below the equilibrium

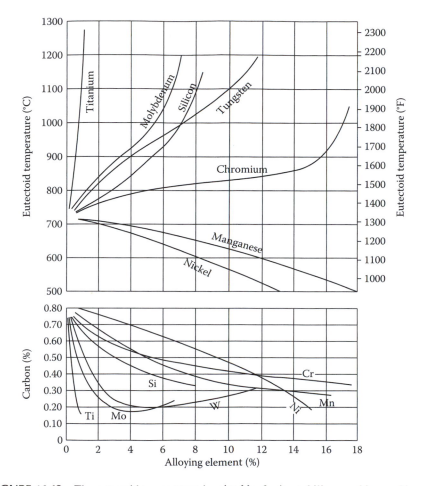

FIGURE 16.12 The eutectoid temperature is raised by ferrite stabilizers and lowered by austenite stabilizers. All alloying elements lower the carbon content at the eutectoid. (From *Making, Shaping and Treating of Steel*, 9th ed., U.S. Steel Corporation, 1971. With permission.)

temperature of 727°C. Since growth depends on diffusion, the growth rate decreases at lower temperatures. As a result, the overall transformation rate first increases as the temperature is decreased below 727°C to about 600°C and then decreases below this.

If austenite is transformed below about 600°C, the resulting structure is called *bainite*. Like pearlite, it consists of ferrite and carbide, but the carbide is not in the form of lamellae. Instead, it consists of very fine isolated particles. Figure 16.19 is an electron photomicrograph of bainite. The size of the carbide particles is finer with lower temperatures of transformation.

As a steel is cooled below a critical temperature, M_s, it transforms very rapidly by shear to a new crystal structure, *martensite*. Diffusion of carbon is not involved, so the amount of this transformation does not increase appreciably with time, but depends only on the temperature. The structure and properties of martensite will be treated later.

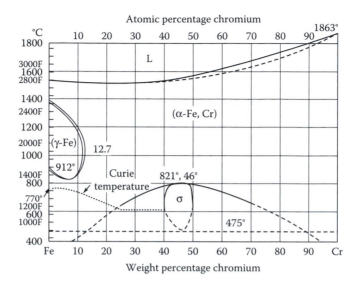

FIGURE 16.13   Iron–chromium phase diagram. (From *Metals Handbook*, 8th ed., Vol. 8, ASM, 1973. With permission.)

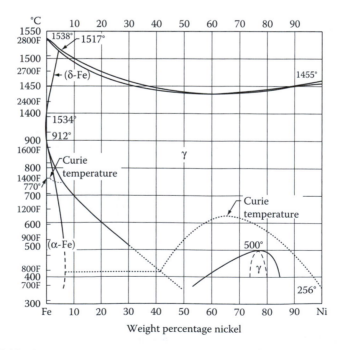

FIGURE 16.14   Iron–nickel phase diagram. (From *Metals Handbook*, 8th ed., Vol. 8, ASM, 1973. With permission.)

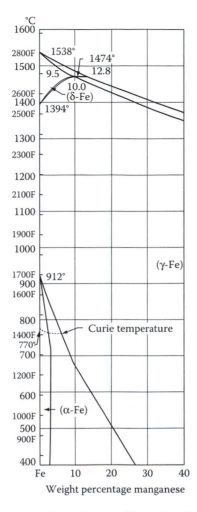

FIGURE 16.15 Iron–manganese phase diagram. (From *Metals Handbook*, 8th ed., Vol. 8, ASM, 1973. With permission.)

Because alloying elements slow the rate of transformation of austenite they shift the pearlite and bainite start times to the right. Chromium and molybdenum delay the transformation austenite to pearlite more than to bainite as shown in Figure 16.20 for a 4340 steel.

The M_s and M_f temperatures are lowered by increased carbon content as shown in Figure 16.21. Increased alloy content has the same effect. The M_f temperature should be regarded as the temperature at which the martensite formation is 99% rather than 100% complete. Increased alloy content also lowers the M_s as shown in Figure 16.22. An empirical relation is

$$M_s(°C) = 539 - 423(\%C) - 30.4(\%Mn) - 12.1(\%Cr) - 17.7(\%Ni) - 7.5(\%Mo). \quad (16.2)$$

With low M_f temperatures, there will be retained austenite at room temperature.

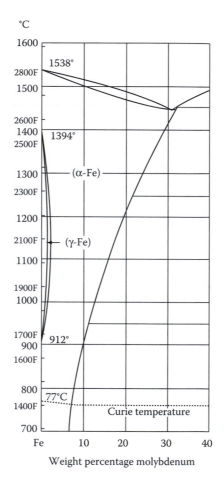

FIGURE 16.16 Iron–molybdenum phase diagram. (From *Metals Handbook*, 8th ed., Vol. 8, ASM, 1973. With permission.)

16.5 CONTINUOUS COOLING DIAGRAMS

For most applications, transformation during continuous cooling is of greater direct interest than isothermal transformation. During continuous cooling, part of the time is spent at high temperatures at which the transformation rates are slow. Therefore, the time to transform is longer for continuous cooling than for during isothermal treatment. The lines representing the start of pearlite and bainite formation are shifted to longer times and lower temperatures. Figure 16.23 shows the continuous cooling transformation diagram for a 4340 steel. Note that because the alloying elements delay the pearlite formation more than the bainite formation, bainite can be formed during continuous cooling.

16.6 MARTENSITE

Martensite has a body-centered tetragonal lattice (Figure 16.24). It can be thought of as supersaturated ferrite, the excess carbon causing the lattice to become elongated

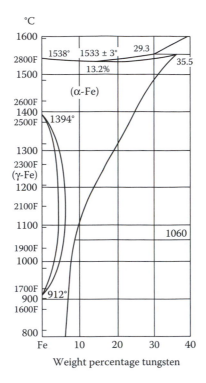

°C

FIGURE 16.17 Iron–tungsten phase diagram. (From *Metals Handbook*, 8th ed., Vol. 8, ASM, 1973. With permission.)

in one direction. Two unit cells of martensite form from one unit cell of austenite (Figure 16.25).

The amount of distortion of the lattice depends on the carbon content (Figure 16.26). The effect of carbon on the lattice parameter of martensite in nanometes (nm) is given by

$$c = 0.2861 + 0.0166x \qquad (16.3)$$

and

$$a = 0.2861 - 0.0013x, \qquad (16.4)$$

where, x is the wt% carbon. The lattice parameter of austenite is

$$a = 0.3548 + 0.0044x. \qquad (16.5)$$

Martensite is very hard, the hardness increasing with carbon content up to a value of about Vickers 850 (Rc 65) at 0.8% carbon (Figure 16.27). The hardness of martensite is independent of the amount of alloying elements present.

16.7   SPECIAL HEAT TREATMENTS

When a steel is quenched, the outside cools much faster than the interior and therefore undergoes the martensitic transformation while the interior is still hot. When the

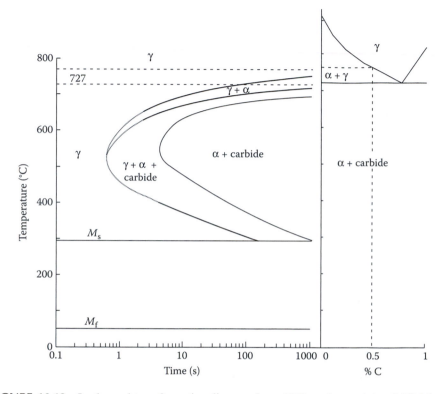

FIGURE 16.18 Isothermal transformation diagram for a 1050 steel containing 0.9% Mn. Note that above about 550°C, ferrite must form before pearlite (From *Metals Handbook*, 8th ed., Vol. 8, ASM, 1973. With permission.).

interior later transforms to martensite, its volume expansion causes residual tensile stresses on the surface. These stresses can be avoided if the surface and interior transform to martensite at the same time. *Marquenching* is the name of a process in which steel parts are quenched into a molten salt bath at a temperature just above the M_s. After thermal equilibrium is reached, but before bainite starts to form, the parts are slowly cooled to form martensite.

Austempering is a process designed to produce bainite. Steel parts are quenched into a molten salt bath at a temperature just above the M_s and held at that temperature until transformation to bainite is complete.

16.8   MISCELLANY

Ancients produced wrought iron directly from the charcoal-reduction of ore. In the production, wrought iron was never completely molten. About the twelfth or thirteenth century higher temperatures were attained by preheating the air for combustion. This led gradually to the development of the modern blast furnace. The product

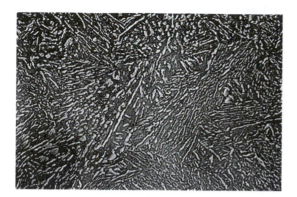

FIGURE 16.19   Electron micrograph of bainite formed at 700°F, 7500X.

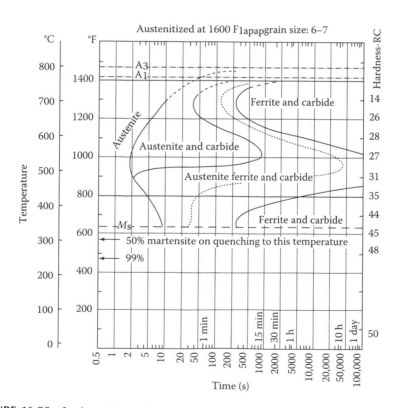

FIGURE 16.20   Isothermal transformation diagram for a 4340 steel containing 0.42% C, 0.78% Mn, 1.79% Ni, 0.80% Cr, and 0.33% Mo. (From *Making, Shaping and Treating of Steels*, 9th ed., U.S. Steel Corporation, 1971. With permission.)

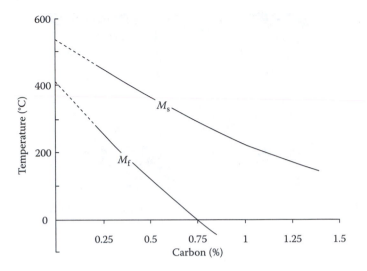

FIGURE 16.21 Increased carbon content lowers the M_s and M_f temperatures. (From *Making, Shaping and Treating of Steels*, 9th ed., U.S. Steel Corporation, 1971. With permission.)

of the blast furnace contains up to 3–4% C and 2–3% Si which makes it very brittle. It is called *pig iron* because the early blast furnaces were tapped into a channel in the sand that led to a number of cavities that could be broken off for sale. As shown in Figure 16.28, the whole arrangement resembled piglets suckling from a sow. Hence the term pig iron.

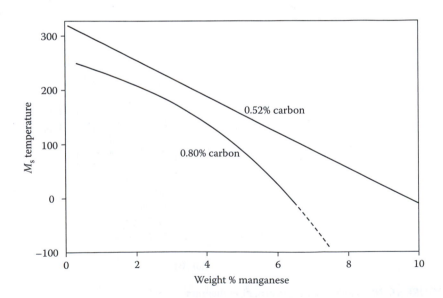

FIGURE 16.22 Increased alloy content lowers the M_s temperature. (From J. V. Russell and F. T. McGuire, *Trans ASM*, 33, 103, 1944. With permission.)

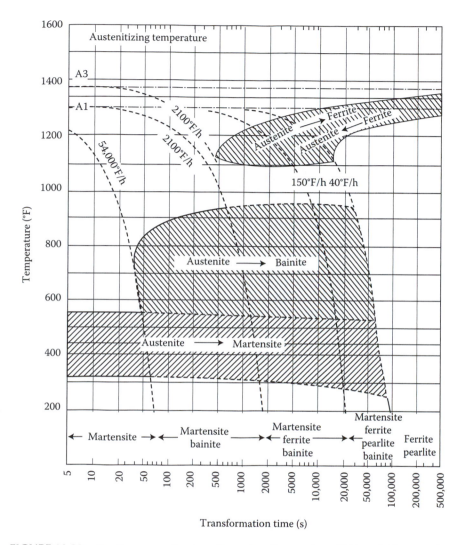

FIGURE 16.23 Continuous cooling transformation diagram for a 4340 steel. (From *Making, Shaping and Treating of Steels*, 9th ed., p. 1096, U.S. Steel Corporation, 1971. With permission.)

Pig iron was often *puddled* under oxidizing conditions which lowered the C and Si contents. As these elements were removed, the melting temperature increases. Slag was entrapped and elongated as the semisolid structure was worked. For centuries wrought iron was the dominant engineering material although steel was made by the crucible process in which pig iron and iron oxide were put in 100 lb crucibles. Heating caused the iron oxide to react with the carbon in the pig iron, lowering its carbon content.

Modern steel making started with development of the Bessemer process in 1856. Air was blown through molten pig iron to reduce the carbon and silicon contents. At about the same time (1858) the open hearth process was developed. In this process,

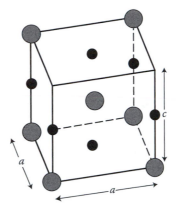

FIGURE 16.24 Martensite unit cell. Black dots indicate possible sites for carbon atoms.

steel was made in a very large but shallow furnace. Carbon reduction was achieved by an oxidizing slag. Although the open hearth process required hours (about 8) in contrast to the short (30 min) cycle of the Bessemer process, its very much larger heat led to it being the dominant process by the mid twentieth century.

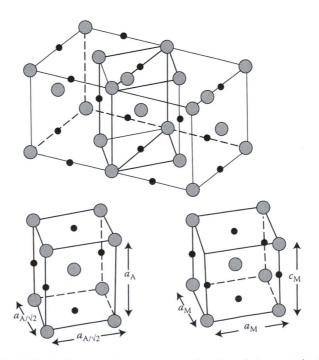

FIGURE 16.25 Relation between the martensite unit cells and the austenite from which it forms. Note that two martensite unit cells are formed for each austenite unit cell and that $c_M > a_A/\sqrt{2}$ and $c_M < a_A$.

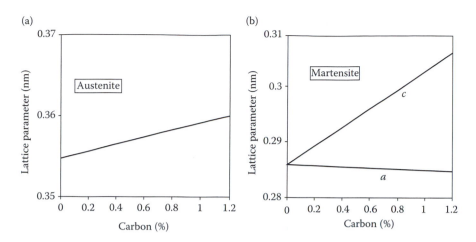

FIGURE 16.26 Changes of the lattice parameters of austenite (a) and martensite (b) with carbon content. Note that as the carbon content of martensite approaches zero, both *c* and *a* approach the lattice parameter of bcc ferrite. (Data from C. S. Roberts, *Trans. AIME*, 197, 1953.)

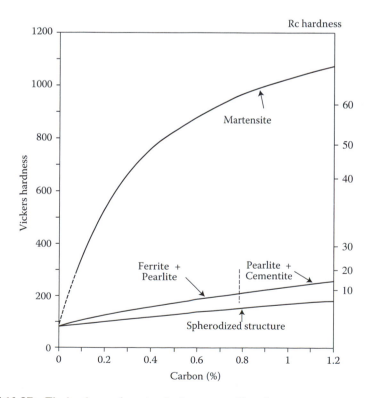

FIGURE 16.27 The hardness of martensite increases with carbon content up to a maximum of about Vickers 1000 at 0.8% C.

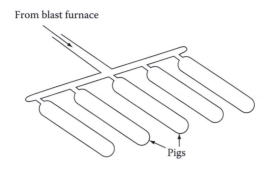

From blast furnace

Pigs

FIGURE 16.28 Sketch showing the tapping of the blast furnace into "pigs."

The basic oxygen process has now largely replaced both processes. Pig iron, together with scrap, is placed in a very large furnace. Carbon is removed by blowing oxygen through a lance into the melt. Today about half of the steel is produced by the basic oxygen process. The remainder is made by arc melting of scrap. In the second half of the twentieth century, continuous casting largely replaced ingot casting.

One may wonder why the Fe–C diagram has phases designated as α, γ, and δ but not β. The reason is that early workers, examining the thermal response of iron interpreted the anomalously high heat capacity near the Curie temperature (770°C) to be a latent heat of transformation (Figure 16.29). As a result, they labeled the structure of iron in the temperature region between 770°C and 912°C as β. Later it was realized that this was not a separate phase, so β was removed from phase diagrams.

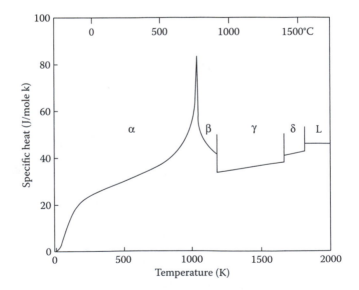

FIGURE 16.29 Temperature dependence of the heat capacity of iron. Early investigators interpreted the anomalously spike in the heat capacity near the Curie temperature to be the latent heat of a phase transformation and called the "phase" between 770°C and 912°C, β. (Adapted from *ASM Handbook*, 9th ed., Vol. 2.)

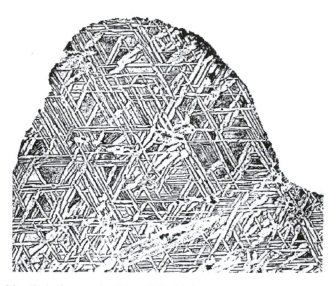

FIGURE 16.30 Etched meteorite. The original is in a book by Widmanstätten and Schriber was printed directly from the etched surface of the meteorite (original size). (From C. S. Smith, *A History of Metallography*, M.I.T. Press, 1960. With permission.)

Examination in 1808 of the structure of iron–nickel meteorites by William Thomson, Alois von Widmanstätten, and Carl von Schriber revealed a very large-scale needle-like structure that has since been called a Widmanstätten structure. In 1820, Widmanstätten and Schriber published a book on meteorites with a print made directly from a heavily etched meteorite (Figure 16.30). This is apparently the first published microstructure.

PROBLEMS

1. a. Use the data below to determine the solubility of carbon in metastable austenite at 620°C. Do this by assuming that the solubility can be expressed by %C = A exp[−Q/(RT)] and finding Q.
   b. In a 0.22% C steel transformed at 620°C, what fraction of the microstructure should be proeutectoid ferrite?
   c. Using your value of Q in part A, find the lowest temperature at which proeutetoid ferrite will form during isothermal transformation of a steel containing 0.50% C. Do this by finding the temperature at which the solubility is 0.50%.

Solubility of C in γ-Fe			
Temperature (°C)	wt% C	Temperature (°C)	wt% C
1100	1.92	1000	1.55
900	1.22	800	0.94
727	0.77		

2. Consider the pearlite formed by isothermal transformation of austenite in a steel containing 0.50% C at 727°C.
    a. What is the composition of this pearlite?
    b. Calculate the ratio of the thicknesses of the α and Fe_3C lamellae. Assume equal densities of the two phases.
    c. What fraction of the microstructure transforms to pearlite?
3. Repeat problem 2 for isothermal transformation at 680°C = 953 K.
4. a. Calculate the % volume change as austenite containing 0.40% carbon transforms to martensite.
    b. Calculate the % change of linear dimensions during the transformation. (Remember one unit cell of austenite transforms to two unit cells of martensite.)
5. Explain why the microstructure of hypoeutectoid steels after slow cooling consists of pearlite regions surrounded by ferrite.
6. An iron–carbon alloy containing 0.50% carbon has a microstructure consisting of 85% pearlite and 15% ferrite. Are these the amounts of the constituents expected after slow cooling? Calculate the expected amounts and offer a reason for the observed amount.
7. The rate of growth, G, of pearlite into austenite depends on the diffusional flux of carbon at the pearlite–austenite interface. Fick's first laws states that the flux, J, is given by $J = -D dc/dx$. The diffusional gradient, dc/dx, is inversely proportional to the pearlite spacing, λ.
    a. Knowing that λ is proportional to $1/\Delta T$, and $D = D_o \exp[-Q/(RT)]$, derive a relationship between G and the transformation temperature, T.
    b. At 675°C, $G = 2 \times 10^{-3}$ mm/s. What is the growth rate at 600°C? Assume $Q = 30,000$ J/mol.
8. a. Estimate the eutectoid temperature and composition for a 4340 steel that contains 0.40% C, 0.80% Mn, 0.22% Si, 0.80% Cr, 1.8% Ni, and 0.20% Mo.
    b. If this steel were transformed at the eutectoid temperature, what fraction of the microstructure would be pearlite?

REFERENCES

J. Chipman, *Met. Trans.*, 3, 55, 1972.

G. Krauss, *Steel–Heat Treatment and Processing Principles*, ASM, 1990.

W. C. Leslie, *The Physical Metallurgy of Steels*, McGraw-Hill, 1981.

D. T. Llewellyn and R. C. Hudd, *Steels: Metallurgy and Applications*, Butterworth Heinemann, 1998.

Making, Shaping and Treating of Steels, U.S. Steel Corporation, 1971.

Metals Handbook, 8th ed., Vol. 7, ASM.

D. D. Pearson and J. D. Verhoeven, *Met. Trans. A*, 15A, 1037, 1984.

C. S. Roberts, *Trans AIME*, 197, 1953.

J. V. Russell and F. T. McGuire, *Trans. ASM*, 33, 103, 1944.

C. S. Smith, *A History of Metallography*, M.I.T. Press, 1960.

A. R. Troyano and A. B. Greninger, *Metals Progress*, Vol. 50, p. 303, 1946.

17 Hardening of Steels

The influence of alloying elements on the rate of pearlite formation influences whether or not martensite will be formed when austenite is quenched since martensite can form only from austenite. If the formation of pearlite is delayed, more austenite will be available at the M_s temperature to transform to martensite. The term *hardenability* is used to describe this effect. We say that alloying elements increase the hardenability of steel, making it possible to harden them to greater depths.

17.1 HARDENABILITY: JOMINEY END-QUENCH TEST

Hardenability may be quantitatively described in several ways. One of the simplest is the Jominey end-quench test. A 4-in long, 1-in diameter bar of the steel in question is austenitized and then placed in a fixture and cooled from one end with a specified water spray (Figure 17.1). The hardness is then measured as a function of distance from the quenched end. Figures 17.2 and 17.3 show the resulting curves for several steels.

Several features should be noted:

1. The cooling rate at the quenched end was fast enough to insure 100% martensite in all the steels, so the hardness at the quenched end depends only on the carbon content.
2. The depth (distance from quenched end) of hardening increases with the amount of alloying addition.
3. The hardenability increases with carbon content.
4. The 1060 steel with larger grain size (ASTM#2) has a higher hardenability. This is because with a larger austenite grain size, there are fewer nucleation sites for pearlite.

Jominey data can be used in several ways. For example, the cooling rates at various positions in round bars during several types of quenches have been experimentally determined (Figure 17.4). For low alloy steels, these cooling rates are independent of the steel composition. Therefore, if the Jominey curve for a steel is known, the hardness distribution in a quenched bar can be predicted. Consider a 2-in diameter bar of 3140 steel quenched in "mildly agitated oil." The surface should cool at the same rate as a spot 5.5/16 in from the end of a Jominey bar and therefore should have a hardness of Rc53. At the mid-radius the equivalent Jominey distance is 10/16 so the

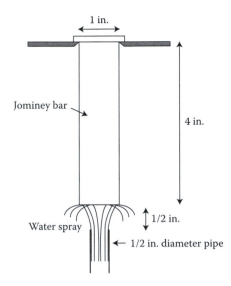

FIGURE 17.1   Jominey end-quench test.

hardness should be Rc43. At the center, the radius of the equivalent Jominey distance is 11/16 so the hardness should be Rc42.

For a part having a complex shape, the hardness at various locations can be predicted by quenching the same shaped part made from known steel. By comparing

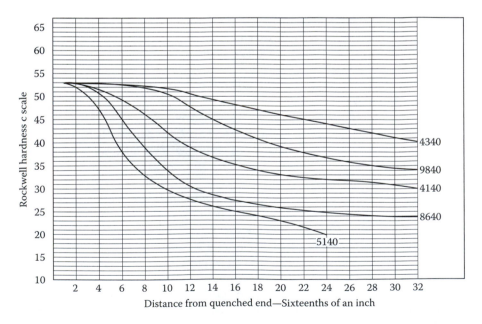

FIGURE 17.2   Hardnesses along Jominey bars of several steels. (From *The Making, Shaping and Treating of Steels*, 9th ed., U.S. Steel Corporation, 1971. With permission.)

Composition of Steels (%)

	C	Mn	Ni	Cr	Mo
4340	0.40	0.70	1.65	0.70	0.20
9840	0.40				
4140	0.40	0.75	—	0.80	0.15
8640	0.40	0.75	0.40	0.40	0.15
5140	0.40	0.70	—	0.70	—

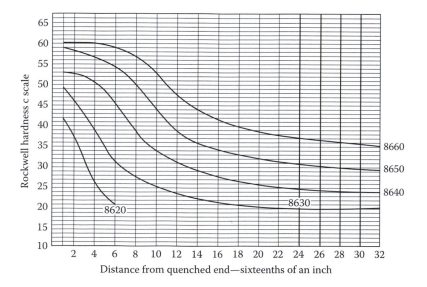

FIGURE 17.3 Hardnesses along Jominey bars of several steels of different carbon contents. All contain 0.70–1.0% Mn, 0.4–0.7% Ni, 0.4–0.5% Cr, and 0.1%–0.25% Mo. (From *The Making, Shaping and Treating of Steels*, 9th ed., U.S. Steel Corporation, 1971. With permission.)

the hardness at a given location with Jominey curve for this steel, the equivalent Jominey distance can be found and this can be used to predict the hardness at this location of a similar part made from any steel.

EXAMPLE PROBLEM 17.1

A test gear was made from a 1060 steel of ASTM GS#2. It was austenitized and quenched in agitated oil. The hardness at a critical spot was found to be Rc35. What would be the hardness if the gear were made of 4140 steel heat treated the same way?

SOLUTION

From Figure 17.2, the hardness of Rc35 in the 1060 steel indicates that the cooling rate is the same as 3/8 in from the end of a Jominey bar. The 4140 steel hardens to Rc53 when cooled at the same rate.

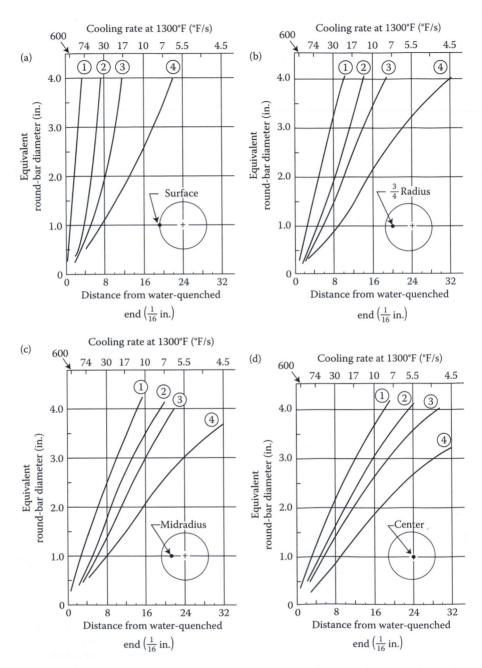

FIGURE 17.4 Cooling rates at different positions in round bars during several quenches related to positions along a Jominey bar. 1 = still water, 2 = mildly agitated oil, 3 = still oil, and 4 = mildly agitated molten salt. (From R. A. Flinn and P. K. Trojan, *Engineering Materials and their Applications*, 4th ed., Houghton Mifflin, p. 400, 1990. With permission.)

EXAMPLE PROBLEM 17.2

Predict the hardness of a 3-in diameter bar of 8660 steel quenched in mildly agitated oil.

SOLUTION

From Figure 17.4, it is seen that in mildly agitated oil, the center of a 3-in diameter bar will cool at the same rate as 20/16 in from the end of a Jominey bar. Figure 17.3 indicates that it will harden to Rc39.

17.2 IDEAL DIAMETER CALCULATIONS

The hardenability of a steel can be predicted from its composition and grain size. The method is as follows: The *critical diameter* and the *ideal diameter* are defined as follows:

The *critical diameter*, D_c, for a steel and quench is the diameter that would harden to 50% martensite at center.

The *ideal diameter*, D_I, is the diameter that would harden to 50% martensite in an ideal quench.

An *ideal quench* is one for which there is no resistance to heat transfer from the bar to the quenching medium ($H = \infty$), so the surface comes immediately to the temperature of the bath.

The values of quench severity for various quenches are given in Table 17.1.

Grossman worked out a scheme for estimating the ideal diameter for a steel. Knowing the carbon content and grain size, one can find the base diameter for a plain-carbon steel (Figure 17.5). Each alloying element has a multiplying effect. The multiplying factors are shown in Table 17.2.

TABLE 17.1
Severity of Quenches

Agitation/Medium	Quench Severity (H)[a]		
	Oil	Water	Brine
None	0.25–0.30	0.9–1.0	2.0
Mild	0.30–0.35	1.0–1.1	2.0–2.2
Moderate	0.35–0.40	1.2–1.3	
Good	0.40–0.50	1.4–1.5	
Strong	0.50–0.80	1.6–2.0	
Violent	0.80–1.1	4.0	5.0

[a] H is defined as the heat transfer coefficient, h in Btu/(h ft² °F) divided by the thermal conductivity of steel = 20 Btu/(h ft °F), so its units are ft⁻¹.

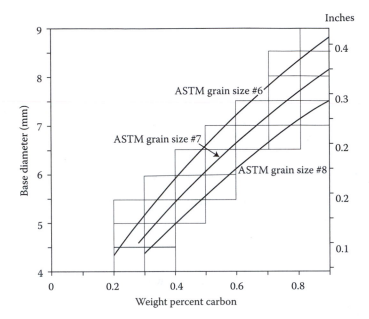

FIGURE 17.5 Dependence of the ideal diameter of a plain-carbon steel on its carbon content and austenitic grain size. (After M. A. Grossman, *Elements of Hardenability*, ASM, 1952.)

TABLE 17.2
Multiplying Factors

Percent	Mn	Mo	Cr	Si	Ni
0.00	1.00	1.00	1.00	1.00	1.00
0.10	1.34	1.31	1.22	1.07	1.04
0.20	1.67	1.62	1.43	1.14	1.08
0.30	2.00	1.93	1.65	1.21	1.11
0.40	2.34	2.24	1.86	1.28	1.15
0.50	2.08	1.35	1.19	2.080	2.50
0.60	3.01		2.29	1.42	1.23
0.70	3.34		2.51	1.49	1.26
0.80	3.68		2.72	1.56	1.30
0.90	4.01		2.94	1.63	1.34
1.00	4.35		3.15	1.70	1.38
1.10	4.78		3.37	1.77	1.41
1.20	5.17		3.58	1.84	1.45
1.30	5.60			1.91	1.49
1.40	6.05			1.98	1.53
1.50	6.6			2.05	1.56
1.60	7.2			2.12	1.60

Source: The Making, Shaping and Treating of Steels,
           9th ed., U.S. Steel, 1971.

EXAMPLE PROBLEM 17.3

Calculate the ideal diameter of a steel containing 0.45% C, 0.5% Mn, 0.2% Si, and 0.35% Cr if the ASTM grain size number is 6.

SOLUTION

For 0.45% C and ASTM grain size 6, Figure 17.4 indicates that the base diameter is 0.25 in. The multiplying factors for 0.5% Mn, 0.2% Si, and 0.35% Cr are 2.667, 1.140, and 1.756, respectively. The ideal diameter $D_I = (0.25)(2.667)(1.140)(1.756) = 1.3347$ in.

However, the critical diameter depends on the quench severity. Figures 17.6 and 17.7 show the relation between the critical and ideal diameters for various quench severities. (In the figure, "D value" means D_c and the values are given in inches.)

EXAMPLE PROBLEM 17.4

Find the critical diameter for the steel in Example problem 17.1, if it were quenched in a moderately agitated oil bath.

SOLUTION

Taking $H = 0.37$ from Table 17.1 and $D_I = 1.33$ in, Figure 17.7 indicates that $D_c = 0.30$ in.

EXAMPLE PROBLEM 17.5

Find the Jominey distance that would harden to 50% martensite for the steel in Example problems 17.3 and 17.4.

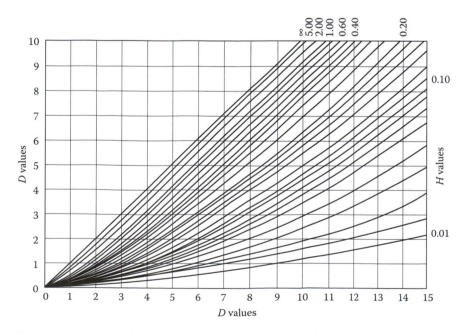

FIGURE 17.6 Relation between critical diameter and ideal diameter for different severities of quench. (From *The Making, Shaping and Treating of Steels*, 9th ed., p. 1098, U.S. Steel Corporation, 1971. With permission.)

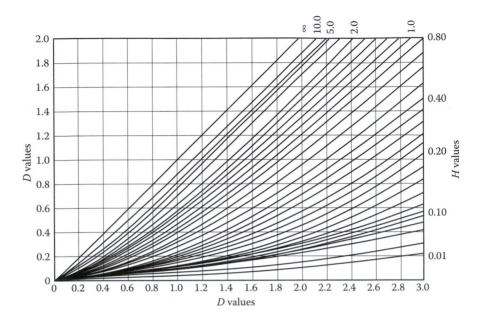

FIGURE 17.7 Enlargement of the low diameter end of Figure 17.5. (From *The Making, Shaping and Treating of Steels*, 9th ed., p. 1098, U.S. Steel Corporation, 1971. With permission.)

SOLUTION

For $D_I = 1.33$ in, Figure 17.8 indicates that the Jominey distance is about 130 mm.

EXAMPLE PROBLEM 17.6

Find the critical diameter for the steel in Example problem 17.1, if it is quenched in a moderately agitated oil bath.

SOLUTION

Taking $H = 0.37$ from Table 17.1 and $D_I = 1.3347$ in, Figure 17.7 indicates that the critical diameter (D_c value) would be 0.30 in.

Figure 17.8 compares the section sizes of bars with square cross sections and plates with round bars with the same of cooling rates at the centers.

There is an empirical relationship between the ideal diameter of a steel and the Jominey distance at which there will be 50% martensite. This is shown in Figure 17.9.

EXAMPLE PROBLEM 17.7

For the steel in Example problems 17.1 and 17.2, find the Jominey distance that would harden to 50% martensite.

SOLUTION

For $D_I = 1.3347$ in $= 33.9$ mm, Figure 17.8 indicates a Jominey distance of about 130 mm.

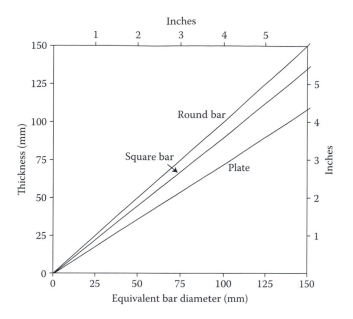

FIGURE 17.8 Comparison of sizes of square bars, round bars, and plates that have the same cooling rate at the center. (Data from W. C. Leslie, *The Physical Metallurgy of Steels*, McGraw-Hill, 1981.)

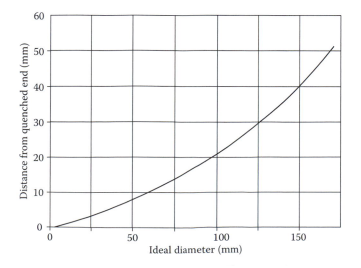

FIGURE 17.9 Relationship between the ideal diameter of a steel and the Jominey position that hardens to 50% martensite. (Data from D. J. Carney, *Trans. ASM*, 46, 822–927, 1954.)

17.3 BORON

Boron in the range of $0.0005 < \text{wt}\% \; B < 0.003$ has a strong effect on hardenability (Figure 17.10). Its effect decreases with increasing % C (Figure 17.11) and depends on austenitizing temperature (Figure 17.12). Boron has very low solubility in austenite and an extremely low solubility in ferrite. It segregates to austenite grain boundaries. Its effect is probably that it lowers austenite grain boundary energy and therefore decreases nucleation rate at grain boundaries.

17.4 MARTENSITE AND RETAINED AUSTENITE

Figure 17.13 shows that the hardness of quenched structures reaches a maximum amount at about 0.8% C and then decreases, even though the hardness of martensite does not drop. The decrease in hardness at high carbon contents is caused by the fact that the amount of retained austenite increases with carbon content. Remember that increased carbon content lowers the M_s temperature and so the amount of retained austenite increases.

Nature of the martensite depends on the carbon content. Lath martensite at low carbon contents is composed of parallel laths separated by high angle boundaries as shown in Figure 17.14. At higher carbon contents the martensite is composed of twinned plates (Figure 17.15). Figure 17.16 shows how the amounts of lath martensite and retained austenite depend on the carbon content.

Hypereutectoid steels are normally austenitized just above the lower critical. In this way, the carbon content of the austenite is about 0.8% C.

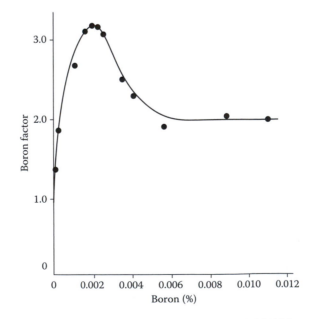

FIGURE 17.10 Multiplying factor for boron in a 0.20% C and 0.55 % Mo steel. (From G. F. Melloy, P. R. Slimmon, and P. P. Podgurski, *Met. Trans.*, 4, 1973. With permission.)

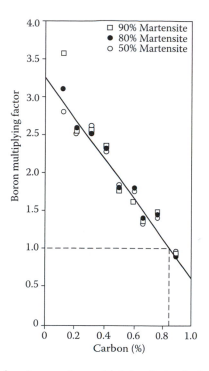

FIGURE 17.11 Effect of carbon on the multiplying factor for boron in 0.8% Mn steels. (From D. T. Llewellyn and W. T. Cook, *Met. Tech.*, 1, 517, 1974. With permission.)

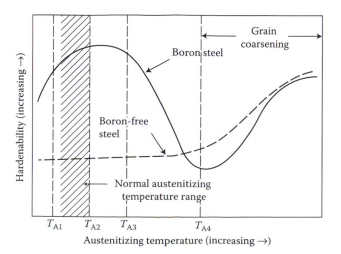

FIGURE 17.12 Effect of boron decreases for high austenitizing temperatures. (From R. A. Grange and J. B. Mitchell, *Trans. ASM*, 53, 157, 1961. With permission.)

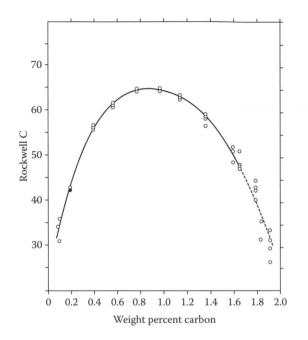

FIGURE 17.13   Above about 0.8% C, the hardnesses of quenched steels decrease with carbon content because of an increased amount of retained austenite. (From A. Lipwinchuk, F. X. Kayser, and H. H. Baker, *J. Mat. Sci.*, 11, 1200, 1976. With permission.)

FIGURE 17.14   Lath martensite. (From G. R. Speich and W. C. Leslie, *Met. Trans.*, 3, 1972. With permission.)

FIGURE 17.15 Twinned plate martensite. (From G. R. Speich and W. C. Leslie, *Met. Trans.*, 3, 1972. With permission.)

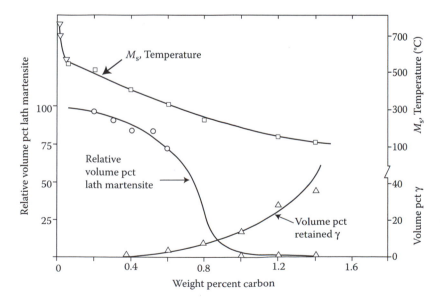

FIGURE 17.16 Amount of lath martensite decreases at high carbon contents. Amount of retained austenite increases as the M_s temperature drops. Data for carbon steels. (After G. R. Speich and W. C. Leslie, *Met. Trans.*, 3, 1972.)

17.5   MISCELLANY

In the period 1932–1936, E. C. Bain and coworkers did pioneering work on the kinetics of decomposition of austenite and presented their results in the form of isothermal transformation diagrams. In 1934, B. F. Shepherd proposed that the hardenability of a steel could be assessed by quenching a bar of steel and measuring the depth of hardening. Walter Jominey and A. L. Boegehold first published the details of the end-quench test in 1938. Also, about the same time, M. A. Grossman and coworkers proposed the concept of an ideal diameter. Robert Mehl explained the physics behind the effects of alloys on hardenability in 1939.

PROBLEMS

1. A test piece of a part made from 4340 steel was austenitized and a good oil quench given. At a critical spot the hardness was measured and found to be Rc50. If the same part were made from 4140 steel and the same heat treatment given, what would be the hardness at the same critical spot?
2. Calculate the critical diameter for a steel containing 0.40% C, 0.20% Si, 0.60% Mn, 0.80% Cr, and 0.15% Mo with an ASTM grain size of 8 quenched in mildly agitated oil ($H = 0.3$).
3. What thickness of plate of the steel in problem 1 would harden to 50% martensite in the center, if quenched in mildly agitated oil ($H = 0.3$)?
4. What diameter of a round bar of 3140 steel would harden to Rc40, if the bar were quenched in mildly agitated oil?
5. A steel contains 0.40% C, 0.25% Si, 0.70% Mn, and 1.0% Cr. This is to be replaced by a second steel containing 0.40% C, 0.25% Si, 0.80% Mn, and 0.5% Cr and enough Mo to achieve the same hardenability.
   A. How much Mo should the steel contain?
   B. Would the alloys in the new steel be more expensive or less expensive?
6. How thick of a plate of 4140 steel (of the composition given in Figure 17.2) could be hardened to 50% martensite at the center, if it were quenched in agitated oil ($H = 0.35$)?
7. Table 17.2 indicates that on a pound-for-pound basis Mo contributes more hardenability than Si, Ni, or Cr, and on an atom-for-atom basis more than Mn. Suggest a possible explanation.

REFERENCES

D. J. Carney, *Trans. ASM*, 46, 822–927, 1954.
R. A. Flinn and P. K Trojan, *Engineering Materials and their Applications*, 4th ed., Houghton Mifflin, p. 400, 1990.
R. A. Grange and J. B. Mitchell, *Trans. ASM*, 53, 157, 1961.
M. A. Grossman, *Elements of Hardenability*, ASM, 1952.
G. Krauss, *Steel—Heat Treatment and Processing Principles*, ASM, 1990.
W. C. Leslie, *The Physical Metallurgy of Steels*, McGraw-Hill, 1981.
A. Lipwinchuk, F. X. Kayser, and H. H. Baker, *J. Mat. Sci.*, 11, 1200, 1976.
D. T. Llewellyn and W. T. Cook, *Met. Tech.*, 1, 517, 1974.
G. F. Melloy, P. R. Slimmon, and P. P. Podgurski, *Met. Trans.*, 4, 1973.
G. R. Speich and W. C. Leslie, *Met. Trans.*, 3, 1972.
The Making, Shaping and Treating of Steels, 9th ed., p. 1098, U.S. Steel Corporation, 1971.

18 Tempering and Surface Hardening

18.1 TEMPERING

Martensite is very brittle. To make it tougher, it is normally heated to *temper* it. Tempering is a complex series of reactions that involve the gradual breakdown of martensite. What happens is usually described by phases. The first phase occurs at the lowest temperature (shortest time) and involves transformation of retained austenite. In the second phase, carbon is redistributed within the martensite to dislocations. Generally, stress relief occurs during this stage. Precipitation of ε carbide ($Fe_{2.4}C$) and η carbide (Fe_2C) from the martensite comprise the third phase. This precipitation lowers the carbon content of the martensite. In stage four, remaining retained austenite decomposes to cementite (Fe_3C) and ferrite. Finally, in stage five, the transition carbides and low-carbon martensite form more ferrite and cementite. These reactions overlap.

There is a gradual loss of hardness throughout tempering (except stage one), at increasing temperatures as a function of the carbon content as shown in Figure 18.1. The amount of tempering depends on time as well as on temperature as indicated in Figure 18.2, although the effect of time is much less than that of temperature.

Tempering can be treated by an Arrhenius equation. The time, t, to reach a given hardness is

$$t = A \exp\left(\frac{+Q}{RT}\right)$$

(18.1)

or the hardness, H, can be expressed as

$$H = f\left[t \exp\left(\frac{-Q}{RT}\right)\right].$$

(18.2)

EXAMPLE PROBLEM 18.1

Figure 18.3 shows how the hardness of a 0.72% C, 0.85% Mn steel depends on time and temperature. Determine the activation energy for tempering by comparing the time–temperature combinations that result in a hardness of Rc 60.

293

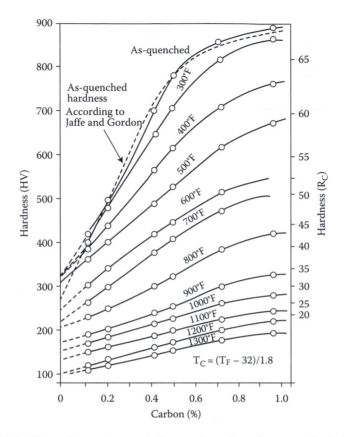

FIGURE 18.1 Tempering martensite in iron–carbon alloys for one and a half hour. The hardness drops with increasing temperature. (From R. A. Grange, C. R. Hribal, and L. F. Porter, *Met. Trans. A*, 8A, 1977. With permission.)

How many months (or years) could this steel be used at 220°C before its hardness falls to Rc 60?

SOLUTION

Equation 18.1 can be expressed as

$$t = A\exp\left(\frac{Q}{RT}\right), \quad \frac{t_2}{t_1} = \exp\left[\left(\frac{Q}{R}\right)\left(\frac{1}{T_2} - \frac{1}{T_1}\right)\right],$$

so

$Q = R\ln(t_2/t_1)/(1/T_2 - 1/T_1)$. Taking points at Rc 60 in Figure 18.3, $t_2 = 800$ min at $T_2 = 250°C = 523°K$ and $t_1 = 0.6$ min at $T_2 = 315°C = 588$ K.

$Q = 7.134 \ln(800/0.6)/(1/523 - 1/588) = 249$ kJ/mol.

Substituting $T_3 = 220°C = 493°K$, into $t_3/t_2 = \exp[(Q/R)(1/T_3 - 1/T_2)] = 35.2$ min.

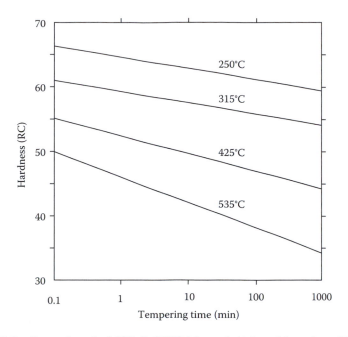

FIGURE 18.2 Tempering of a 0.82% C, 0.75% Mn steel. (Adapted from data of E. C. Bain, *Functions of Alloying Elements in Steel*, ASM, 1969.)

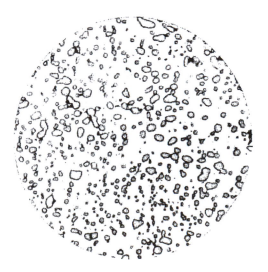

FIGURE 18.3 Spheroidite produced by tempering just below the lower critical temperature. (From *Making, Shaping and Treating of Steel*, 9th ed., U.S. Steel Corporation, 1977. With permission.)

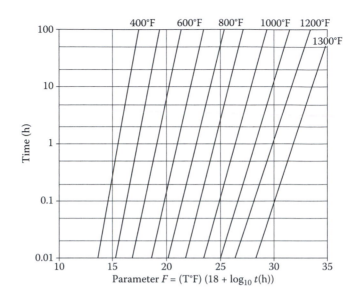

FIGURE 18.4 Time–temperature combinations that result in the same parameter describing the extent of tempering. (Adapted from R. A. Grange and R. W. Baughman, *Trans. ASM*, 48 165–197, 1956.)

At very high tempering temperatures, the carbides spheroidize, producing a product called spheroidite* as shown in Figure 18.3.

Different times at different temperatures will result in equivalent amounts of tempering. Grange and Baughman showed that the degree of tempering is related to a parameter,

$$F = T(18 + \log_{10} t), \tag{18.3}$$

where T is the temperature in °F plus 480 and t is the time in hours. Figure 18.4 shows this relation.

Alloying elements slow the tempering rate and lessen the decrease of hardness. Figure 18.5 illustrates this. This retardation of tempering by alloying elements is partially a result of the fact that substitutional elements diffuse slower than carbon. Many of the alloying elements also form harder carbides than iron.

It is interesting to compare tempering of steel with precipitation hardening of nonferrous alloys. Both processes involve the precipitation of fine particles from a supersaturated solid solution (martensite can be thought of as a supersaturated solution of carbon in ferrite). The precipitates cause a hardening but the loss of solid solution hardening causes an accompanying softening. In the case of precipitation

---

* Spheroidite is not normally produced by tempering of martensite. Hot-rolled structures of medium carbon steels can be spheroidized by heating to just below the lower critical for several hours. Alternatively, they can be heated into the α + γ range and then cooled to just below the lower critical and held for a number of hours. The former way is much faster.

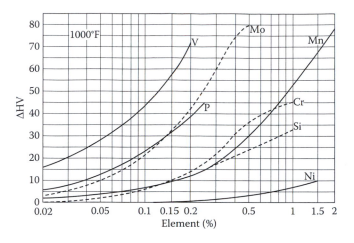

FIGURE 18.5 Effect of alloying elements on the hardness of martensite after tempering at 1000°F (538°C) for 1 h. (From R. A. Grange, C. R. Hribal, and L. F. Porter, *Met. Trans.*, 8A, 1775–1785, 1977. With permission.)

hardening, the increase of hardness from the precipitates more than makes up for the loss of solid solution strengthening. In the case of tempering, the loss of solution hardening is much greater than the precipitation hardening effect.

In alloy steels, some of the retained austenite may transform to martensite on cooling from tempering. This untempered martensite will lower the toughness. To alleviate this problem, sometimes steels are double tempered, the second tempering treatment to temper the martensite formed on cooling from the first tempering.

18.2 SECONDARY HARDENING

There is a strong effect of alloying elements on how much softening occurs. With large amounts of Cr, Mo, and W, there may be secondary hardening. Steels containing these alloys are resistant to tempering. When carbides do precipitate at high temperatures, they have a very strong precipitation-hardening effect and are resistant to coarsening. Figure 18.6 shows the effect of molybdenum. This effect is very important in high-speed tool steels.

18.3 TEMPER EMBRITTLEMENT

There are two forms of temper embrittlement: One is the 500°F (350°C) embrittlement that occurs in low alloy steels after tempering in the 250–450°C temperature range (Figure 18.7). The fractures are intergranular. This is particularly troublesome if the combination of strength and ductility obtainable in this temperature range is needed. Here bainite is useful.

The other form of temper embrittlement (two-step embrittlement) is reversible. It occurs only in alloy steels tempered in the range of 600–700°C and slowly cooled through 600–350°C. It is caused by Sb and P (and secondarily by Sn and As). Ni,

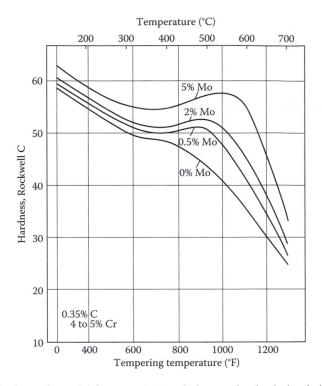

FIGURE 18.6 Increasing molybdenum content results in secondary hardening during tempering. (From *Making, Shaping and Treating of Steel*, U.S. Steel Corporation, 1973. With permission.)

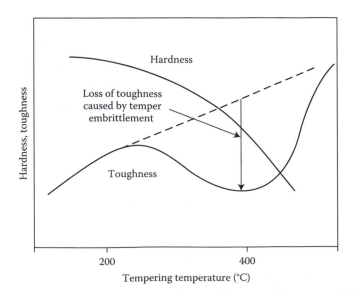

FIGURE 18.7 Hardness and toughness changes on tempering. The low toughness associated with tempering in the 250–450°C range is called 500°F embrittlement.

Mn, Cr, and Si aggravate the embrittlement. Mo, Ti, and Zr delay its onset. Heating to over 600°C followed by a rapid cooling reverses the embrittlement.

18.4 CARBURIZING

There are several techniques for hardening the surfaces of steels. These include diffusing carbon C (carburizing) and nitrogen (nitriding) into the surface and heat treating the surface without heat treating the core (case hardening).

In carburizing, the steel is heated into the austenite region (900–930°C) and subjected to a carburizing atmosphere. This is usually a mixture of carbon monoxide and water vapor. Carbon is deposited in the steel by the reaction, $CO + H_2 \rightarrow \underline{C} + H_2O$. The CO/H_2 ratio is controlled to give carbon potential in equilibrium with 0.9% C in the austenite. Often a two-step process, which initially uses higher C potential than desired in the final product followed by period with a lower carbon potential, which allows time for diffusion. If % C is too high hardness falls because of retained austenite, as shown in Figure 17.12.

The high temperatures required for carburizing coarsen the austenite grain size of the core. This can be alleviated with a second lower temperature austenitization treatment as illustrated in Figure 18.8. Figure 18.9 is the cross section of a carburized gear.

The depths of hardening (Figure 18.10) typically vary from 0.020–0.060 in (0.5–1.5 mm). The depth is controlled by the diffusion of carbon into the surface. All solutions to Fick's second law predict that concentration is a function of $x/\sqrt{Dt}$ where x is the distance at which a certain concentration occurs, D is the diffusivity, and t is the time. Hardness is a function of carbon concentration, so the depth, x, at which a given hardness will be found depends only on D and t.

For a fixed hardness,

$$x = A\sqrt{Dt}, \qquad (18.4)$$

where A is a constant. To double the depth of hardening at fixed carburizing conditions, the time must be increased fourfold.

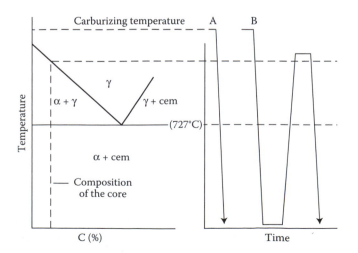

FIGURE 18.8 Effect of heat treatment on the core and case of a carburized part.

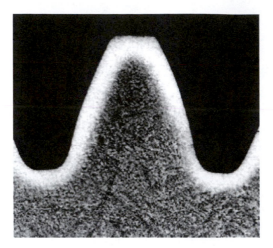

FIGURE 18.9 Cross section of a gear of a steel containing 0.016% C, 0.6% Mn, 1.65% Cr, and 3.65% Ni carburized at 920°C, austenitized at 830°C, quenched and tempered at 150°C. The case depth is 0.8–1.0 mm. (From *ASM Handbook*, 9th ed., Vol. 9, 1985. With permission.)

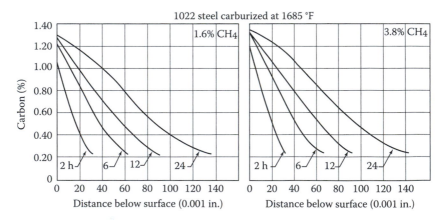

FIGURE 18.10 Depth of carburization at several times. Note that the depth at a particular level of hardness is proportional to √(time). (From *Metals Handbook*, 8th ed., Vol. 2, ASM, 1964. With permission.)

Alternatives to carburizing in a mixed CO/H_2 atmosphere include mixed CO/CO_2, low-pressure methane atmosphere, and pack carburizing. Pack carburizing involves heating the part to be carburized while it is packed in coke or charcoal. In pack carburizing, almost all of the transport of carbon from the coke to the steel is by CO.

18.5 NITRIDING

In nitriding, nitrogen is diffused into the steel at temperatures below the lower critical (500–570°C). The sources of nitrogen include NH_3, NH_3–N_2 mixtures, NH_3 mixed

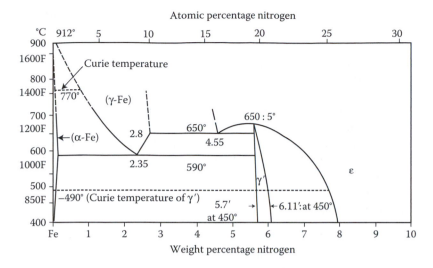

FIGURE 18.11 The iron–nitrogen phase diagram. (From *Metals Handbook*, 8th ed., Vol. 8, ASM, 1973. With permission.)

with endothermic gas (40% N_2, 20% CO, and 40% H_2), and NH_3–N_2–CO_2 mixtures. Diatomic nitrogen, N_2, will not dissolve in steel. Figure 18.11 shows the Fe/N phase diagram. The depths of nitrided layers are less than the depths of carburized layers. A depth of 0.01–0.015 in (0.25–0.37 mm) can be obtained in 48 h. In unalloyed steels, the hardness is a result of ε carbide (Fe_3N). Some steels used for nitriding usually contain aluminum, chromium, or molybdenum. A typical nitriding composition is 1.0% Al, 0.25% C, 0.5% Mn, 0.25% Si, 1.0% Cr, and 0.2% Mo.

A process called carbonitriding is done in molten cyanide salts, which allow diffusion of both carbon and nitrogen into the steel. It is basically a carburizing treatment in which some nitrogen enters the steel. The nitrogen acts mainly to increase the hardenability. With carbonitriding, there is a greater problem of retained austenite than with carburizing.

18.6 CASE HARDENING WITHOUT COMPOSITION CHANGE

If a very strong heat source is applied to a part, the surface can be heated into the austenite temperature range while the interior remains cold. As soon as the heat source is removed, the unheated interior provides the quench necessary for martensite formation. The key is a rapid heating of the surface. With high frequency induction, heating is restricted to the surface. Surface heating may also be accomplished with a laser, or an intense flame.

18.7 FURNACE ATMOSPHERES

Different gases have different characteristic. Oxygen is oxidizing. It reacts with dissolved carbon to form carbon monoxide, $O_2 + 2\underline{C} \rightarrow 2CO$. It must be excluded for bright anneals. CO_2 is also oxidizing and decarburizing.

CO is a source and carrier of carbon. It is reducing. The CO_2/CO ratio controls whether the atmosphere is oxidizing or reducing. A ratio of 0.6 is oxidizing at 800°C; a ratio of 0.4 is not oxidizing but will decarburize a 1% C steel. For low-carbon steels ratio of 0.5 is used for bright anneals. Lower ratio may carburize.

Nitrogen is inert. It is used for ion nitriding. Hydrogen, H_2, is highly reducing: Water, H_2O, is oxidizing. Hydrocarbons and proprietary mixtures are source of carbon. These include methane (glow discharge for carburizing) forms C^+ and "endothermic gas": which is a mixture of 20% CO, 40% H_2O, 40% N_2, and methane and is used for carburizing. Ammonia, NH_3, when decomposed in an arc is used as a source of nitrogen for nitriding. A mixture of ammonia and endothermic gas can be used for carbonitriding.

18.8 MISCELLANY

The word "tempering" has different meanings when applied with relation to different materials. With most nonferrous alloys, "temper" refers to the degree of hardness produced by cold work. The higher the temper number, the *harder* the alloy is. With steels, it refers to the softening that occurs when martensite is heated. The more the "tempering," the *softer* the steel is. This is in general agreement that to temper something is to moderate or mitigate it. When the term tempered steel is used in the popular press it implies "hardened and tempered." Finally, when the term "tempering" is applied to glass, it refers to the process of rapidly cooling the glass to strengthen it by imparting residual compression stresses in the surface.

Tempering is usually done in air. Oxidation of the surface of steel develops colors that depend on the thickness of the oxide, and hence the time and temperature of tempering. In the past, heat treaters used these temper colors to gauge whether tempering was complete. Gun barrels were tempered to a light blue. Hence the term "gun barrel blue." Today certain chemical agents may be used to hasten the formation of a color.

PROBLEMS

1. Figure 18.12 shows how the hardness of a 1080 steel depends on time and temperature.
   a. Determine the activation energy for tempering by comparing the time–temperature combinations that result in a hardness of Rc 50.
   b. How many months (or years) could this steel be used at 200°C before its hardness falls to Rc 50?

2. The ABC Company carburizes 1 in diameter shafts of an 8220 steel at 885°C for 8 h. The case depth at which the hardness drops to Rc 50 is 0.030 in. The boss thinks that the carburization time could be shortened if the temperature were raised to 920°C. What time would be required to obtain the same depth of carburization? For diffusion of carbon in α-iron, $D_o = 220 \times 10^{-6}$ m²/s and $Q = 122.5$ kJ/mol and for diffusion of carbon in γ-iron, $D_o = 20 \times 10^{-6}$ m²/s and $Q = 142$ kJ/mole.w

   Solution:
   In both cases, the diffusion is in austenite. $t_2/D_1 = t_1/D_2$ so $t_2 D_1 = t_1 (D_1/D_2) = $ (8 h)exp$[(-Q/R)(1/T_1 - 1/T_2)] = $ (8)exp$[-(122,500/8.314)(1/1028 - 1/1193)] = 1.1$ h.

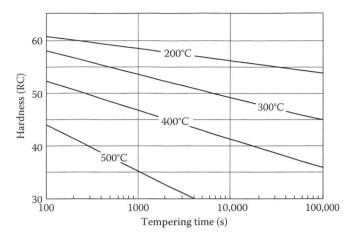

FIGURE 18.12   Tempering kinetics of a 1080 steel.

3. a. Calculate the diffusivity of carbon in γ-iron at 915°C using the data in problem 2.
   b. Calculate the diffusivity of carbon in α-iron at 720°C.
   c. Why carburizing is not done at 720°C rather than at 900°C?
4. Rapid tempering of parts is often done on a continuous conveyer line. What tempering temperature is required to achieve the same degree of tempering in 15 s as achieved in 30 min at 550°C?
5. Using the data in Figures 18.1 and 18.5, estimate the hardness of a steel containing 0.4% C, 0.2% Mo, 0.7% Mn, 0.5% Cr, and 0.5% Ni after hardening and tempering for 1 h at 538°C (1000°F).
6. Figure 18.5 indicates that Ni and Si have very much less effect on tempered hardness than Mn and that Mo and V have a much greater effect than Mn. Suggest an explanation.
7. Would an austenite surface in equilibrium with graphite contain more, less, or the same amount of carbon as an austenite surface in equilibrium with cementite? Explain your answer.

REFERENCES

ASM Handbook, 9th ed., Vol. 9, 1985.
E. C. Bain, *Functions of Alloying Elements in Steel*, ASM, 1969.
R. A. Grange and R. W. Baughman, *Trans. ASM*, 48 165–197, 1956.
R. A. Grange, C. R. Hribal, and L. F. Porter, *Met. Trans.*, 8A, 1775–1785, 1977.
G. Krauss, *Steel–Heat Treatment and Processing Principles*, ASM, 1990.
W. C. Leslie, *The Physical Metallurgy of Steels*, McGraw-Hill, 1981.
Making, Shaping and Treating of Steel, U.S. Steel Corporation, 1973.
Making, Shaping and Treating of Steel, 9th ed., U.S. Steel Corporation, 1977.
Metals Handbook, 8th ed., Vol. 2, ASM, 1964.
Metals Handbook, 8th ed., Vol. 8, ASM, 1973.

19 Low-Carbon Sheet Steel

Low-carbon sheet steel may be finished by hot rolling or by cold rolling. The *hot-rolled* steel has a rougher surface finish that limits its use to applications where surface appearance is not important (e.g., auto underbodies and firewalls). *Cold-rolled* steels are almost always recrystallized before sale to fabricators. They are therefore softer than hot-rolled steels, and have a much better surface finish.

19.1 STRAIN AGING

Low-carbon steels (%C ≈ 0.06 or less) are usually finished by cold rolling and annealing except in heavy gauges (2 mm thick or less). They are marketed after annealing at 600–700°C. Historically, they were produced by casting into ingots and were classified as either *rimmed steel* or *aluminum-killed steel*. A rimmed steel was one that was not deoxidized before ingot casting. During freezing, dissolved oxygen and carbon react to form CO. Violent evolution of CO bubbles threw sparks into the air. The bubbles stirred the molten metal breaking up boundary layers. This allowed segregation of carbon to the center of the ingot producing a very pure iron surface. In contrast, killed steels are deoxidized with aluminum, so the violent reaction is "killed." The solidification is quiet and so a boundary layer forms and prevents surface-to-center segregation.

Today continuous casting has almost completely replaced ingot casting. As a result, almost all steels are killed. There has been a trend to casting thinner sections, which require less rolling. This saves money but does not refine the grain structure as much.

Figure 19.1 shows a tensile stress–strain curve of an annealed low-carbon steel. Loading is elastic until yielding occurs (point A). Then the load suddenly drops to a lower yield stress (point B). Continued elongation occurs by propagation of the yielded region at this lower stress until the entire specimen has yielded (point C). During the extension at the lower stress, there is a sharp boundary or *Lüder's band* between the yielded and unyielded regions as shown in Figure 19.2. Behind this front, all of the material has suffered the same strain. The Lüder's strain or yield point elongation is typically from 1% to 3%. Only after the Luder's band has traversed the entire specimen does strain hardening occur. Finally, at point E the specimen necks.

If the specimen in Figure 19.1 were unloaded at some point, D, after the lower yield region and immediately reloaded, the stress–strain curve would follow the original curve. However, if the steel were allowed to strain age between unloading and reloading, a new yield point would develop. During strain aging interstitially

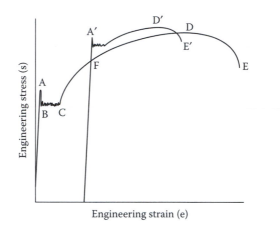

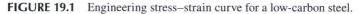
Engineering strain (e)

FIGURE 19.1   Engineering stress–strain curve for a low-carbon steel.

dissolved nitrogen and carbon segregate to dislocation lowering their energy. A higher stress is required to move the dislocations away from these interstitials than to continue their motion after they have broken free.

Aluminum-killed steels are much more resistant to strain aging than rimmed steels, but strain aging will occur at the temperatures of paint baking. At this point the strain aging is beneficial because the parts have already been formed and the

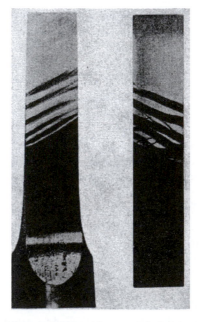

FIGURE 19.2   Tensile specimen of a low-carbon steel during extension. Deformation occurs by movement of a Lüder's band through the specimen. (From F. Körber, *J. Inst. Metals*, 48, 317–342, 1932. With permission.)

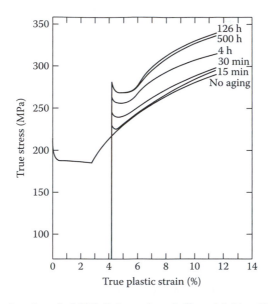

FIGURE 19.3   Strain aging of a 0.03% C rimmed steel. (From *Making, Shaping and Treating of Steel*, 9th ed., U.S. Steel Corporation, 1971. With permission.)

strain aging adds strength (Figure 19.3). Strain aging is much slower in steels that have been aluminum-killed than those that have not.

The yield point and stretcher strains can be removed by causing a small amount of deformation in the sheet by roller leveling or by temper rolling. In roller leveling the sheet is bent back and forth by a series of small diameter rolls. This causes local regions to yield without much change in thickness. In temper rolling the sheet is subjected to a very light reduction (usually 0.5–1.5%). This is also called a pinch pass or skin rolling. In both cases the overall strain is less than the yield-point elongation. Only small regions need deform as shown in Figure 19.4.

Usually aluminum is added to molten low-carbon steel as it is poured. Without the addition of aluminum, dissolved oxygen would react with dissolved carbon to form CO. This reaction is violent, the CO bubbles causing steel droplets to fly into the air where they ignite. The addition of aluminum reacts with the dissolved oxygen so it is not available to react with carbon. This process is called *killing* and the steel called

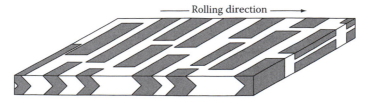

FIGURE 19.4   Deformed regions in roller-leveled material. (Adapted from D. S. Blickwede, *Metals Progress*, July 1969.)

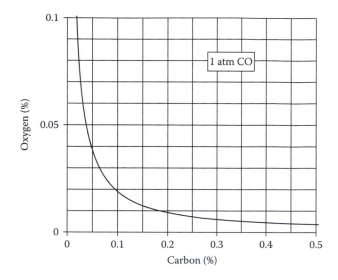

FIGURE 19.5 Equilibrium between oxygen and carbon in molten steel at 1540°C. With high carbon contents, the oxygen solubility is low and vice versa.

an AKDQ (aluminum-killed, drawing quality) is the designation for most of the steel sheet used in forming operations. The amount of oxygen that can dissolve in molten steel decreases with increasing carbon contents, as shown in Figure 19.5. Therefore killing is not required for higher carbon contents. Today almost all low-carbon steel is continuously cast. AKDQ sheets usually contain 0.03% carbon or less.

The aluminum that does not react to form Al_2O_3 can react after solidification with dissolved nitrogen to form AlN.

19.2 QUALITY

Low-carbon sheet is formed by manufacturers into a variety of shapes including automotive bodies, appliances, and cans. Formability and surface appearance are vital in most applications. The six important properties of sheet steels are

1. Surface finish
2. Strain hardening
3. Strain rate sensitivity
4. Anisotropy
5. Freedom from yield point effect
6. Yield strength (YS).

Surface finish: Cold-rolled steels have a much better surface finish than hot-rolled steels. The term *cold rolled* means cold rolled and recrystallized (unless otherwise stated). Stretcher strains, resulting from strain aging (Figure 19.6) cannot be tolerated for most applications.

FIGURE 19.6 Stretcher strains on a formed part. (From *Metals Handbook*, 8th ed., Vol. 7, p. 14, ASM, 1973. With permission.)

Large grain sizes, resulting from high annealing temperatures, cause an orange peel effect as shown in Figure 19.7.

Strain hardening: The strain hardening is usually expressed by

$$\sigma = K\varepsilon^n, \tag{19.1}$$

where the strain hardening exponent, n, describes the persistency of hardening and is equal to the uniform elongation in a tension test. Figure 19.8 illustrates the relation between n and the shape of the stress–strain curve. A high n is indicative of a high stretchability.

Yield strength: Higher carbon and alloy content promote higher yield strength (YS), but lower n (Figure 19.9). *R*-value is affected by the rolling and annealing cycles. With increasing interest in reducing the weight of automobiles, the use of higher YS sheet has increased.

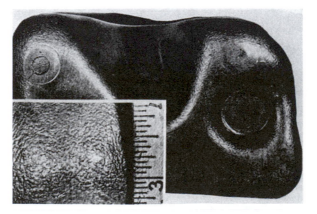

FIGURE 19.7 Orange peel. The surface rumpling is on the scale of the grain size. (Courtesy of American Iron and Steel Institute.)

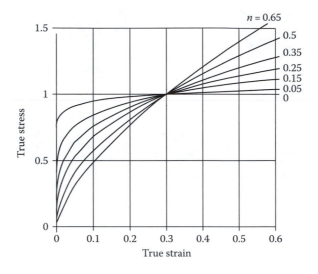

FIGURE 19.8 True stress–strain curves approximated by $\sigma = K\varepsilon^n$, with several values of n.

An empirical relation between YS and n is

$$YS = \frac{69\ \text{MPa}}{n}. \tag{19.2}$$

Strain rate sensitivity: The effect of the strain rate on the stress–strain curves is usually expressed by

$$\sigma = C\dot{\varepsilon}^m \tag{19.3}$$

for a constant level of strain. In steels with a high m, necks tend to localize less rapidly.

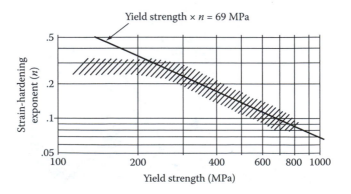

FIGURE 19.9 Relation between yield strength and strain-hardening exponent, n. (Adapted from S. P. Keeler and W. G. Brazier, in *Microalloying 75*, Union Carbide, 1977.)

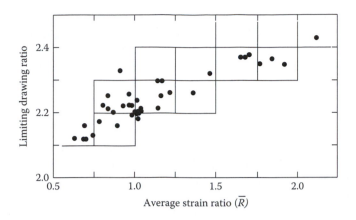

FIGURE 19.10 Limiting drawing ratio increases with the *R*-value. An increased limiting drawing ratio allows deeper cups to be drawn in a single operation. (Adapted from R. W. Logan, D. J. Meuleman, and W. F. Hosford, in *Formability and Metallurgical Structure*, A. K. Sachdev and J. D. Embury, Eds, TMS, 1987.)

Anisotropy: This is usually expressed by the plastic strain ratio. Generally, high *R* leads to increased drawability (Figure 19.10) and less wrinkling. Aluminum-killed steels usually have a strong texture with {111} parallel to the sheet. This leads to *R*-values of about 1.6–1.8.

19.3 GRADES OF LOW-CARBON STEELS

There are a number of grades of low-carbon steels. Among these are

Interstitial-free (IF) steels: These are steels from which carbon and nitrogen have been reduced to extremely low levels (<0.005%). After vacuum degassing, titanium is added to react with any carbon or nitrogen in solution. Titanium reacts preferentially with sulfur, so the stoichiometric amount of titanium that must be added to eliminate carbon and nitrogen is

$$\%\text{Ti} = \left(\frac{48}{14}\right)(\% \text{ N}) + \left(\frac{48}{32}\right)(\% \text{ S}) + \left(\frac{48}{12}\right)(\% \text{ C}). \tag{19.4}$$

A typical composition is 0.002% C, 0.0025% N, 0.025% Ti, 0.15% Mn, 0.01% Si, 0.01% P, 0.04% Al, and 0.016% Nb. Niobium may be used instead of titanium but it is more expensive. IF steels typically have very low strengths. The advantage of these steels is that they are very formable. Control of crystallographic texture is also fundamental in producing exceptional deep drawability. Typically, $\bar{R} = 2.0$ (*R*-values of traditional aluminum-killed steels rarely exceed 1.8). Formability is also enhanced by high *n*-values ($\approx$2.5).

Fine grain sizes and higher strengths can be achieved by alloying with Nb. High-strength IF steels are solution hardened with small amounts of Mn, Si, and P. The tensile strength is increased 4 MPa by 0.1% Mn, 10 MPa by 0.1% Si, and 100 MPa by 0.1% P in solution. The presence of titanium reduces the phosphorus in solution by forming FeTiP. The increased strength comes at the expense of a somewhat reduced formability. A composition of 0.003% C, 0.003% N, 0.35% Mn, 0.05% P, 0.03% Al, 0.035% Nb, 0.2% Ti, and 0.001% B has the following properties: YS = 220 MPa, TS = 390 MPa, elongation = 37%, $\bar{R}$ = 1.9, and n = 0.21.

AK steels: Dissolved carbon and oxygen in steels having carbon contents 0.05–0.10% will react to form CO on freezing. This causes a violent rimming action as the CO bubbles are emitted, causing tiny drops of iron to burn in the air. Aluminum is added in the ladle to react with oxygen removing it in the form of Al_2O_3, which rises to the surface and is scraped off. This "kills" the rimming action. Aluminum also ties up dissolved nitrogen as AlN. Because of the removal of nitrogen, strain aging only occurs at elevated temperatures. Typically, an aluminum-killed steel will have an n-value of about 0.22 and an R-value of 1.8.

Bake-hardenable steels: Steels having enough carbon and/or nitrogen in solution to strain age at the temperatures for paint baking are termed bake-hardenable steels. The low yield strength without a yield point prior to forming and the high strength caused by strain aging are useful for producing dent resistance or downsizing the thickness for weight reduction. Bake-hardenable steels are used in hoods, quarter-deck panels, roofs, doors, and fenders. With a low carbon level they have good weldability.

HSLA steels (high-strength low-alloy): These are much stronger than plain-carbon steels. They are used in cars, trucks, cranes, bridges, and other structures where stresses may be high. A typical HSLA steel may contain 0.15% C, 1.65% Mn, and low levels (below 0.035%) of P and S. They may also contain small amounts of Cu, Ni, Nb, N, V, Cr, Mo, and Si. The term "microalloying" is often used because of small amounts of alloying elements. As little as 0.10% niobium and vanadium can have profound effects on the mechanical properties of a 0.1% C and 1.3% Mn steel. The Mn provides a good deal of solid solution strengthening. The other elements form a fine dispersion of precipitated carbides in an almost pure ferrite matrix. Rapid cooling produces a fine grain size that also contributes to the strength. Yield strengths are typically between 250 and 590 MPa (35,000–85,000 psi). The ductility and n values are lower than in plain-carbon steels. HSLA steels are also more rust resistant than most carbon steels, due to their lack of pearlite. An alloy with small amounts of Cu, called Cor-ten, forms a rust that is quite adherent and is used architecturally.

Dual-phase steels: These have been heat treated to obtain 5–15% martensite in a ferrite matrix replacing many HSLA grades. Uses include front and rear rails, bumpers, and panels designed for energy absorption.

Trip steels (transformation-induced plasticity): The microstructure of Trip steels consists mainly of ferrite but there is also martensite, bainite, and retained austenite. The various levels of these phases give Trip steels their unique balance of properties. During forming, the retained austenite transforms to martensite. This results in a high rate of work hardening that persists to higher strains, in contrast to that of dual-phase steels which decreases at high strains. This causes enhanced formability. The carbon content controls the strain level of retained austenite-to-martensite transformation. With low carbon levels, transformation starts at the beginning of forming, leading to excellent formability and strain distribution at the strength levels produced. With high carbon levels, retained austenite is more stable and persists into the final part. The transformation occurs at strain levels beyond those produced during stamping and forming. Transformation to martensite occurs during subsequent deformation, such as a crash event, and provides greater crash energy absorption. Spot welding of Trip steels is made more difficult by the alloying elements.

Complex-phase (CP) steels: These have a very fine microstructure of ferrite with martensite and bainite. They are further strengthened by precipitation of niobium, titanium, or vanadium carbonitrides. They are used for bumpers and B-pillar reinforcements because of their ability to absorb energy.

Martensitic grades: The microstructures of martensitic grades are completely martensite. Tensile strengths vary between 900 and 1500 MPa (130 and 220 ksi). These grades can be made directly at the steel mill by quenching after annealing or by heat treating after forming. Mill-produced material has a very low ductility so it is typically roll formed.

The carbon content controls the strength level. The tensile strength in MPa is approximately

$$TS = 900 + 2800 \times \% \, C. \tag{19.5}$$

Manganese, silicon, chromium, molybdenum, boron, vanadium, and nickel are used in various combinations to increase hardenability. Typical applications for martensitic steels usually are those requiring high strength and good fatigue resistance, with relatively simple cross sections, including door intrusion beams, bumper reinforcement beams, side sill reinforcements, and belt line reinforcements.

Properties of some grades of low-carbon steels are given in Table 19.1.

TABLE 19.1
Properties of Some Grades of Low-Carbon Sheet Steels

Steel	YS (MPa)	TS (MPa)	El (%)	n	R	m
IF	150	300	45	0.28	2+	0.015
IF w/P	220	390	37	0.21	1.9	0.015
AKDQ		350	32–40	0.20–0.22	1.4–2	0.015
BH210/340	210	340	34–39	0.18	1.8	
BH260/370	260	370	29–34	0.13	1.6	
DP280/600	280	600	30–34	0.21	1.0	
DP300/500	300	500	30–34	0.16	1.0	
DP350/600	350	600	24–30	0.14	1.0	
DP400/700	400	700	19–25	0.14	1.0	
DP500/800	500	800	14–20	0.14	1.0	
DP700/1000	700	1000	12–17	0.09	1.0	
HSLA350/450	350	450	23–27	0.14	1.1	0.005–0.01
TRIP450/800	450	800	26–32	0.24	0.9	
Mart950/1200	950	1200	5–7	0.07	0.9	
Mart1250/15201250		1520	4–6	0.065	0.9	

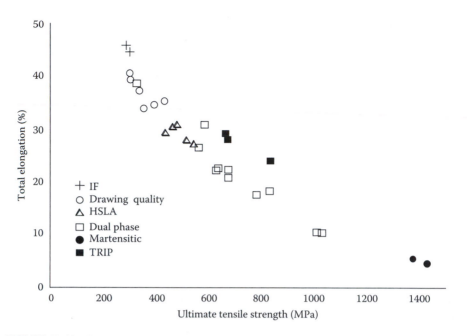

FIGURE 19.11   Strong correlation between the strengths and ductilities of various grades. (From *Making, Shaping and Treating of Steel*, 9th ed., U.S. Steel Corporation, 1971. With permission.)

Figure 19.11 shows a strong correlation between the strengths and ductilities of various grades.

Constructional steels: These are sold in the hot-rolled condition for use as beams, plates, rods, and rerods. They are not heat treated.

Rusting steels: Steel containing 0.02–0.04% copper form an adherent rust that slows atmospheric corrosion. These steels may also contain small amounts of nickel or chromium. Such rusting steels are often used without painting in sculpture and exteriors of buildings.

Steel sandwiches: Panels consisting of two steel sheets with a polymeric core of polyethylene or polyvinyl acetate are used by the automotive, railroad, and ship-building industries. Sandwich panels reduce noise transmission and reduce weight for the same stiffness.

19.4 COATINGS

Galvanized grades: Galvanized means zinc plated. Because zinc is anodic to iron, galvanizing protects the steel from corrosion. Zinc may be applied either by dipping in molten zinc or by electroplating. Figure 19.12 shows the surface produced by dip galvanizing. There is also a *galvaneal* process in which the zinc coating is diffused into the surface of the steel.

Terne plate: Steel coated with lead containing 3–15% tin has an excellent corrosion resistance. It is often used in automobile gas tanks and applications requiring corrosion resistance. The plating is applied by hot dipping.

FIGURE 19.12   Typical spangles on a sheet of dip-galvanized steel (about 75% size).

Tin plate: Steel is normally plated with tin by electrolytic deposition although application by hot dipping is possible. The sheet used for "tin cans" have either no tin (*tinless tin plate*) or extremely thin tin coatings because of the cost of tin.

Phosphate coatings are commonly applied to steel that is to be cold formed or painted. The phosphate coatings are less reactive than the steel and absorb lubricants and paint. They are applied by dipping the steel in dilute phosphoric acid.

19.5   SPECIAL CONCERNS

Inclusion shape control: Splitting of steels parallel to the rolling direction is sometimes a problem during fabrication or in service. Usually the cracks form along elongated MnS particles. This causes the ductility measured in the transverse and through-thickness directions to be much lower than that in the rolling direction. HSLA steels are much more sensitive to this than AKDQ steels. One common remedy is *inclusion shape control.* Additions of Ce and rare earths to the steel form sulfides that are much more resistant than MnS to elongation during hot rolling.

Hot shortness: Copper and tin are called tramp elements. They enter steel through recycling of scrap. Unlike most other alloying elements, they are not oxidized during steel making. Their concentrations in steel are increasing from year to year because of repeated recycling. Trace amounts in steels can cause hot shortness. They are much less oxidizable than iron; so when the steel surface is oxidized during hot rolling, the concentrations of Cu and Sn can increase enough to melt.

Machinability of steel is improved by intentional additions of sulfur (resulfurized steels), which forms MnS inclusions. Resulfurized steels must have increased manganese contents. Sometimes lead is added to hardenable steels for increased machinability.

19.6 MISCELLANY

Before the advent of the Bessemer and open-hearth processes, wrought iron was a very important material. It was used for nails, horse shoes, chains, plows, hinges, and a wide variety of products now made from low-carbon steel. Molten pig iron was exposed to air and stirred (puddle) to reduce the carbon content. As the carbon content fell, the solidus and liquidus temperatures increased, until the material was semisolid. The final product contained very little carbon but did have slag inclusions, entrapped during the puddling. Wrought iron was very ductile so the blacksmith could shape it into useful forms. No wrought iron has been produced in the United States since the middle to the twentieth century. What is often called wrought iron today is really low-carbon steel.

PROBLEMS

1. *Making, Shaping, and Treating of Steels* gives the following table of combination of aging times and temperatures that cause the same amount of strain aging.

Temperature (°C)	0	21	100	120	150
	1 year	6 months	4 h	1 h	10 min
	6 months	3 months	2 h	30 min	5 min
	3 months	6 weeks	1 h	15 min	2.5 min

   a. Make a plot of time (log scale) or log(time versus $1/T$ for equivalent aging.
   b. Find the apparent activation energy from the straight-line portion of the plot.
   c. Offer an explanation of why the 0°C data do not fall on the line.
2. Dual-phase steels are low-carbon steels that are quenched from a two-phase $\alpha + \gamma$ field. When quenched the austenite transforms to martensite. Consider a steel containing 0.10% C, 1.5% Mn, and 0.53% Si. For this composition, the eutectoid temperature and composition are approximately 725°C and 0.65% C.
   a. This steel was held at 750°C until equilibrium was reached and then quenched. Determine what fraction of the microstructure is martensite. What is the rest of the microstructure?
   b. Using Figure 16.27, assuming the rule of mixtures, predict the Vickers hardness of the steel after quenching.
3. Show whether more CO would be formed during solidification of a liquid steel containing 0.10% C or 0.05% C if the liquid steel were in equilibrium with 1 atm CO. Assume that the solid contains no dissolved CO.
4. Calculate the volume of CO that would be released during solidification of 10 tons of a liquid steel containing 0.05% C in equilibrium with 1 atm CO. Assume that all of the oxygen reacts to form CO.
5. The use of higher temperature recrystallization temperatures for low-carbon steels produce higher values of n and R. Why are they not always used?

6. Hot-rolled steels are cheaper than cold-rolled steels. Why are not they used for exterior auto body panels?
7. One engineer specified that a part be made from an extra-low-carbon grade of steel. Although it costs more than the usual grade, he thought that with the usual grade there might be an excessive scrap rate.
   a. How could you determine whether the cheaper, usual grade could be used?
   b. Would the substitution of a cheaper grade result in an inferior product?

REFERENCES

D. S. Blickwede, *Metals Progress*, July 1969.

Flat Rolled Products III, Metallurgical Society Conference, Vol. 16, Interscience Publishers, Wiley, 1962.

Daniel J. Schaeffler, *Stamping Journal*, 2004.

S. P. Keeler and W. G. Brazier, in *Microalloying 75*, Union Carbide, 1977.

F. Körber, *J. Inst. Metals*, 48, 317–342, 1932.

W. C. Leslie, *The Physical Metallurgy of Steels*, McGraw-Hill, 1981.

D. T. Llewellyn and R. C. Hudd, *Steels: Metallurgy and Applications*, Butterworth-Heinemann, 1998.

R. W. Logan, D. J. Meuleman, and W. F. Hosford, in *Formability and Metallurgical Structure*, A. K. Sachdev and J. D. Embury, Eds, TMS, 1987.

Making, Shaping and Treating of Steel, 9th ed., U.S. Steel Corporation, 1971.

Metals Handbook, 8th ed., Vol. 7, p. 14, ASM, 1973.

20 Special Steels

20.1 STAINLESS STEELS

Stainless steels are characterized by a very good aqueous corrosion resistance and by a very good resistance to oxidation at high temperatures. All stainless steels contain at least 11.5% Cr. Many contain nickel as well. For the aqueous corrosion resistance, the steels must contain a minimum of 12% chromium. With 12% chromium they become passive in oxidizing solutions. Even more chromium is required for passivity in nonoxidizing solutions. Unless the chromium content is sufficient for passivity, their corrosion resistance is similar to steels without any chromium. Table 20.1 is a galvanic series of alloys. It shows that stainless steels may occupy two different positions corresponding to the active and passive conditions.

There are five major types of stainless steels. They are ferritic, martensitic, austenitic, duplex, and precipitation hardenable.

20.2 FERRITIC STAINLESS STEELS

These contain 11.5–30% chromium with minor amounts of silicon and manganese. The carbon content is kept as low as possible. There are several grades with a 4xx designation. As Figure 20.1 shows, with Cr > 12.7%, pure Fe–Cr alloys are ferritic (bcc) at all temperatures. Because they are bcc, ferritic stainless steels undergo a ductile–brittle transition at low temperatures. If C + N are kept below 0.015%, the transition temperature is below room temperature.

The mechanical properties are similar to low-carbon steels. The strain-hardening exponent is about 0.20 and the R-value is about 1. Major uses include architectural and automotive trim. A serious surface appearance may develop when sheets are stretched. Hills and valleys may form parallel to the prior rolling direction seen in Figure 20.2. This phenomenon known as *roping* or *ridging* is a result of bands of grains of two different crystallographic textures as illustrated in Figure 20.3. When a sheet is extended, grains with a $\{111\}\langle112\rangle$ orientation contract less laterally than grains with a $\{001\}\langle110\rangle$ orientation and therefore buckle.

If carbides precipitate during heat treatment, the stainless steels will become *sensitized* and susceptible to intergranular corrosion. This will be discussed in detail later. In ferritic stainless steels with very high chromium contents, sigma phase may form during heat treatment or in service at high temperatures. Figure 20.4 is an isothermal diagram for sigma phase formation.

TABLE 20.1

Galvanic Series in Sea Water

Most anodic magnesium

Magnesium alloys

Zinc

Aluminum

Aluminum alloys

Low-carbon steels

Austenitic stainless steel (active)

Lead

Tin

Muntz metal

Nickel

Brass

Copper 70-30 cupro-nickel

Austenitic stainless steel (passive)

Most cathodic titanium

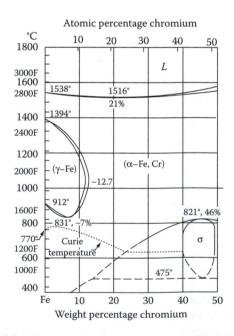

FIGURE 20.1   Iron-rich end of the Fe–Cr phase diagram. With chromium contents over
12.7%, stainless steels are ferritic at all temperatures. Sigma phase may develop at very high
chromium contents, but the reaction is sluggish. (From *Metals Handbook*, 8th ed., Vol. 8,
ASM, 1973. With permission.)

FIGURE 20.2   Ridging on a ferritic stainless steel hubcap. (From *The Making, Shaping and Treating of Steels*, 9th ed., U.S. Steel Corporation, 1971. With permission.)

In ferritic stainless steels with very high chromium contents, sigma phase may form during heat treatment or in service at high temperatures (see Figure 20.1). This is regarded as undesirable because of the brittleness of this phase. Figure 20.4 is an isothermal transformation diagram for sigma phase formation.

20.3   MARTENSITIC STAINLESS STEELS

These contain 12–17% Cr and 0.1–1.0% C. These, too, have (4xx) designations. The effect of C on the phases in a 12% Cr steel is shown in Figure 20.5. With 0.1%

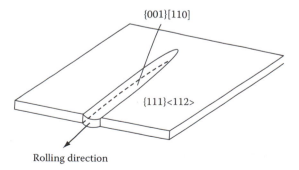

$\{001\}[110]$

$\{111\}<112>$

Rolling direction

FIGURE 20.3   Combination of two textures that leads to ridging in ferritic stainless steels.

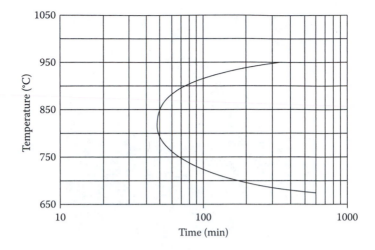

FIGURE 20.4 Isothermal transformation of sigma phase in a steel containing 25% Cr, 3% Mo, and 4% Ni. (Adapted from E. L. Brown, M. E. Burnett, P. T. Purtscher, and G. Krauss, *Met. Trans.*, 14A, 791–800, 1983.)

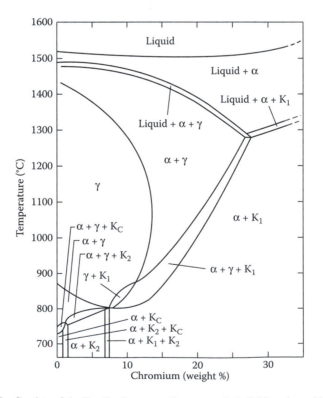

FIGURE 20.5 Section of the Fe–Cr–C ternary diagram at 10% C. Note that with 12% Cr, the steel can be heated into the austenite region. This makes it possible to form martensite. (From *Making, Shaping and Treating of Steels*, 9th ed., U.S. Steel Corporation, 1971. With permission.)

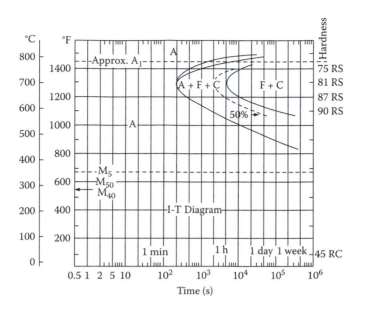

FIGURE 20.6 Isothermal transformation diagram for a 12% Cr, 0.10% C steel. (From *Making, Shaping and Treating of Steels*, 9th ed., U.S. Steel Corporation, 1971. With permission.)

carbon, it is possible to form austenite. With 12% Cr, the hardenability is so great that rapid quenching is not necessary. The isothermal transformation diagram for a 12% Cr, 0.10% C steel (Figure 20.6) shows that about 3 min at 700°C are required to form the first pearlite. Martensitic stainless steels are used largely for razor blades, knives, and other cutlery.

20.4 AUSTENITIC STAINLESS STEELS

These contain 17–25% Cr and 8–20% Ni. The amount of carbon is kept very low. Austenitic stainless steels form the (2xx and 3xx) series. Figure 20.7 is the isothermal section of the Fe–Cr–Ni ternary phase. The tendency of nickel to stabilize austenite is sufficient to overcome the ferrite-stabilizing tendency of chromium. It should be noted that 18% Cr, 8% Ni falls just inside the γ region. At lower temperatures the γ should transform to α, but the transformation is very sluggish so a steel containing 18% Cr, 8% Ni will be austenitic at all temperatures. However, cold working may cause an austenitic stainless steel to transform by a martensitic reaction.

Other elements affect the occurrence austenite versus ferrite in stainless steels. Carbon and manganese tend to stabilize austenite while Cr, Mo, Si, and Nb stabilize ferrite. The effects of these are characterized by nickel and chromium *equivalents*, as shown in Figure 20.8. Nitrogen is a strong austenite stabilizer and has been used to substitute for a portion of the nickel.

EXAMPLE PROBLEM 20.1

Figure 20.8 shows that a stainless steel containing 14% Cr requires 14% Ni to be sure that no martensite can be formed. In a steel containing 12% Cr, 1.0% Si, and

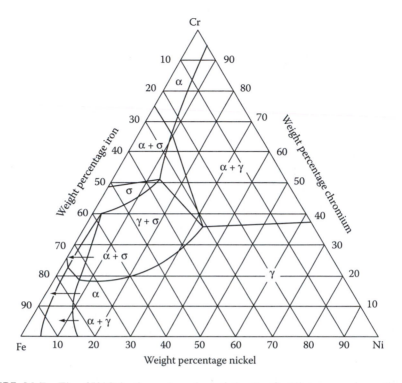

FIGURE 20.7 The 650°C isothermal section of the Fe–Cr–Ni ternary phase diagram. (From *Metals Handbook*, 8th ed., Vol. 8, ASM, 1973. With permission.)

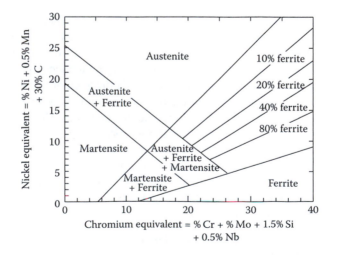

FIGURE 20.8 Nickel and chromium equivalents of several elements and their effects on the phases present. (Adapted from *Metals Progress Data Book*, Vol. 112, ASM, 1977.)

0.05% C, how much nickel is necessary to be sure that no martensite can be formed?

SOLUTION

The chromium equivalent of the steel is $12 + 1.5 \times 1 = 13.5$. Figure 19.8 indicates that the nickel equivalent should be 13. The 0.05% carbon contributes $0.05 \times 30 = 1.5$ to the nickel equivalent, so the needed amount of nickel is $13–1.5 = 11.5\%$ Ni.

Austenitic stainless steels are resistant to aqueous corrosion over a larger range of conditions than ferritic stainless steels. Their mechanical properties include a very high strain-hardening exponent, n (in the range of 0.5–0.65) and a low R-value, which is typical of fcc metals. The high n can be attributed to the low stacking fault energy. Because they are fcc, they are not embrittled at low temperatures. One of the uses is in cryogenics. They are not nonferromagnetic, unless cold worked, which is useful in instruments that might be affected by magnetic fields.

20.5　OTHER STAINLESS STEELS

Precipitation hardening stainless steels have an austenitic or a martensitic base with additions of Cu, Ti, Al, Mo, Nb, or N. Duplex stainless steels have mixed austenite–ferrite microstructure. They contain 23–30% Cr, 2.5–7% Ni, with small amounts of Ti or Mo.

20.6　SENSITIZATION

The corrosion resistance of both austenitic and ferritic grades may be lost by heating to 600–650°C or slowly cooling through that temperature range. Carbon solubility of carbon in an 18% Cr, 8% Ni austenitic stainless is shown in Figure 20.9. Unless the carbon content is less than about 0.03%, precipitation of $(Fe,Cr)_{23}C_6$ in the grain boundaries depletes chromium from the adjacent regions, lowering the % Cr below the critical 12% (Figure 20.10). This sets up active–passive cells with the unaffected grain interiors being the cathodes and the grain-boundary regions being the anodes. Rapid intergranular corrosion can result. This is a particular problem in welded structures because the regions just outside of the weld will be heated into the sensitizing temperature range. While heating the entire welded structure above 800°C to dissolve the carbides and quenching would relieve the problem, this is usually not practical. The common remedy is to use special grades of stainless steels where welding is required. These are either extra-low-carbon grades or are grades containing Ti or Nb, which are such strong carbide formers that they remove the carbon from solution so it cannot react with the chromium.

20.7　OXIDATION RESISTANCE

The resistance to oxidation is provided by an adherent Cr_2O_3 oxide film, which protects the underlying steel from oxidation. Oxidation requires diffusion of Cr^{+3} ions through the Cr_2O_3 film, which has few ionic defects. Oxidation resistance increases with increasing Cr content of the steel.

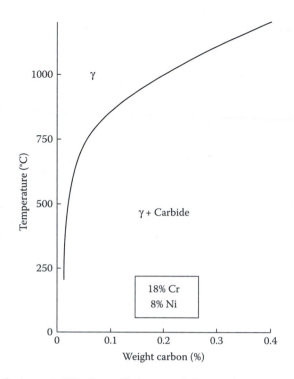

FIGURE 20.9   Carbon solubility in an 18–8 type stainless steel.

20.8   HADFIELD AUSTENITIC MANGANESE STEEL

Manganese is a powerful austenite stabilizer as indicated by the Fe–Mn phase diagram (Figure 16.15). Hadfield manganese steels containing 10–14% Mn and 1–1.4% C are austenitic at all temperatures. They are extremely wear resistant, and find use

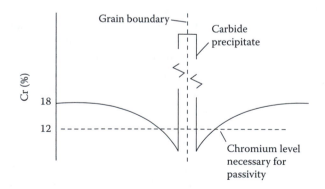

FIGURE 20.10   Precipitation of chromium carbide at the grain boundaries depletes the chromium in the adjacent regions to less than the critical 12% needed for passivity.

in ore grinding and teeth on earth moving equipment. These steels work harden very rapidly and as a consequence are very difficult to machine. Parts are almost always cast to final shape.

20.9 MARAGING STEELS

Maraging steels develop high hardness by precipitation hardening of a very low-carbon martensite. They contain about 18% Ni, 8–12% Co, and 4% Mo with very low carbon (<0.03%) and titanium (0.20–1.80%) and aluminum (0.10–0.15%). Their primary use is in tools and dies. Because the carbon content is so low, the martensite, which is formed by austenitizing at 850°C and cooling, is soft enough to be machined. Finished tooling can then be hardened by aging at 480°C for 3 h. Whereas tools made from conventional tool steels have to be austenitized, quenched, and tempered after machining, maraging steels need only be heated to a moderate temperature to age and can be furnace cooled. This avoids oxidation, distortion, and cracking that often occurs during conventional heat treatment. The main disadvantage is the high cost that results from the high Ni, Mo, and Co contents.

20.10 TRIP STEELS

Trip is an acronym for *transformation-induced plasticity*. A typical composition is 0.25% C, 2% Mn, 2% Si, 10% Cr, 9% Ni, and 5% Mo. A high combination of strength and ductility can be achieved by (1) solution treating at 1200°C and quenching, (2) deforming about 80% above the M_s temperature which induces a fine precipitate, and (3) cold forming below the M_d to transform some of the austenite.

20.11 TOOL STEELS

Most steel tools are not made from tool steels—rather carbon or low-alloy steels. Tool steels are used for shaping and cutting metal or rocks, wood or concrete. The various classes are

W—Water hardening. They are used for files, wood-working tools, drill bits, axes, and taps. They contain 0.6–1.4% C. Some have small amount of Cr and V.

S—Shock resisting. Their uses include chisels, punches, and riveting tools. Typical compositions are 0.5% C with 1 to 3% Si and smaller amounts of Cr, Mo, or W.

P—Mold steels. The carbon contents are lower so they can be machined and then carburized. Typical compositions are 0.10–0.30% C, 1.5–5% Cr, and 0.0–4% Ni. The lower carbon core gives them thermal shock resistance.

D—Cold work. Applications include cold-forming dies and thread-rolling dies. They have high carbon contents (1.5–2%) with typically 12% Cr and 1% Mo. They are air or oil hardening and have good dimensional stability.

H—Hot working. These are used for forging dies, extrusion dies, and die casting molds. Compositions range from 0.35% to 0.40% C, 3% to 5% Cr, 0.4% to 2% V, and either 1.5–2.5% Mo or 1.5–4% W. The latter elements assure hot hardness.

M—Molybdenum-type high-speed steels. These contain 0.8–1.1% C, 4–9% Mo, 1–2% W, 4% Cr, 1–2% V, and often Co.

T—Tungsten-type high-speed steels. These contain 0.7–1.8% C, 18% W, 4% Cr, 1–2% V, and often 5% or more Co.

Both molybdenum and tungsten high-speed tool steels rely on secondary hardening to retain a high hardness during use at the high temperatures that cutting tools experience during high-speed machining. A number of different carbides are precipitated during secondary hardening (Figure 20.11) and most of these are harder than the martensite from which they precipitate as indicated in Figure 20.12.

20.12 SILICON STEELS

Steels containing 3–3.5% Si and very low carbon are used for soft magnetic applications such as motors, generators, and transformers. They will be discussed in detail in Chapter 25.

20.13 MISCELLANY

The original austenitic manganese steel, containing about 1.2% C and 12% Mn, was invented by Sir Robert Hadfield in 1882.

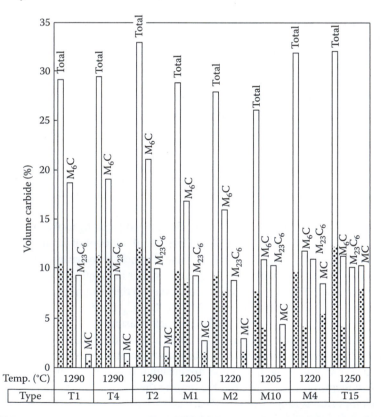

FIGURE 20.11 Carbides formed in T- and M-type tool steels after being annealed at the temperatures indicated. (From F. X. Kayser and M. Cohen, *Metals Progress*, Vol. 61. With permission.)

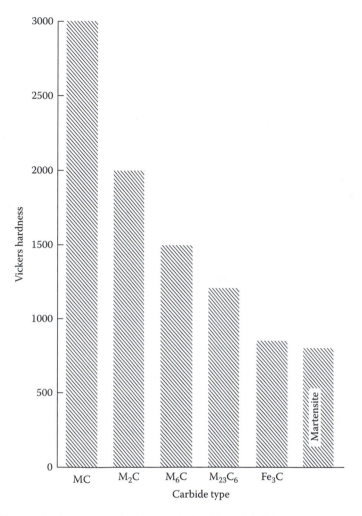

FIGURE 20.12 The hardnesses of various carbides. (From W. G. Moffatt, G. W. Pearsall, and J. Wulff, *Structure and Properties of Metals*, vol. 1, p. 195, Wiley, 1964. With permission.)

Carl Zapffe's *Stainless Steel* (ASM 1949) gives a review of the development of stainless steels. He states that it is remarkable that they were not discovered much earlier than the first decade of the twentieth century. In 1821, Berthier, in France, found that iron "highly alloyed with chromium was more resistant to acids than chromium-free iron." Maullet confirmed that chromium imparted corrosion resistance to iron but then dismissed the importance of this by proposing that chromium would leach out of the steel leaving it less corrosion resistant. Hadfield in 1892 studied steels containing up to 16.76% chromium, but concluded that chromium *impaired* the corrosion resistance. There are three reasons for his conclusions. The first is that he concentrated his efforts on compositions containing 1–9% Cr (less than the critical for passivity). The second is that his alloys contained from 1–2% carbon and were therefore probably sensitized. The last is that he chose to 50% sulfuric acid as his corrosive agent.

Between 1904 and 1906, Leon Guillet studied metallography and mechanical behavior of compositions with sufficient chromium and low enough carbon to be considered ferritic, austenitic, and martensitic stainless steels. He did not however study their corrosion resistance.

It was P. Monnartz in Germany who first disclosed in works between 1908 and 1911 the phenomenon of passivity, the remarkable corrosion resistance. Commercial application of stainless steels follows shortly.

Tool steels were developed earlier. In 1868 Robert Mushet promoted a steel containing tungsten for its hardness. However, 25 years passed before it was recognized that the steel retained its hardness when red hot. It contained about 2% C, 2.5% Mn, and 7% W. In the 1890s chromium was used to replace much of the manganese. After 1900, many new compositions were introduced, containing vanadium and cobalt. Because of the shortage of tungsten during World War II, molybdenum was substituted for tungsten. Because it is cheaper than tungsten, molybdenum-containing high-speed steels are a major part of the market.

PROBLEMS

1. True or false:
   a. In order to be passive, a stainless steel requires the presence of oxygen.
   b. The corrosion resistance of a stainless steel increases with chromium content up to about 12% Cr.
   c. An austenitic stainless steel contains nickel as well as chromium.
   d. The sensitization of stainless steels during welding occurs because chromium carbides in the grain boundaries are anodic to the rest of the material.
2. Manganese and nitrogen are sometimes used in austenitic stainless steels. What purpose might they serve?
3. Which type of stainless steel would be specified for each of the following uses:
   a. Kitchen knife.
   b. Automobile trim.
   c. Kitchen sink.
   d. Carving knife.
   e. Low-temperature cryogenics.
4. According to Figure 20.8, for a steel containing 19.5% Cr, 0.3% Si, 0.5% Mn, and 0.05% C, how much Ni must it contain to be austenitic?
5. What effect do Cr, Ni, and C have on the melting point of iron?
6. Sometimes martensitic stainless steels are subjected to a subzero heat treatment. What purpose would this serve?
7. Why are tool steels sometimes tempered a second time after the initial quenching and tempering?
8. Microscopic examination of an M-2 tool steel shows that the volume fractions of the various carbides are M_6C (0.15), $M_{23}C_6$ (0.06), and MC (0.02). The rest of the microstructure is ferrite. Estimate the hardness of this steel. Assume that the hardness of the ferrite is 150 Vickers.

REFERENCES

ASM Specialty Handbook: Stainless Steels, 1994.

E. L. Brown, M. E. Burnett, P. T. Purtscher, and G. Krauss, *Met. Trans.*, 14A, 791–800, 1983.

C. Zapffe, *Stainless Steels*, ASM, 1949.

E. Haberling, A. Rose, and H. H.Weigand, *Stahl und Eisen*, 93, 645, 1973.

F. X. Kayser and M. Cohen, *Metals Progress*, Vol. 61.

G. Krauss, *Steel–Heat Treatment and Processing Principles*, ASM, 1990.

W. C. Leslie, *The Physical Metallurgy of Steels*, McGraw-Hill, 1981.

Metals Handbook, 8th ed., Vol. 8, ASM, 1973.

Metals Progress Data Book, Vol. 112, ASM, 1977.

W. G. Moffan, G. W. Pearsall, and J. Wulff, *Structure and Properties of Metals*, vol. 1, p. 195, Wiley, 1964.

R. L. Rickett, W. F. White, C. S. Walton, and J. C. Butler, *Trans. ASM*, 44, 1952.

The Making, Shaping and Treating of Steels, 9th ed., U.S. Steel Corporation, 1971.

21 Cast Irons

21.1 GENERAL

In the iron–carbon system, true equilibrium is between iron and carbon. Fe_3C is not an equilibrium phase. In steels the carbon content is low enough so that graphite rarely forms. In contrast, most cast irons have sufficient carbon that graphite does form. Figure 21.1 is the iron–carbon phase diagram. Note that the solubility of carbon in α and γ is lower when they are in equilibrium with graphite than in metastable equilibrium with cementite. The reason is illustrated in Figure 21.2.

Cast irons contain 2–4% C and 0.5–3% Si. Silicon shifts the eutectic to lower carbon contents. The percent carbon at the eutectic composition can be approximated by

$$\% \, C = 4.3 - 0.3\% \, Si. \tag{21.1}$$

The formation of graphite instead of Fe_3C is favored by

1. High carbon contents.
2. High silicon contents.
3. Long times at high temperatures or slow cooling.
4. The absence of carbide formers (Cr, Mo, etc.) is also important.
5. There are four basic types of cast iron: white, gray, ductile, and malleable.

21.2 WHITE IRONS

White cast irons tend to have less carbon and silicon than the other grades (typically 2.5% C and 0.5% Si), so graphite does not form. The microstructure (Figure 21.3) contains massive carbides. White cast irons are very brittle, fracture occurring along paths of carbides. Their use is limited to special applications requiring high wear resistance as tappets for automobile engines. Another use is as a precursor for the formation of malleable cast iron.

21.3 GRAY IRONS

These have sufficient carbon and silicon (typically 3.2% C and 2.5% Si) that graphite forms during freezing. It occurs in the form of flakes as shown in Figure 21.4.

333

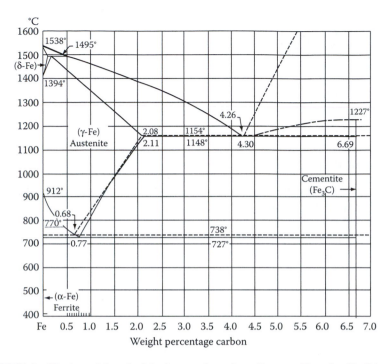

FIGURE 21.1 The iron-rich end of the iron–carbon phase diagram. Note that Fe₃C is a metastable phase. True equilibrium exists between γ and graphite and between α and graphite. The dashed lines are the true equilibrium with graphite whereas the solid lines are the metastable equilibrium with cementite. (From J. Chipman, *Metals Handbook*, 8th ed., Vol. 8, ASM, 1973. With permission.)

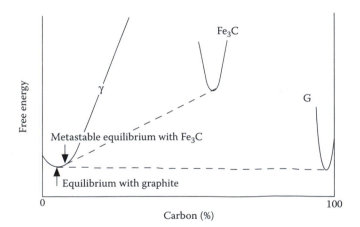

FIGURE 21.2 Solubility of carbon in austenite corresponds to the point of tangency of lines drawn between the free energy curves of austenite and either graphite or cementite. Since the graphite curve is lower, the point of tangency is at a lower carbon content. (From W. G. Moffatt, G. W. Pearsall, and J. Wulff, *Structure and Properties of Metals*, vol. 1, p. 195, Wiley, 1964. With permission.)

FIGURE 21.3 Microstructure of a white cast iron. Primary dendrites of austenite surrounded by an austenite–carbide eutectic formed during solidification. Later on cooling the austenite is transformed to pearlite. (From *Metals Handbook*, 8th ed., Vol. 7, ASM, 1972. With permission.)

FIGURE 21.4 Microstructure of a gray cast iron, unetched. The flakes of graphite are apparent. (From *Metals Handbook*, 8th ed., Vol. 7, ASM, 1972. With permission.)

FIGURE 21.5   Electron micrograph of a gray iron that has been deeply etched to remove the ferrite and pearlite. The three-dimensional structure of the graphite flakes is apparent. (From *Metals Handbook*, 8th ed., Vol. 7, ASM, 1972. With permission.)

The three-dimensional nature of the graphite flakes is apparent in Figure 21.5. Gray cast irons are not very ductile. One percent elongation in a tension test is typical. Fracture follows a path through graphite flakes so that the fracture surface is almost entirely graphite. The fracture appearance is the origin of the name, *gray iron.*

Gray cast iron is a very cheap material. It is much easier to cast than steel because of its lower melting point and because there is almost no liquid-to-solid shrinkage. It is easy to machine because the graphite flakes allow chips to break into small pieces. Gray cast iron has a very high capacity to dampen vibrations because of the graphite flakes. This is desirable in applications such as bases for lathes, milling machines, and other types of machinery that are sensitive to vibrations. Because of its low cost it finds wide use for manhole covers, fire hydrants, sewer gratings, drainage pipe, and lamp posts.

The tensile strength of gray iron decreases with the amount and size of graphite flakes. Because of this increased carbon and silicon contents the tensile strength lowers. On the other hand, most alloying elements increase the tensile strength. Slower cooling promotes larger graphite flakes. Therefore the tensile strength decreases with increased section size. See Problem 21.8.

Compact graphite iron has properties and a microstructure intermediate between gray iron and ductile iron. It is sometimes called seminodular or vermicular graphite. The graphite occurs as blunt flakes. This microstructure results in properties that are intermediate between gray and ductile irons. Production of compact graphite iron is similar to that of ductile iron requiring close metallurgical control and rare earth element additions. An alloying element like titanium is necessary to prevent the formation of spheroidal graphite. Compact graphite iron has much of the castability of gray iron, but with a higher tensile strengths.

21.4 DUCTILE CAST IRON

Ductile cast iron has about the same compositions as gray iron except that the sulfur content is very low. Magnesium (or Ce and rare earths) is added to the melt just before pouring to further reduce the sulfur. The result is that the graphite forms as spheroids or nodules as shown in Figure 21.6. The final product has considerable toughness and ductility. It is sometimes called semisteel. It has replaced forged steel in applications such as automobile crankshafts. There is a form of cast iron in which the graphite structure is between that of gray cast iron and ductile cast iron. This is referred to as *compacted graphite*. Its properties are between those of gray iron and ductile iron. Its formation is favored by a low Mg content and a small amount of Ti.

21.5 MALLEABLE CAST IRON

Heating a white cast iron to just below the eutectic temperature allows the iron carbide to decompose, $Fe_3C \rightarrow \alpha +$ graphite. The graphite forms as clumps or rosettes called *temper carbon*. Figure 21.7 is a micrograph of this structure. The properties are similar to ductile cast iron. The heat treatment required to form malleable iron is long (2–5 days) and expensive. For this reason, it has largely been replaced by ductile cast iron.

FIGURE 21.6 Ductile cast iron, unetched. Note that the graphite is in the form of spheroids, in contrast to the flake form in gray iron. (From *Metals Handbook*, 8th ed., Vol. 7, ASM, 1972. With permission.)

FIGURE 21.7   Malleable cast iron, unetched. Note that the graphite is in the form of rosettes, in contrast to the spheroidal form in ductile iron. (From *Metals Handbook*, 8th ed., Vol. 7, ASM, 1972. With permission.)

21.6   MATRICES

The matrix of gray, ductile, and malleable cast iron is essentially steel. Cast irons are described by the structure of the matrix, for example, by the terms, ferritic, pearlitic, martensitic, bainitic, or even austenitic. The nature of the matrix depends on the composition and the rate of cooling through the eutectoid temperature. It may be altered from the original cast structure by subsequent heat treatment. Figures 21.8 through 21.11 show several cast iron microstructures.

Austempering of ductile cast iron produces a microstructure of acicular ferrite in a carbon-stabilized austenite matrix rather than the bainite that is found in austempered steels. The properties can be varied with heat treatment. Very high tensile strengths can be achieved along with very good toughness.

Figure 21.12 summarizes the effects of composition on the microstructure for cast irons in the sand-cast condition. Silicon lowers the composition of the eutectic according to the equation

$$\% \ C = 4.3 - 0.3 \, (\% \ Si). \tag{21.1}$$

Figure 21.13 gives a summary of how composition and heat treatment control the phases in cast irons.

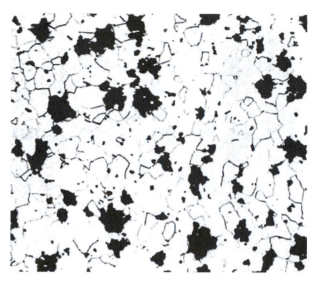

FIGURE 21.8 Ferritic malleable iron. (From *Metals Handbook*, 8th ed., Vol. 7, ASM, 1972. With permission.)

Successful welding of gray cast iron is difficult. The rapid cooling usually leads to the formation of martensite, further reducing ductility. The stresses set-up during cooling cause cracking. Very slow cooling after welding is required. The usual alternative to welding is braze welding in which a copper-base material that has a melting temperature under the lower critical is used to join the two pieces of cast iron.

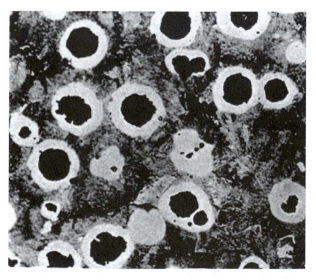

FIGURE 21.9 Ductile cast iron with a mixed ferritic–pearlitic matrix. Note that the ferrite is adjacent to the graphite. (From *Metals Handbook*, 8th ed., Vol. 7, ASM, 1972. With permission.)

FIGURE 21.10 Pearlitic gray iron. (From *Metals Handbook*, 8th ed., Vol. 7, ASM, 1972. With permission.)

21.7 MISCELLANY

The names gray cast iron and white cast iron come from the appearance of their fracture surfaces. Fractures in gray cast iron run through the graphite flakes so most of the surface is graphite. Hence the gray color. On the other hand, fractures in white cast iron run through the white cementite.

FIGURE 21.11 Martensitic ductile iron. (From *Metals Handbook*, 8th ed., Vol. 7, ASM, 1972. With permission.)

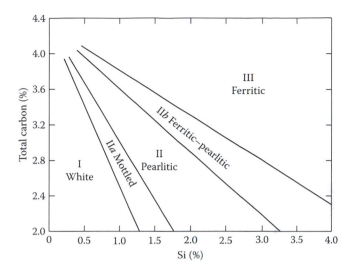

FIGURE 21.12 Effect of composition on the microstructure of sand-cast bars. (From J. Wulff, H. F. Taylor, and A. J. Shaler, *Metallurgy for Engineers*, Wiley, 1952. With permission.)

Ductile cast iron was first developed in 1943 by Keith Millis and L. Gagnebine at the laboratories of the International Nickel Company. They were trying to find a substitute for chromium as a carbide stabilizer in white cast iron. They tried most of the elements in the periodic table. When they added magnesium, there was a great deal of fireworks as magnesium vapor burned, but they found they had a product in which the carbon was presenting spheroids. They coined the term *ductile cast iron* to describe its toughness. (The term *malleable cast iron* was already in use.) In 1948, Henton Morrogh announced the successful production of spherical graphite in hypereutectic gray iron by the addition of small amounts of cerium on October 25, 1949. It took years of development to make ductile cast iron commercially viable. Real growth occurred in the 1960s. Today ductile cast iron accounts for about 40% of the cast iron market. This development ranks as one of the most important inventions of the twentieth century. Unlike most inventors, Gagnebine later became the CEO of INCO.

PROBLEMS

1. In Figure 21.9, the black spheres are graphite, the gray areas are pearlite, and the white regions are ferrite.
   a. Determine the volume fractions of ferrite, pearlite, and graphite.
   b. Calculate the % carbon in the cast iron. Assume densities of 7.87 mg m^{-3} for ferrite and pearlite and 2.25 mg m^{-3} for graphite.
2. The graphite flake size in a class 20 gray cast iron is about nine times as large as in a class 60 gray cast iron. Predict the fracture strength of a class 60 cast iron knowing that the fracture strength of a class 20 cast iron is about 20,000 ksi. The fracture strength is given by $\sigma_f = C/\sqrt{a}$, where C is a constant and a is half of the length of a preexisting crack.
3. Explain why graphite flakes in gray cast iron and/or spheroids in ductile cast iron are surrounded by ferrite rather than pearlite.

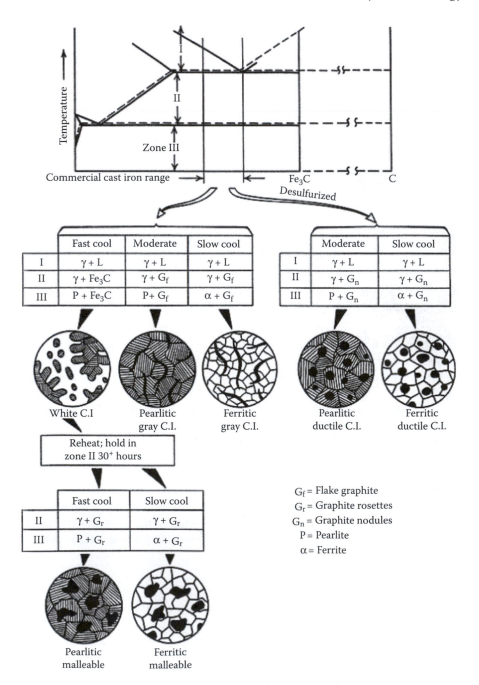

FIGURE 21.13 Effect of composition and heat treatment on the structure *of cast irons.* (From W. G. Moffatt, G. W. Pearsall, and J. Wulff, *Structure and Properties of Metals*, Vol. 1, p. 195, Wiley, 1964. With permission.)

4. Would a cast iron containing 3.2% C and 3.0% Si be hyper- or hypoeutectic?

5. A gray cast iron containing 3.2% C and 2.1% Si has a microstructure consisting of ferrite and graphite.
   a. What is the volume fraction of graphite? Assume that all of the Si is in the ferrite.
   b. In view of your answer to A, why does a machined surface of the cast iron appear metallic while a fracture surface has the color of the end of a lead pencil?

6. Calculate the density of a ferritic gray cast iron containing 3.5% carbon. Assume that the density of the ferrite is 7.87 mg m^{-3} and the graphite is 2.25 mg m^{-3}.

7. Tensile strength of gray iron depends on the diameter, the carbon equivalent, CE, and alloy content. A formula for estimating the tensile strength is

$$TS = (20,000 \text{ psi})(AF)(5.7d^{-.09} - CE), \tag{21.2}$$

where d is the diameter in inches, and CE is given by

$$CE = \% \text{ C} + \% \text{ Si}/3 + \% \text{ P}/4 \tag{21.3}$$

and the alloy factor, AF, depends on alloying elements. Figure 21.14 shows how AF depends on the composition.
   a. Calculate the tensile strength of a 1-in diameter bar containing 3.5% C, 1.4% Si, and 0.25% Cr.
   b. Explain why the effects of copper and nickel content are so much less than that of V, Cr, and Mo.

8. In the production of ductile cast iron, magnesium is added in the form of a nickel–magnesium alloy, rather than as pure magnesium. Why?

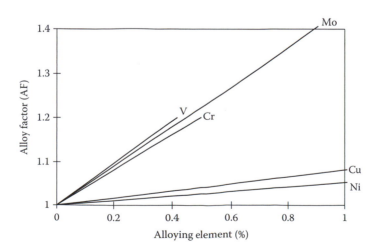

FIGURE 21.14 Dependence of the alloy factor on composition.

REFERENCES

J. Chipman, *Metals Handbook*, 8th ed., Vol. 8, ASM, 1973.

G. Krauss, *Steel—Heat Treatment and Processing Principles*, ASM, 1990.

Metals Handbook, 8th ed., Vol. 7, ASM, 1972.

W. G. Moffatt, G. W. Pearsall, and J. Wulff, *Structure and Properties of Metals*, Vol. 1, p. 195, Wiley, 1964.

J. Wulff, H. F. Taylor, and A. J. Shaler, *Metallurgy for Engineers*, Wiley, 1952.

22 Welding, Brazing, and Soldering

Welding is joining of two pieces of metal—usually of the same chemical composition. If a filler metal is used, it is of substantially the same chemical composition as the pieces to be joined. There may be melting or not. *Brazing* is joining of two metal pieces with a nonferrous filler that melts at temperatures below that of the metals being joined, but above 535°C (1000°F), *Soldering* is like brazing but with a filler that melts below 535°C (1000°F).

22.1 WELDING WITHOUT MELTING

Whenever there is a sufficient area of metal-to-metal contact, welding will occur. However, oxides usually prevent metal-to-metal contact. Metals that do not form oxides will weld when in contact. Powder of pure gold cannot be kept long before the individual particles weld (sinter). Snowflakes sinter to form hard ice. Blacksmiths of the past often welded iron to iron by hammering two hot pieces together after sprinkling their surfaces with sand. The sand formed a low melting eutectic with the iron oxide that was liquid at the forging temperature. If two sheets of titanium in contact are heated in an oxygen-free atmosphere, they will weld. Titanium dissolves its own oxide so there will be metal-to-metal contact.

Aluminum is often pressure welded. Two pieces are placed together and indented generating new oxide-free surface. The oxide-free area increases with the amount of deformation (Figures 22.1 and 22.2). Friction or inertia welding involves rotating one piece of steel or aluminum at a high speed and bringing it into contact with a stationary piece. The oxide is rubbed off as the rotating piece is brought to rest.

22.2 RESISTANCE WELDING

Other forms of welding involve heating to melt the metals. Often the heat is supplied electrically by passing a current through the pieces to be welded. Among the processes are butt welding (Figure 22.3), spot welding (Figure 22.4), and seam welding (Figure 22.5). Projection welding (Figure 22.6) is similar to spot welding.

FIGURE 22.1 Metal-to-metal contact will occur where the oxide films are broken.

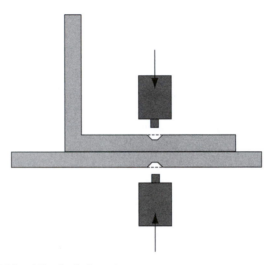

FIGURE 22.2 Cold welding by indentation.

22.3 ARC WELDING

There are a number of different arc welding processes. In tungsten-inert gas (TIG) (Figure 22.7) and atomic hydrogen welding (Figure 22.8), the electrode is not consumed and a filler rod is used.

Consumable electrode processes include AC and DC with coated electrodes (Figure 22.9). Electrode coatings serve several functions. They may provide a

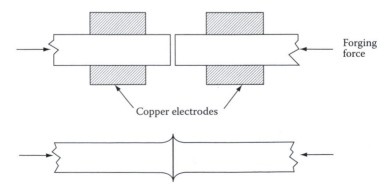

FIGURE 22.3 Butt welding.

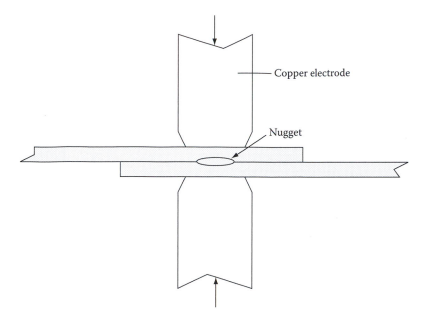

FIGURE 22.4 Spot welding setup. Electric current between electrodes melts the material in what is called a nugget.

protective atmosphere around the molten weld metal; they provide a glassy slag that protects the hot weld and they may have additives to stabilize the weld. For DC welding coatings contain cellulose, $(C_6H_{10}O_5)_n$ which decomposes by the reaction

$$(C_6H_{10}O_5)_n + \left(\frac{n}{2}\right)O_2 \rightarrow 6nCO + 5nH_2. \tag{22.1}$$

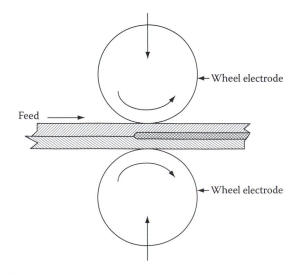

FIGURE 22.5 Seam welding.

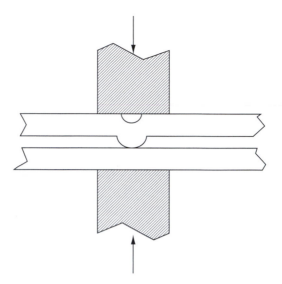

FIGURE 22.6 Projection welding. (From H. Udin, E. R. Funk, and J. Wulff, *Welding for Engineers*, Wiley, 1954. With permission.)

The carbon monoxide and hydrogen form a protective atmosphere. Oxides in the coating form a protective slag. Clay helps in the extrusion of the coatings on the electrodes and provides Al_2O_3 and SiO_2 for the slag. CaO, FeO, and MgO are frequently added to lower the melting point of the slag. Often bare (uncoated) electrodes are used.

With DC arcs, 70% heat is generated at the anode. With normal polarity, work is anode. Therefore DC polarity is desirable for nonconsumable electrodes. However, with consumable electrodes, heat would be transferred from electrode to work by molten metal. Reverse polarity is desirable for welding of aluminum and magnesium

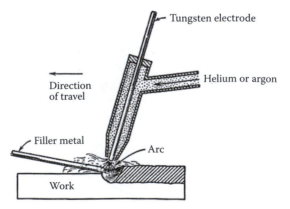

FIGURE 22.7 TIG welding. (From J. Wulff, H. Taylor, and J. Shaler, *Metallurgy for Engineers*, Wiley, 1952. With permission.)

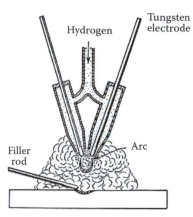

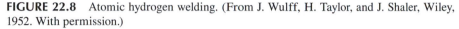

FIGURE 22.8 Atomic hydrogen welding. (From J. Wulff, H. Taylor, and J. Shaler, Wiley, 1952. With permission.)

because it helps to break up oxides. AC arcs are more stable and wander less than that of DC arcs.

Other arc welding processes include metal-inert gas (MIG) welding that uses equipment like the TIG welding (Figure 22.7) except that the electrode is the filler metal and must be fed continually into the torch. It is used widely for welding aluminum alloys. In the submerged arc welding process (Figure 22.10), the arc passes through a thick slag deposit formed from powdered flux. The welding current and speed are much faster than ordinary arc welding processes.

22.4 HEAT DISTRIBUTION

In arc welding, heat is generated by an electric arc between an electrode and the work piece or between two electrodes. As the arc move along the pieces to be welded, a heated region is left behind. The isotherms around a welding electrode are shown in Figure 22.11.

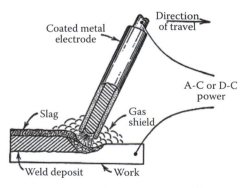

FIGURE 22.9 Coated electrode welding. (From J. Wulff, H. Taylor, and J. Shaler, Wiley, 1952. With permission.)

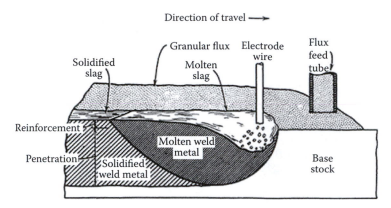

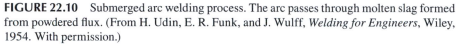

FIGURE 22.10 Submerged arc welding process. The arc passes through molten slag formed from powdered flux. (From H. Udin, E. R. Funk, and J. Wulff, *Welding for Engineers*, Wiley, 1954. With permission.)

The width of the heated zone around the weld depends on the welding current. With greater currents, the weld speed must be increased to maintain the same temperature at the weld. The faster speed results in a narrower heat-affected zone. The rate of cooling, |dT/dt|, behind the weld increases with welding speed. If a slow cooling rate is required, preheating or postheating can be used.

22.5 TORCH WELDING

Figure 22.12 illustrates torch welding. Acetylene (C_2H_2) is the most common fuel for torch welding. Depending on the oxygen/acetylene ratio in oxyacetylene welding, the reaction may be either

$$(C_2H_2) + \left(\frac{5}{2}\right)O_2 \rightarrow 2CO_2 + H_2O \qquad (22.2)$$

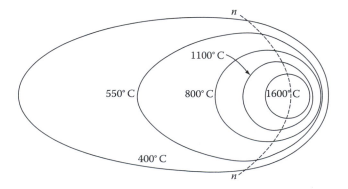

FIGURE 22.11 Isotherms calculated for 2 mm below the surface of a steel plate welded at 2.4 mm/s. The line *n-n* is locus of maximum temperatures. (From H. Udin, E. R. Funk, and J. Wulff, *Welding for Engineers*, Wiley, 1954. With permission.)

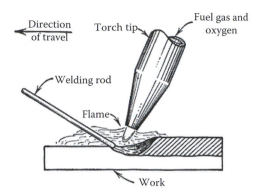

FIGURE 22.12 Torch welding. (From H. Udin, E. R. Funk, and J. Wulff, *Welding for Engineers*, Wiley, 1954. With permission.)

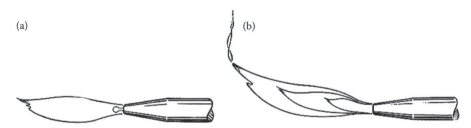

FIGURE 22.13 Oxyacetylene flames. Oxidizing flame (a) and reducing flame (b). (From H. Udin, E. R. Funk, and J. Wulff, *Welding for Engineers*, Wiley, 1954. With permission.)

or

$$C_2H_2 + O_2 \rightarrow 2CO + H_2. \tag{22.3}$$

The former flame is much hotter, but it is oxidizing. Figure 22.13 shows the appearances of the oxidizing and reducing flames. Oxygen–hydrogen flame welding is sometimes used.

Other sources of heat for welding include atomic hydrogen, electron beams, lasers, and the thermit. The thermit depends on the reaction between powders of aluminum and iron oxide to provide both heat and molten iron for the filler material.

$$8Al + 3Fe_3O_4 \rightarrow 9Fe + 4Al_2O_3. \tag{22.4}$$

Figure 22.14 shows a typical setup. Thermit welding is particularly useful in remote areas where electric power is not available.

22.6   METALLURGY OF WELDING

The freezing of weld metal is basically the same as freezing of any casting. Cooling is provided by the colder metal adjacent to the weld. Columnar grains grow toward

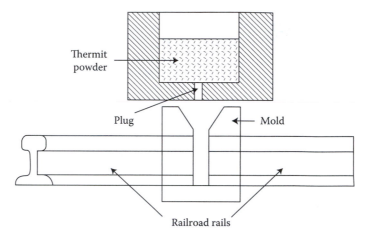

FIGURE 22.14   Thermit welding.

the center of the weld. Shrinkage may result in pipe. The material in the heat-affected zone around the weld is annealed. If it has been cold worked it softens because of recrystallization. Near the weld grain growth will occur. Figure 22.15 illustrates this. In the case of steel, austenite will form near the weld as illustrated in Figure 22.16. If there is sufficient hardenability, the austenite will transform to martensite, leaving the weld and the region around it harder and more brittle than the base metal.

Formation can be prevented by preheating the plates to be welded. This slows down the cooling rate after the heat source passes. If martensite does form, it can be tempered by postheating.

22.7   RESIDUAL STRESSES AND DISTORTION

During welding, the different temperatures in different regions cause residual stresses. The largest stresses are parallel to the weld. Because the metal in the weld

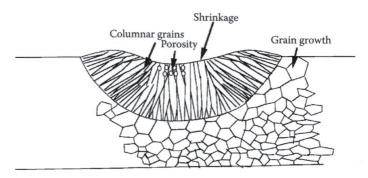

FIGURE 22.15   Columnar grains in weld and grain growth in the heat-affected zone near a weld.

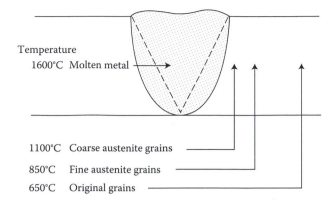

1100°C   Coarse austenite grains

850°C   Fine austenite grains

650°C   Original grains

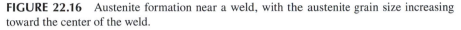

FIGURE 22.16 Austenite formation near a weld, with the austenite grain size increasing toward the center of the weld.

is hotter than the surrounding material, it would contract more on cooling if it were unattached. However, the dimensions of the plate and weld metal must match after cooling. The dimensional mismatch is relieved by an elastic tension in the weld and a lower elastic compression outside the weld region as illustrated in Figure 22.17. If the plates are restrained from contracting in the direction perpendicular to the weld, there will also be a lateral tensile stress.

Distortion will occur where there is no constraint. Because the top surface is heated more than the bottom, it will contract more on cooling leaving a bent plate. T-joints will warp. These distortions are illustrated in Figure 22.18.

22.8   BRAZE-WELDING OF CAST IRON

Cast iron is particularly difficult to weld, because the stresses built up during welding are sufficient to crack the material. In addition, the high carbon and silicon contents

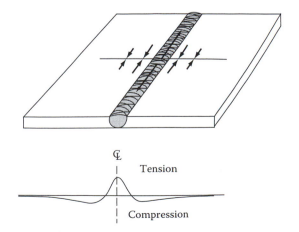

FIGURE 22.17 Pattern of residual stresses in a welded plate.

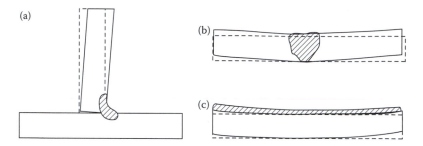

FIGURE 22.18 Distortions caused by welding. T-joint (a). Warping of a plate (b) and (c).

impart a high hardenability, so martensite usually forms on cooling, adding to the magnitude of the stresses. For this reason, broken cast iron machinery is usually repaired by braze-welding. This process involves welding with a bronze that melts at a temperature well below the melting temperature of the cast iron. Braze-welding differs from true brazing in that sufficient filler material is used that capillarity plays no role in making the joint. A copper-base alloy (39% Zn, 1% Sn, balance Cu) is frequently used. With a moderate preheating, martensite can be avoided. Other metals can be braze-welded, but the technique is most often applied to cast irons.

22.9 BRAZING AND SOLDERING

Brazing and soldering are welding processes with liquid nonferrous metal. Melting of base does not occur. The joints are thin and capillarity plays an important role in filling of joints. The difference between brazing and soldering is merely a question of definition. The process is called brazing if the filler metal melts above 535°C (1000°F) and soldering if the filler metal melts below 535°C. Brazing and soldering may be done by heating with a torch. Alternatively, the parts to be joined together with the filler metal may be placed in a furnace.

Many filler materials are used. Pure copper is used in for furnace brazing of steel and stainless steel. Many of the brazing alloys are copper alloyed with zinc and silicon, tin and phosphorous, to lower the melting point. Silver-base alloys (called silver solders) are usually alloys of silver, copper, and zinc. Aluminum can be brazed with an aluminum alloy containing 5–13% silicon, but this is tricky because the melting point is only about 55°C below that of aluminum.

Brazing joints are thin. Capillarity plays an important role in filling the joints. A simple tube-to-plate joint is illustrated in Figure 22.19. Figure 22.20 illustrates the role of the surface tension in filling a vertical joint between two plates. The gravitational force per length of joint is $\rho g h t$, where ρ is the density of the liquid, g is the acceleration of gravity, t is the thickness of the joint, and h is the capillary rise. This is balanced by the surface tension force, $2\gamma \cos \theta$, where θ is the equilibrium wetting angle. The capillary rise, h, is

$$h = \frac{2\gamma \cos \theta}{\rho g t}. \tag{22.5}$$

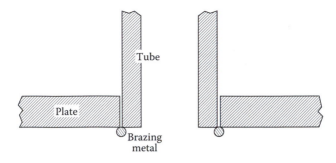

FIGURE 22.19   Tube-to-plate brazing joint.

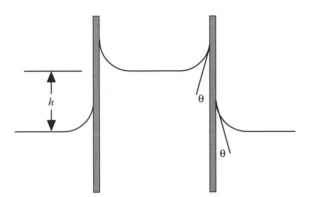

FIGURE 22.20   Schematic of the capillary rise between two plates.

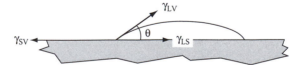

FIGURE 22.21   Liquid droplet on a solid surface.

For a high capillary rise, the joint thickness, t, and the wetting angle, θ, should be small. Figure 21.21 illustrates the equilibrium wetting angle, θ, between a liquid droplet and a solid surface. A balance of forces gives,

$$\gamma_{SV} = \gamma_{LS} + \gamma_{LV} \cos \theta. \tag{22.6}$$

For complete wetting, $\theta = 0$, so

$$\gamma_{SV} \geq \gamma_{LS} + \gamma_{LV}. \tag{22.7}$$

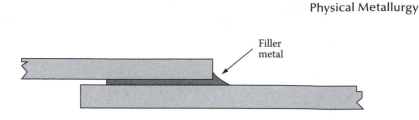

FIGURE 22.22 Lap joint.

Often a flux is added which aids the wetting. Borax is often used as a flux. The most common soft solders are lead–tin alloys. The greater the tin content, the more expensive the solder is. Because of the toxicity of lead, great effort has been expended in recent years to find lead-free solders.

22.10 JOINT DESIGN

For lap joints, the strength depends on amount of overlap (Figure 22.22). For tensile joints, the strength increases as the ratio of the joint thickness to diameter decreases as indicated in Figure 22.23. The reason for this is that the softer filler material is constrained from deforming by the material being brazed.

22.11 HARD FACING

Sometimes hard facings are put on dies tools by welding to improve wear resistance. Typical applications are rock-crushing equipment, tractor wheels, and shears. The weld material normally has a higher alloy content than the base material. During welding some dilution will occur from the melted base material. Figure 22.24 illustrates hard facing of the lips of a shear.

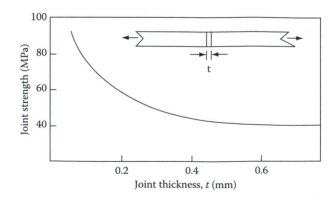

FIGURE 22.23 Effect of joint thickness on the strength of a silver-soldered butt joint between two bars of stainless steel bars.

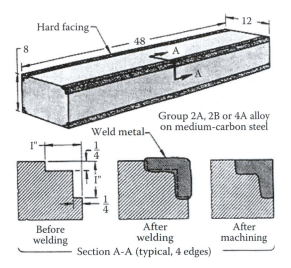

FIGURE 22.24 Hard facing of a lip of a shear. (From *Metals Handbook*, 8th ed., Vol. 6, p. 157, ASM, 1971. With permission.)

22.12 MISCELLANY

In the eighteenth and nineteenth centuries, blacksmiths used a form of cold welding to joint pieces of wrought iron and to close links of forged chain. The smith would sprinkle sand on the surfaces of the hot iron to be joined. Silica would combine with iron oxide to form a liquid flux. The flux would flow away when the two surfaces were hammered together, allowing metal-to-metal contact.

Welding defects were responsible for failures of a large number of natural gas pipelines. Most occurred during testing. Crack lengths ranged from 180–3200 ft. The fact that brittle cracks propagate faster than the speed of sound in methane meant that the stress at the crack tip could not be relived. Figure 22.25 shows such a failure.

During World War II, there was a rapid expansion of welded ship construction. Over 2000 liberty ships and about 500 T-2 tankers were produced. Of these 145 liberty ships and 53 T-2 tankers failed by brittle fracture. One of the reasons for the alarming number of ship failures has been attributed to inferior welding procedures. In the rush to get ships built some operators were poorly trained. Others tried to speed up the welding by filling the joints with unmelted welding rods. Finally, ship design at first was not modified from that of riveted ships to account for the greater rigidity of the welded structures. Figure 22.26 shows the failure of a ship that failed in port.

PROBLEMS

1. True or false:
    a. During welding, the width of the heat-affected zone increases as the welding rate decreases.

FIGURE 22.25   A natural gas pipe line that failed during field testing. (From E. Parker, *Brittle Fracture of Engineering Structures*, p. 269, Wiley, 1957. With permission.)

FIGURE 22.26   A ship that fractured while in port. (From C. F. Tipper, *The Brittle Fracture Story*, Cambridge University Press, 1963. With permission.)

    b.   Preheating decreases the rate of cooling in the heat-affected zone.

    c.   One of the important functions of an electrode coating is to provide a protective atmosphere.

    d.   After welding there is residual tension in the weld parallel to the weld.

2. A liquid between two plates will undergo a capillary rise if the wetting angle is <90°. Make a force balance to relate the capillary rise, h, to the liquid density, ρ, the liquid surface tension, γ, the wetting angle, θ, the separation of the two plates, t, and the gravitational constant, g.

    a.   How high would liquid copper rise between two steel plates separated by 0.10 mm? The surface tension of copper is 1300 dynes/cm = 13 mN/m. Copper completely wets steel, so $\theta = 0$.

3. From what you know, speculate as to how bimetallic strips are manufactured.

4. Assuming that 2 cm in Figure 22.11 represents 1 actual cm, calculate the average cooling rate (°C/s) along the centerline between 800°C and 550°C.

REFERENCES

Metals Handbook, 8th ed., Vol. 6, p. 157, ASM, 1971.

E. Parker, *Brittle Fracture of Engineering Structures*, p. 269, Wiley, 1957.

C. F. Tipper, *The Brittle Fracture Story*, Cambridge University Press, 1963.

H. Udin, E. R. Funk, and J. Wulff, *Welding for Engineers*, Wiley, 1954.

J. Wulff, H. Taylor, and J. Shaler, *Metallurgy for Engineers*, Wiley, 1952.

23 Powder Processing

23.1 GENERAL

Many parts are made by powder metallurgy, which involves pressing powder to a desired shape and sintering. Most nonclay, nonglass ceramics are processed in this way. Teflon is consolidated by powder processing because it cannot be melted without decomposing. Our interest here however is powder metallurgy. Reasons for making items by powder metallurgy instead of by casting or mechanical working include the following:

1. Powder processing is economical for high production of small items of simple shape such as lock parts, electrical contacts, magnets, and gears.
2. Preforms for forging can be made more accurately by powder processing than by cutting bar stock.
3. Powder processing can consolidate materials that are difficult to melt such as tungsten, molybdenum, and carbides.
4. It may be the only way of processing obtaining mixtures that cannot be cast together (e.g., high friction materials).

Porous parts (e.g., filters and oilless bearings) can be produced by powder metallurgy.

23.2 POWDERS

Powders are produced by direct reduction of oxides, precipitation from solution, electrolytic deposition, atomization, grinding, and crushing to name a few. Powder shape, size, and size distribution are important. Screening is used to separate powders by size. Very fine powder is highly combustible because of high surface area. (The surface area-to-volume ratio is inversely proportional to the particle diameter.) The amount of gas adsorbed depends on the surface area, so fine powders adsorb more gas.

23.3 POWDER PRESSING

Usually powder compaction is done with a double-action die. Figure 23.1 shows typical punch and die set and Figure 23.2 illustrates the operation. Because of friction between the powder and the die walls, the pressure decreases with distance from the

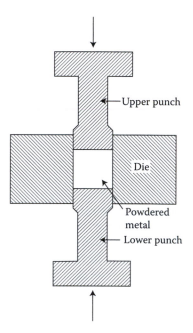

FIGURE 23.1   Punch and die set for compaction.

punch. Figure 23.3 illustrates this for both double-action and single-action pressing. The variation of pressure and therefore pressed density with double-action pressing is about half as much as the variation of density with single-action pressing. The ratio of the part's height to diameter is very important. The variation of pressure is large with double-action pressing if the height-to-diameter ratio is greater than 1 and in single-action pressing if the height-to-diameter ratio is greater than one half. Pressure variation is important because it causes a variation in density after compaction. This variation in density in turn causes cracking or distortion during later sintering.

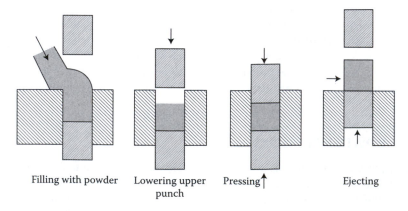

Filling with powder   Lowering upper   Pressing↑   Ejecting
                            punch

FIGURE 23.2   Operation of a punch and die set.

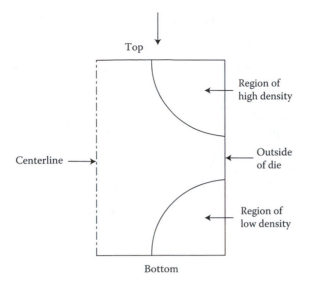

FIGURE 23.3 Density variation after compacting with a single-action pressing. A double-action pressing would produce a region of high density in the lower right-hand side.

However, the dependence of density on pressure decreases at high pressure, so high pressure compaction lessens the problem. There is a danger of overpressing. With too much compaction, gas may become entrapped. If this occurs, gas expansion during sintering can rupture the part.

23.4   SINTERING: EARLY STAGES

After pressing, parts are sintered. Diffusion increases the bonding between particles, which in turn improves mechanical properties. Figure 23.4 is an electron micrograph showing the neck formed between two nickel spheres during sintering. The driving force for sintering is the lowering of surface area. As an atom moves to neck between the two particles, it lowers the total surface area. There are two important parameters that indicate the degree of sintering. One is the ratio of radius of contact between the two spheres to the neck radius, X. The other is the change of distance between their centers, ΔL. These are illustrated in Figure 23.5. There are a number of mechanisms of sintering. Evaporation–condensation, surface diffusion, and volume diffusion from the surface to the base of the neck increase the radius of contact, X, but do not change the distance between the centers of the spheres. Hence they do not contribute to densification. On the other hand, plastic deformation and diffusion of atoms from the area of contact to the base of the neck by either grain boundary or volume diffusion decrease the distance between the centers as well as neck increases the radius of contact. Hence they do cause densification. The two equations that describe the extent of the initial stage of sintering are

$$X = ct^m \, R^{m/n} \tag{23.1}$$

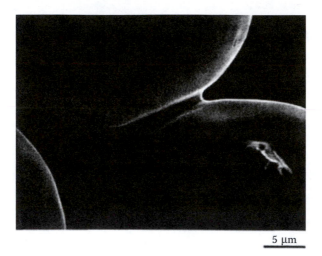

FIGURE 23.4 The neck formed between two spheres of nickel. The marker is 5 mm. (From R. M. German, *Powder Metallurgy Science*, Metal Powder Industries Federation, 1984. With permission.)

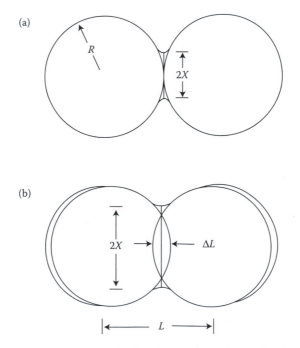

FIGURE 23.5 Important parameters for describing the extent of early sintering. (a) Surface diffusion and vapor transport increase the contact, X, but do not increase density. (b) Diffusion in the bulk and in boundary between particles decreases, L, causing density increase as well as increase contact, X. (From R. M. German, *Powder Metallurgy Science*, Metal Powder Industries Federation, 1984. With permission.)

TABLE 23.1
Time Exponents for Various Sintering Mechanisms

Mechanism	Example	m	n
Viscous flow	Glass	1/2	1
Vapor	NaCl	1/3	0
Volume diffusion	Metals	1/5	2/5
Surface diffusion	Ice, metals	1/7	0
Grain boundary diffusion (vac form)	Metals	1/4	1/2
Grain boundary diffusion (vac motion)	Metals	1/6	1/8

and

$$\frac{\Delta L}{L} = ct^n, \tag{23.2}$$

where n and m depend on mechanism. The time dependence can aid in identifying the dominant mechanism. Table 23.1 lists the values of m and n for various mechanisms.

The mechanism may change during sintering. Figure 22.8 is a map showing the dominant mechanism as a function of the particle size and temperature.

23.5 SINTERING: LATER STAGE

Figure 23.6 is a sketch of the initial pore shape. Figure 23.7 illustrates schematically how the porosity changes during the course of sintering. Surface diffusion can play an important role in densification as long as there are continuous surface paths. However, as the channels close, pores become isolated as indicated in Figure 23.8a.

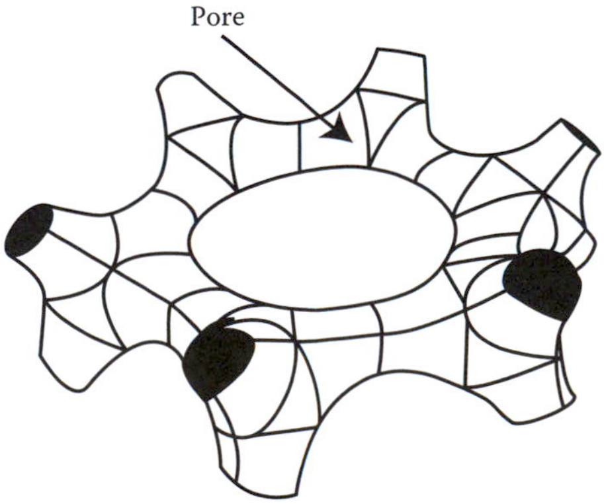

FIGURE 23.6 Sketch of the initial pore shape. The pores form continuous channels. (From R. M. German, *Powder Metallurgy Science*, Metal Powder Industries Federation, 1984. With permission.)

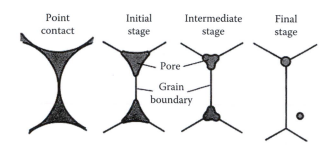

FIGURE 23.7 Changes of pore morphology during sintering. (From R. M. German, *Powder Metallurgy Science*, Metal Powder Industries Federation, 1984. With permission.)

As long as the pores are on grain boundaries, grain boundary diffusion can play a role in densification. However, grain growth can cause grain boundaries to leave pores (Figure 23.8b) causing further slowing of densification. Figure 23.9 illustrates pore isolation by grain boundary migration.

Often powders of more than one material are mixed. Alloying occurs during sintering if the materials are soluble. However diffusion can cause Kirkendall porosity, which acts to slow densification and may even cause a volume increase.

23.6 LIQUID PHASE AND ACTIVATED SINTERING

In two-phase systems, if one phase melts during sintering, it can form a continuous liquid film around the solid phase. Some of the two-phase systems are W–Cu, W–Ni–Fe, Cu–Sn, Fe–Cu, TiC–Ni, and WC–Co. Complete densification is possible in a reasonably short time. Figure 23.10 is the cobalt–tungsten carbide phase diagram. Figure 23.11 shows the microstructure of a 94%WC–6%Co alloy that was liquid-phase sintered. The molten cobalt sintered the carbide particles.

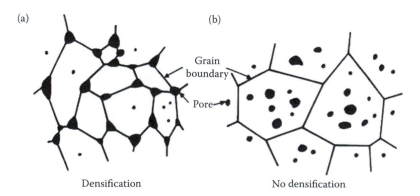

FIGURE 23.8 As long as the pores are on grain boundaries (a) densification can occur by grain boundary diffusion. Once the boundaries migrate away from pores (b), grain boundary diffusion stops and sintering becomes exceedingly slow. (From R. M. German, *Powder Metallurgy Science*, Metal Powder Industries Federation, 1984. With permission.)

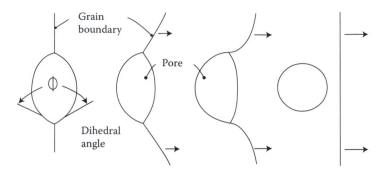

FIGURE 23.9 Grain boundary migrating away from a pore. (From R. M. German, *Powder Metallurgy Science*, Metal Powder Industries Federation, 1984. With permission.)

Most carbide tools contain 75–95% WC with cobalt (25–5%) as a binder. The Co–W phase diagram shown in Figures 23.10 and 23.11 is a typical microstructure. Carbide tools are made by pressing mixed powders of cobalt and tungsten carbide and sintering at 1350–1500°C. Sintering is rapid because a liquid phase is formed. Sintered carbide tools retain a high hardness at very high temperatures and are therefore useful in high-speed machining. For machining steel, mixed titanium carbide or tungsten carbide is preferred because of lower friction. Often an alloy of nickel and cobalt is used as a binder instead of pure cobalt.

There are other means of increasing the speed of sintering. Very minor additions of Ni, Pt, and Pd to tungsten have been shown to increase the sintering rate. It is

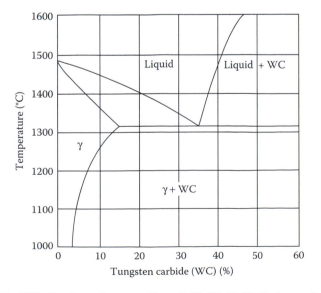

FIGURE 23.10 WC–Co phase diagram. (From J. Wulff, H. W. Taylor, and A. J. Schaler, *Metals Handbook*, 8th ed., Vol. 7, p. 127, 1972. With permission.)

FIGURE 23.11 Microstructure of sintered WC–Co tool material. (From J. Wulff, H. W. Taylor, and A. J. Schaler, *Metals Handbook*, 8th ed., Vol. 7, p. 127, 1972. With permission.)

possible that the effect is due to thin liquid films. Irradiation and cycling iron through its α–γ transformation temperature are other means of increasing the sintering rate.

23.7 HOT PRESSING

Hot pressing combines the compaction and sintering steps. At high temperatures, the compaction is increased because of the lower flow stresses. The biggest disadvantage of hot pressing is the greatly increased time spent in the compaction press. This results in a greatly decreased rate of production.

Sometimes powder compacts are subjected to hot isostatic pressing (*hipping*). Powder are initially pressed to the desired shape and coated with an impervious material. They are then placed in a pressurized chamber that is heated. Very low porosities can be achieved in this way. Hipping is also used to close porosity in castings.

23.8 STRENGTH AND DUCTILITY

Powder performs for forging offer better control over properties than performs cut from rod.

23.9 MISCELLANY

Metals are not the only materials formed from powder by pressing and sintering. Most ceramics, other than clay products and glasses, are formed into useful shapes from powder by pressing and sintering. Teflon (polytetrafluoroethylene) cannot be melted without it being decomposed. Teflon objects are therefore made by pressing and sintering. When children make snowmen, they are practicing powder processing. They first compact the powder (snow), and then allow it to sinter. (Ice will sinter at −7°C, 20°F.)

PROBLEMS

1. Stainless steel powder with a mean particle diameter of 50 mm has been compacted to a green density of 58% and sintered in pure H_2. The resulting shrinkage measurements are given below. Published diffusion data for this stainless steel show that the activation energies are 225 kJ/mol for surface diffusion, 200 kJ/mol for grain boundary diffusion, and 290 kJ/mol for volume diffusion. Use the data below to determine the mechanism.

Temperature (°C)	Time (h)	Shrinkage (%)	Temperature (°C)	Time (h)	Shrinkage (%)
1050	2.0	0.62	1200	1.5	1.63
1100	2.0	0.91	1200	2.0	1.82
1150	2.0	1.31	1250	2.0	2.49
1200	0.5	1.05	1300	2.0	3.33
1200	1.0	1.38			

2. What is the surface area of 1 g of copper powder, 50 μm diameter? What is the total energy of the surface? ($\gamma_{SV} = 1700$ ergs cm^{-2})
3. The density of a powder after compaction is 85% and after sintering is 99%. What diameter die and punch should be used to make a cylinder 25 mm in diameter and 22 mm tall?

REFERENCES

R. M. German, *Powder Metallurgy Science*, Metal Powder Industries Federation, 1984.

R. M. German, *Powder Metallurgy of Iron and Steel*, Wiley, 1998.

F. V. Lenel, *Powder Metallurgy: Principles and Applications*, Metal Powder Industries Federation, 1980.

F. Thümmler and R. Oberacker, *Introduction to Powder Metallurgy*, The Institute of Materials, 1993.

G. S. Upadhyaya, *Powder Metallurgy Technology*, Cambridge International Science Publishing, 1997.

J. Wulff, H. W. Taylor, and A. J. Schaler, *Metals Handbook*, 8th ed., Vol. 7, p. 127, 1972.

24 Magnetic Materials

24.1 GENERAL

Magnetism seems to be a mysterious phenomenon. The discovery of lodestone (Fe_3O_4) has led to many myths (Figure 24.1). When we speak of "magnetic behavior," we usually mean ferromagnetic behavior. Actually there are two other types of magnetic behavior: Diamagnetic behavior, which is a weak repulsion of a magnetic field, and diamagnetism, which is a weak attraction of a magnetic field.

Ferromagnetism, in contrast, is a very strong attraction of a magnetic field. There are only a few ferromagnetic elements. The important ones are iron, nickel, and cobalt. A few rare earths are ferromagnetic at low temperatures. Table 24.1 lists all of the ferromagnetic elements and the temperature above which they cease to be ferromagnetic (Curie temperature).

Atoms of other transition elements may act ferromagnetically in alloys where the distance between atoms is different than in the elemental state. These include the manganese alloys Cu_2MnAl, Cu_2MnSn, Ag_5MnAl, and $MnBi$.

The physical basis for ferromagnetism is an unbalance of electron spins in the 3-d shell of the transition elements and the 4-f shell of rare earths. An unbalanced spin causes a magnetic moment. In metals with valences of 1 or 3 (e.g., Cu or Al), each atom has an unbalance of spins, but the unbalance is random so there is no net effect. With the transition elements, the 3-d and 4-s energy bands overlap at the distance between atoms in the crystal (Figure 24.2).

For some, the total energy is lowered in a magnetic field, if one half of the 3-d band is completely full, causing an unbalance of electron spins as shown schematically in Figure 24.3. This results in a strong magnetic effect. The field caused by neighboring atoms is strong enough to cause this shift.

This lowering of energy caused by alignment of the unbalanced spins with that of the neighboring atoms is called the exchange energy. It depends on the interatomic distance. For example, bcc iron is ferromagnetic but fcc iron is not. Figure 24.4 illustrates this. Figure 24.5 shows how the maximum number of unbalanced spins per atom (number of Bohr magnetons) depends on number of 3-d electrons.

Ferromagnetic behavior depends on four important energy terms:

1. *Exchange energy*, which has already been discussed
2. *Magnetostatic energy*
3. *Magnetocrystalline energy*
4. *Magnetostrictive energy*

FIGURE 24.1 Typical legend about lodestone, the first permanent magnet known to man. (From E. A. Nesbitt, *Ferromagnetic Domains*, Bell Telephone Laboratories, 1962. With permission.)

TABLE 24.1
Ferromagnetic Elements and Curie Temperatures

Metals	Curie Temperature (°C)
Iron	1044
Cobalt	1121
Nickel	358
Gadolinium	16
Terbium	−40
Dysprosium	−181

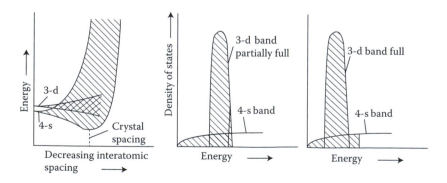

FIGURE 24.2 For the transition elements, the 3-d and 4-s energy levels overlap. (From R. E. Smallman, *Modern Physical Metallurgy*, 2nd ed., Butterworths, 1963. With permission.)

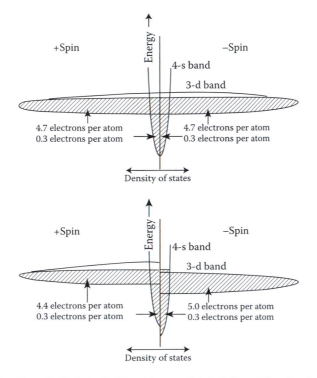

FIGURE 24.3 If one half of the 3-d band is completely full and the other half is partially full, there is a strong unbalance of electron spins causing a strong magnetic effect.

The exchange energy is minimized and neighboring atoms are magnetized in the same direction. This causes the formation of magnetic domains in which all of the neighboring atoms are magnetized in the same direction. These may contain 10^{15} atoms. Figure 24.6 schematically shows parts of two domains.

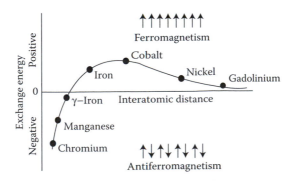

FIGURE 24.4 Dependence of exchange energy saturation on atomic separation. (From J. K. Stanley, *Electrical and Magnetic Properties of Metals*, ASM, 1963. With permission.)

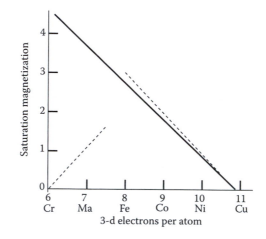

FIGURE 24.5 Variation of the magnetization on the number of 3-d electrons.

24.2 MAGNETOSTATIC ENERGY

However, incomplete magnetostatic circuits raise the total energy because the circuits must be completed externally (Figure 24.7). The reason why horseshoe magnets attract iron is to complete magnetostatic circuits in iron (Figure 24.8). A typical domain structure is composed of domains, which form complete circuits as shown in Figure 24.9.

When there are equal numbers of domains aligned in opposing directions, their magnetic fields cancel and externally the material appears not to be magnetized. However,

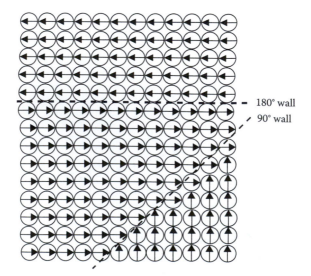

FIGURE 24.6 Ferromagnetic domains are regions in which unbalanced electron spins are aligned. Parts of three domains are indicated. The dashed lines are 180° and 90° domain walls.

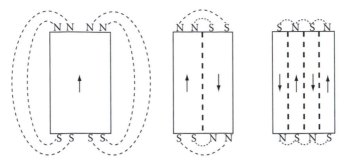

FIGURE 24.7 Incomplete magnetostatic circuits raise the energy. (From E. A. Nesbitt, *Ferromagnetic Domains*, Bell Telephone Laboratories, 1962. With permission.)

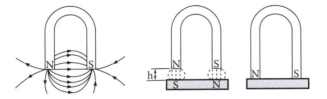

FIGURE 24.8 A horseshoe magnet attracts iron to complete a magnetostatic circuit. (From E. A. Nesbitt, *Ferromagnetic Domains*, Bell Telephone Laboratories, 1962. With permission.)

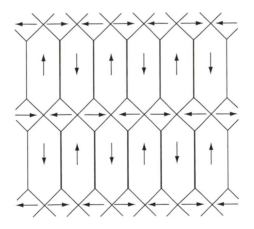

FIGURE 24.9 Typical domain structure composed of complete magnetostatic circuits.

when an external magnetic field is imposed, those domains most nearly aligned with the field will grow at the expense of those aligned in the opposite direction.

24.3 MAGNETOCRYSTALLINE ENERGY

Each of the ferromagnetic materials has a specific crystallographic direction in which it is naturally magnetized. Figure 24.10 shows the B–H curves for iron crystals of different orientations. Table 24.2 lists the directions of easy magnetization.

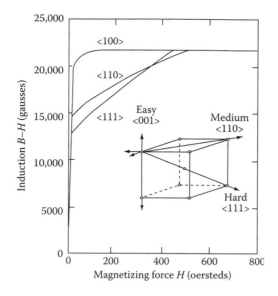

FIGURE 24.10 *B–H* curves for several directions in iron. (From J. K. Stanley, *Electrical and Magnetic Properties of Metals*, ASM, 1963. With permission.)

TABLE 24.2
Directions of Easy Magnetization

Iron	⟨100⟩
Nickel	⟨111⟩
Cobalt	[0001]

24.4   MAGNETOSTRICTION

When a material is magnetized, it undergoes a small dimensional change in the direction of magnetization. This phenomenon is called magnetostriction. Figure 24.11 shows the magnetostriction in the three common metals.

24.5   PHYSICAL UNITS

Further discussion of magnetic behavior requires definition of some terms. The units used to describe these are listed in Table 24.3, which include both cgs and mks units and their relation. The intensity of the magnetic field or magnetizing force, *H*, is measured in henrys (or oersteds). The magnetic induction, *B*, is measured in teslas (or gauss).

24.6   THE *B–H* CURVE

When a magnetic field is imposed on a ferromagnetic material, the domains most nearly aligned with the field will grow at the expense of others as illustrated in

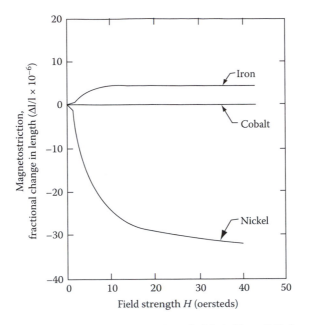

FIGURE 24.11 Magnetostriction of iron, cobalt, and nickel. (From J. K. Stanley, *Electrical and Magnetic Properties of Metals*, ASM, 1963. With permission.)

Figure 24.12. As they do, the material's magnetic induction will increase as shown in Figure 24.13. At first favorably aligned domains grow. Final induction occurs by rotation of the direction of magnetization out of the easy direction to be aligned with the field. Figure 24.14 shows an entire *B–H* curve.

On removal of the field, there is a *residual magnetization* or *remanence*, B_r. A reverse field, H_c, (*coercive force*) is required to demagnetize the material. The area enclosed by the *B–H* curve (*hysteresis*) is the energy loss per cycle. The *permeability* is defined as $\mu = B/H$. The initial permeability μ_0 and the maximum permeability, μ_{max}, are material properties.

24.7 CURIE TEMPERATURE

The Curie temperature is the temperature above which a material ceases to be ferromagnetic. Figure 24.15 shows the decrease of saturation magnetization, B_{max}, with temperature.

TABLE 24.3
Units

Units	mks	cgs
H	henry = A/m	$4\pi \times 10^{-3}$ oersted
B	tesla = weber/m^2	10^4 gauss
μ_0	henry/m	$(10^7/4\pi)$ gauss/oersted

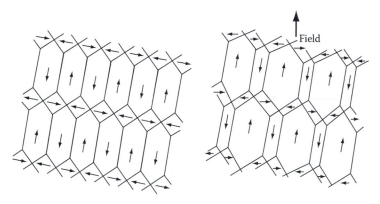

FIGURE 24.12 Imposition of an external field causes domains most nearly aligned with the field to grow at the expense of those antialigned.

24.8   BLOCH WALLS

The boundaries between domains are regions where there is a gradual change in the direction of magnetization. The width of these (perhaps 20 atoms) is a compromise between the magnetocrystalline and exchange energy terms. A wider boundary would require more atoms to be magnetized out of the direction of easy magnetization. The exchange energy is minimized if the boundary is very wide so the direction of magnetization changes very little between neighboring atoms. There are two possibilities. In Bloch walls, the direction of magnetization rotates in a plane parallel to the wall as illustrated schematically in Figure 24.16 for 180° and 90° domain walls. In Neel walls, the axis of rotation is in the plane of the wall.

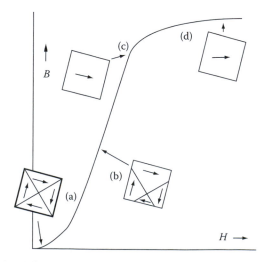

FIGURE 24.13   Magnetization of a material.

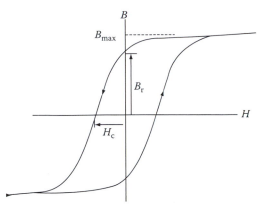

FIGURE 24.14 A typical *B–H* curve. Initially, magnetization increases by growth of favorably oriented domains. At high fields, the direction of magnetization rotates out of the easy direction.

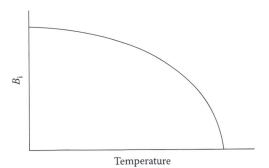

FIGURE 24.15 Decrease of saturation magnetization with temperature up to the Curie temperature.

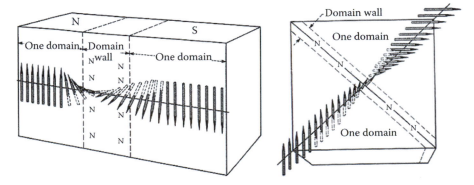

FIGURE 24.16 180° and 90° domain walls. (From E. A. Nesbitt, *Ferromagnetic Domains*, Bell Telephone Laboratories, 1962. With permission.)

24.9 SOFT VERSUS HARD MAGNETIC MATERIALS

There are two main types of magnets: hard and soft. Hard magnetic materials are permanent magnets. They are difficult to demagnetize. The hysteresis is very large. The remanence, B_r, and coercive force, H_c, are high. Soft magnetic materials are easily demagnetized. The terms soft and hard are historic. The best permanent magnets in the 1910s were made of martensitic steel, which is very hard and the best soft magnets were made from pure annealed iron. The differences of the B–H curves are shown in Figure 24.17. Table 24.4 shows the extreme differences.

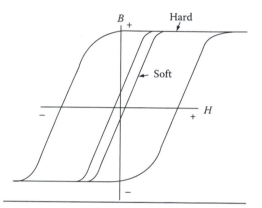

FIGURE 24.17 A hard magnetic material has a much greater hysteresis than a soft magnetic material. The differences are much greater than shown in this figure.

TABLE 24.4
Coercive Forces of Several Materials

Coercive Forces in Ferromagnetic Materials		
Material	**Composition**	**Coercive Force H_c, from Saturation (Oersteds)**
Supermalloy	79% Ni, 5% Mo	0.002
Oriented silicon steel	3.25% Si	0.1
Hot-rolled silicon steel	4.5% Si	0.5
Mild steel (normalized)[a]	0.2% C	4.0
Carbon magnet steel	0.9% C, 1% Mn	50
Alnico V	24% Co, 14% Ni, 8% Al, 3% Cu	600
Alnico VIII	35% Co, 14.5% Ni, 7% Al, 5% Ti, 4.5% Cu	1450
Barium ferrite (oriented)	$BaO \cdot 6Fe_2O_3$	1900
Bismanol	MnBi	3650
Platinum–cobalt	PtCo (77% Pt)	4300

Source: Adapted from J. K. Stanley, *Electrical and Magnetic Properties of Metals*, ASM, 1963.
[a] Heating to above the transformation range followed by cooling to room temperature in still air.

24.10 SOFT MAGNETIC MATERIALS

For a material to be soft magnetically, its domain walls must move easily. The principal obstacles to domain wall movement are inclusions and grain boundaries. Low dislocation contents and residual stresses are also important. A low interstitial content is also important.

Inclusions are important obstacles to domain wall movement because the energy of the system is lowered when a domain wall passes through an inclusion than when the boundary has separated from the inclusion. This is illustrated in Figure 24.18.

Uses of soft magnetic materials include transformers, motor and generator cores, solenoids, relays, magnetic shielding, and electromagnets for handling scrap. Many of these applications employ silicon iron (usually 3–3.5% Si). Alloys containing 3% or more silicon are ferritic at all temperatures up to the melting point (Figure 24.19). Silicon increases the electrical resistance of iron. A high electrical resistance is desirable for transformers because eddy currents are one of the principal power losses in transformers. Remember that power loss is inversely proportional to resistance ($P = EI = E^2/R$). Use of thin sheets also minimizes eddy current losses.

It is possible to control the crystallographic texture of silicon iron sheet by controlling the rolling, and heat-treating schedules. The usual texture for the transformer sheets is {110}<001>, which is called the Goss texture. Because this texture has the <001> easy direction of magnetization aligned with the prior rolling direction, transformers can be made so that they will be magnetized in a <001> direction. The cube texture, {100}<001>, is even more desirable, but it is more difficult to produce. Both are illustrated schematically in Figure 24.20.

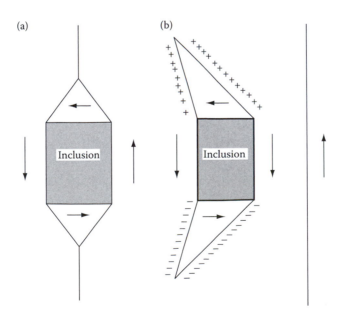

FIGURE 24.18 Difference between the domain boundaries at an inclusion depending on whether the inclusion lies on a boundary (a) or not (b). The total length of boundary is lowered by the inclusion.

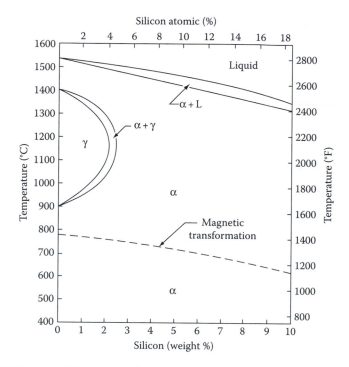

FIGURE 24.19   Iron–silicon phase diagram.

Core losses decrease with increasing silicon content and increase with increasing frequency.

For very soft magnetic magnets, magnetostriction should be minimized. The reason is that magnetostriction causes dimensional incompatibilities at 90° domain boundaries that must be accommodated by elastic straining of the lattice. This is illustrated in Figure 24.21.

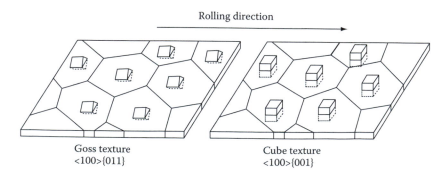

FIGURE 24.20   Textures in silicon. In both the Goss and the cube textures the $\langle 100 \rangle$ direction is aligned with the rolling direction. The {011} is parallel to the sheet in the Goss texture and the {001} is parallel to the sheet in the cube texture.

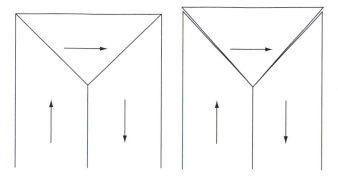

FIGURE 24.21 In iron, magnetostriction causes an elongation in the direction of magneti-zation. This creates a misfit along 90° domain boundaries, which must be accommodated elastically.

In iron–nickel alloys, the magnetostriction and the magnetocrystalline anisotropy are very low at about 78% Ni. Iron–nickel alloys have very high initial permeabili-ties. Mu metal (75% Ni), permalloy (79% Ni), and supermalloy (79% Ni, 4% Mo) are examples that find use in audio transformers. By heat treating in a magnetic field, a texture can be formed that has a square-loop hysteresis curve (Figure 24.22). This is useful in logic circuits where the material is magnetized in one direction or the other. For high fidelity transformers, linearity (constant μ) is needed.

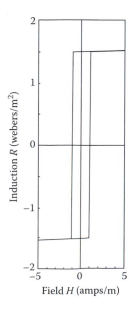

FIGURE 24.22 A square-loop hysteresis curve for permalloy can be obtained by heat treat-ing in a magnetic field.

24.11 HARD MAGNETIC MATERIALS

For hard magnets, a high H_c coercive force is desirable, but most important is a high $B \times H$ product. The second quadrant of the $B-H$ curve (Figure 24.23) is most important. Often the maximum $B \times H$ product (Figure 24.24) is taken as a figure of merit.

1. Small, isolated particles that are single domains.
2. Elongated particles.
3. A high magnetocrystalline energy.

In a microstructure consisting of small isolated particles surrounded by a nonferromagnetic phase, there are no domain walls that can move. The direction of magnetization can be changed only by rotating the magnetization out of the easy direction into another equivalent easy direction. If there is a high magnetocrystalline energy, this will require a high field. Hexagonal structures are useful here because there are only two easy directions, [0001] and [000$\bar{1}$], which differ by 180°. With elongated ferromagnetic particles, the intermediate stage will have a high magnetostatic energy. Figure 24.25 illustrates this.

The most popular magnetic materials are aluminum–nickel–cobalt–iron alloys called alnico. Alnico V contains 8% Al, 14.5% Ni, 23% Co, 3% Cu, and 0.5% Ti with the balance Fe. At very high temperatures, it is a single bcc phase but it decomposes into two bcc phases below 800°C. The phase high in Co and Fe is ferromagnetic and it precipitates as fine particles. If the precipitation occurs in a magnetic field, the particles are elongated (Figure 24.26a) whereas they are equiaxed in the absence of a field (Figure 24.26b). The difference in the $B-H$ curves is shown in Figure 24.27.

Some of the best hard magnetic materials are those with a hexagonal structure. In these, there are only two possible domains, differing by 180°. Table 24.5 lists the maximum $B-H$ product for several alloys.

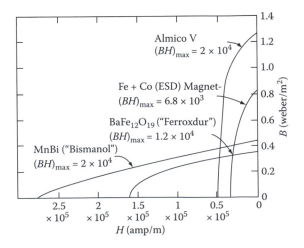

FIGURE 24.23 Second quadrant of $B-H$ curves for selected alloys. (From R. M. Rose, L. A. Shepard, and J. Wulff, *The Structure and Properties of Materials Vol. IV, Electronic Properties*, Wiley, 1966. With permission.)

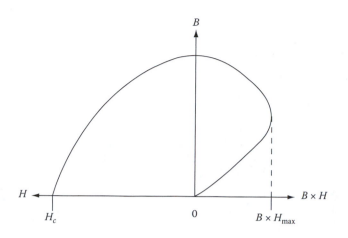

FIGURE 24.24 Second quadrant of a B–H curve (left) and the corresponding $B \times H$ product (right). The maximum $B \times H$ product is a figure of merit.

Cheap permanent magnets can be made by aligning fine iron powder in a magnetic field while it is being bonded by a rubber or a polymer.

24.12 SUMMARY

Exchange energy—near neighbors tend to be magnetized in the same direction because of the magnetic field set up by neighbors. This is the reason for domains.

Magnetostatic energy—lower energy with complete magnetostatic circuits. This is the reason why permanent magnets attract and the reason for normal domain structure in soft magnetic materials.

Magnetocrystalline energy—energy is lowest when the magnetization direction is parallel to a characteristic crystalline direction. To change the direction, a high field is required.

Magnetostriction—dimensional change when magnetization occurs. Magnetostriction makes movement of domain walls more difficult.

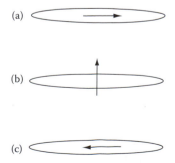

FIGURE 24.25 As the direction of magnetization of an elongated particle is reversed (from a to c) it must be magnetized in a direction that increases its magnetostatic energy (b).

(a)   (b)

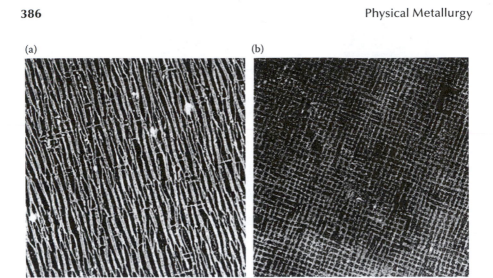

FIGURE 24.26 Microstructure of alnico V (approximately 50,000X) after precipitation in a magnetic field (a) and in the absence of a magnetic field (b). (From R. M. Rose, L. A. Shepard, and J. Wulff, *The Structure and Properties of Materials Vol. IV, Electronic Properties*, Wiley, 1966. With permission.)

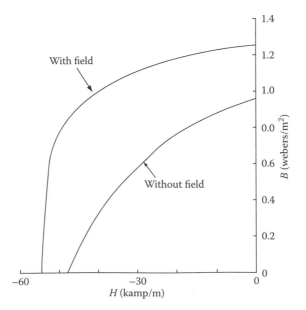

FIGURE 24.27 The effect of heat treating in a magnetic field on the demagnetization curves for alnico V.

24.13 MISCELLANY

The magnetic characteristics of lodestone, Fe_3O_4, were known to the ancients. There are myths about its power to pull nails out of ships. Probably, the first real use of the

TABLE 24.5
Maximum *B–H* Products for Several Alloys

Samarium–Cobalt	120,000 A-Webers/m³
Platinum–cobalt	70,000
Alnico	36,000
Carbon steel	1500

magnetic phenomenon should be attributed to the Vikings. Their development of magnetic compasses enabled then to travel far at sea, even in foggy conditions.

A magnet provides a simple way of distinguishing austenitic stainless steels from the other grades.

PROBLEMS

1. In iron, the magnetic permeability, μ, is highest in the $\langle 100 \rangle$ directions. With the Goss texture in sheets of silicon iron, most grains are oriented so that $\{110\}$ planes are nearly aligned with the rolling plane and $\langle 100 \rangle$ directions with the rolling direction. This $\{110\}\langle 001 \rangle$ texture is called a "cube on edge" texture. It is also possible to produce a "cube" or $\{100\}\langle 001 \rangle$ texture in which $\{100\}$ planes are nearly aligned with the rolling plane and $\langle 001 \rangle$ directions with the rolling direction. Sketch plots of how μ varies in the sheet with the angle, θ, from the rolling direction. Do this for
   a. The Goss texture.
   b. The cube texture. Put both plots on the same axes and plot for $0 \leq \mu \leq 180°$.
2. Figure 24.28 shows the core losses of transformers made from oriented and unoriented silicon steel. Consider the operation at 12 kG. Calculate the temperature rise after 30 min if the transformer is not cooled if
   a. The transformer is made from randomly oriented steel.
   b. The transformer is made from grain-oriented steel (M-6).
   The heat capacity of iron is 34 J/mol-K.
3. Consider the *B–H* curves in Figure 24.29 below.
   a. Which alloy has the highest coercive force?
   b. Which alloy has the highest residual magnetization (remanence)?
   c. Which alloy has the greatest hysteresis loss?
   d. Which alloy would make the best permanent magnet?
   e. Which alloy would be preferred for a memory device?
   f. In which alloy are the highest fractions of the domains aligned with the direction in which the field had been applied after the field is removed?
   g. Which alloy probably has the highest mechanical hardness?
4. A steel containing 18% Cr, 8% Ni, and 0.08% C is nonmagnetic after fast cooling from 900°C. However, it becomes magnetic when it is deformed at −200°C. Explain.
5. Sketch the (110) and (100) pole figures for the Goss texture with the sheet normal in the center and the rolling direction at the top.

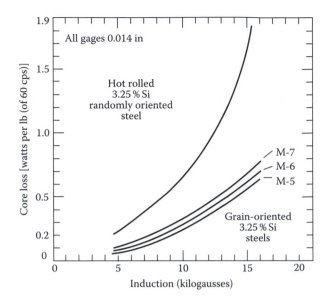

FIGURE 24.28 Hysteresis losses in several silicon steels. (From J. K. Stanley, *Electrical and Magnetic Properties of Metals*, p. 246, ASM, 1963. With permission.)

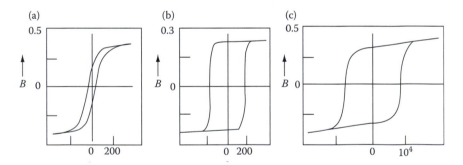

FIGURE 24.29 *B–H* curves for three magnetic alloys. The units of *B* are Wb/m² and the units of *H* are A/m.

REFERENCES

E. A. Nesbitt, *Ferromagnetic Domains*, Bell telephone Laboratories, 1962.

R. M. Rose, L. A. Shepard, and J. Wulff, *The Structure and Properties of Materials Vol. IV Electronic Properties*, Wiley, 1966.

R. E. Smallman, *Modern Physical Metallurgy*, 2nd ed., Butterworths, 1963.

J. K. Stanley, *Electrical and Magnetic Properties of Metals*, ASM, 1963.

25 Corrosion

25.1 CORROSION CELLS

Corrosion is the opposite of electroplating (Figure 25.1). There always is a cathode and an anode. (The anode is where ions go into solution $M \rightarrow M^{+n} + ne^-$. The cathode is where electrons are consumed.) Possible cathode reactions are as follows:

$M^{+n} + ne^- \rightarrow M$ (this can occur only if there is a high concentration of M^{+n} ions)

$2H^+ + 2e^- \rightarrow H_2$ (this can occur only if the solution is acid)

$O_2 + 2H_2O + 4e^- \rightarrow 4(OH)^-$ (there must be O_2 in the solution. This is the most common cathode reaction)

$O_2 + 4H^+ + 4e^- \rightarrow 2H_2O$ (there must be O_2 in an acid solution).

The Daniel cell is a battery involving zinc and copper, which illustrates corrosion. Zinc goes into the solution, $Zn \rightarrow Zn^{+2} + 2e^-$, and copper plates out, $Cu^+ + e^- \rightarrow Cu$. This creates a voltage in an external circuit (Figure 25.2).

Table 25.1 is the electromotive series. It gives the electrode potentials for various possible half reactions for solutions with 1 M concentration. If the solution is not 1 M, the potential must be adjusted by the Nernst equation

$$\varepsilon = \varepsilon_o + \left(\frac{0.00257V}{n}\right)\ln C. \tag{25.1}$$

The voltages for the various reactions are measured against a hydrogen electrode ($H_2 \rightarrow 2H^+ + 2e^-$), which is defined as zero (Figure 25.3).

Seawater is a common corrosive environment. The Galvanic series ranks alloys according to their cathode potential in seawater (Table 25.2).

The Nernst equation predicts that there can be concentration cells as a result of different concentrations in different places. Oxygen concentration cells are very common. Figure 25.4 illustrates some sources of oxygen concentration cells.

Corrosion cells can arise from two-phase microstructures, from cold work (Figure 25.5) and from grain boundaries (Figure 25.6).

25.2 CURRENT DENSITY

Current density controls the corrosion current. As corrosion occurs, the anode and cathode polarities change as indicated in Figure 25.7. The corrosion current

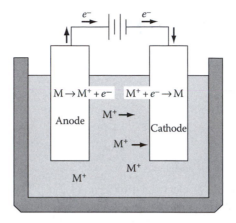

FIGURE 25.1 Corrosion cell. The anode is where electrons are generated in the external circuit and the cathode is where they are consumed.

corresponds to the intersection of the two. It is much better to have a large anode area and a small cathode area than a large cathode area and a small anode area. A large cathode area will lead to a greater corrosion current and a greater loss of metal. See Figure 25.8. Furthermore, with the same corrosion current, the depth of corrosion on a small anode area will be greater than that on a large anode area.

25.3 POLARIZATION

Accumulation of positive ions near the anode and negative ions near the cathode (Figure 25.9) will polarize the cell, decreasing the voltage (Figure 25.10). With polarization the corrosion current decreases as indicated in Figure 25.8.

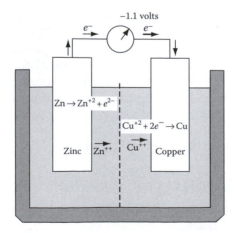

FIGURE 25.2 Daniel cell.

TABLE 25.1
Electrode Potentials (25°C; Molar Solutions)

Anode Half-Cell Reaction (the Arrows are Reversed for the Cathode Half-Cell Reaction)	Electrode Potential Used by Electrochemists and Corrosion Engineers, Volts	
$Au \rightarrow Au^{3+} + 3e^-$	+1.50	
$2H_2O \rightarrow O_2 + 4H^+ + 4e^-$	+1.23	
$Pt \rightarrow Pt^{4+} + 4e^-$	+1.20	Cathodic (noble)
$Ag \rightarrow Ag^+ + e^-$	+0.80	
$Fe^{2+} \rightarrow Fe^{3+} + e^-$	+0.77	
$4(OH)^- \rightarrow O_2 + 2H_2O + 4e^-$	+0.40	
$Cu \rightarrow Cu^{2+} + 2e^-$	+0.34	
$H_2 \rightarrow 2H^+ + 2e^-$	0.000	Reference
$Pb \rightarrow Pb^{2+} + 2e^-$	−0.13	
$Sn \rightarrow Sn^{2+} + 2e^-$	−0.14	
$Ni \rightarrow Ni^{2+} + 2e^-$	−0.25	
$Fe \rightarrow Fe^{2+} + 2e^-$	−0.44	
$Cr \rightarrow Cr^{2+} + 2e^-$	−0.74	
$Zn \rightarrow Zn^{2+} + 2e^-$	−0.76	Anodic (active)
$Al \rightarrow Al^{3+} + 3e^-$	−1.66	
$Mg \rightarrow Mg^{2+} + 2e^-$	−2.36	
$Na \rightarrow Na^+ + e^-$	−2.71	
$K \rightarrow K^+ + e^-$	−2.92	
$Li \rightarrow Li^+ + e^-$	−2.96	

Source: From L. H. Van Vlack, *Elements of Materials Science and Engineering,* 3rd ed., Addison–Wesley, p. 414, 1974. With permission.

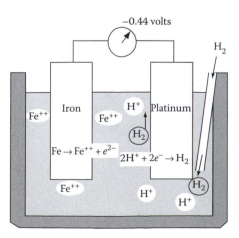

FIGURE 25.3 Hydrogen cathode.

TABLE 25.2

Galvanic Series of Common Alloys[a]

	Cathodic		
Graphite			Nickel—A
Silver	↑		Tin
12% Ni, 18% Cr, 3% Mo steel—P			Lead
20% Ni, 25% Cr steel—P			Lead–tin solder
23 to 30% Cr steel—P			12% Ni, 18% Cr, 3% Mo steel—A
14% Ni, 23% Cr steel—P			20% Ni, 25% Cr steel—A
8% Ni, 18% Cr steel—P			14% Ni, 23% Cr steel—A
7% Ni, 17% Cr steel—P			8% Ni, 18% Cr steel—A
16 to 18% Cr steel—P			7% Ni, 17% Cr steel—A
12 to 14% Cr steel—P			Ni-resist
80% Ni, 20% Cr—P			23 to 30% Cr steel—A
Inconel—P			16 to 18% Cr steel—A
60% Ni, 15% Cr—P			12 to 14% Cr steel—A
Nickel—P			4 to 6% Cr steel—A
Monel metal			Cast iron
Copper–nickel			Copper steel
Nickel–silver			Carbon steel
Bronzes			Aluminum alloy 2017-T
Copper			Cadmium
Brasses			Aluminum, 1100
80% Ni, 20% Cr—A			Zinc
Inconel—A	↓		Magnesium alloys
60% Ni, 15% Cr—A		Anodic	Anodic Magnesium

Source: From L. H. Van Vlack, *Elements of Materials Science and Engineering,* 3rd ed., Addison–Wesley, p. 430, 1974. With permission.

[a] A signifies active state and P signifies the passive state. The subject of passivity is covered later.

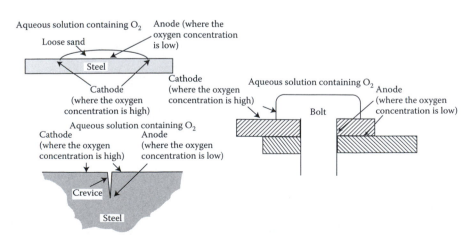

FIGURE 25.4 Oxygen concentration cells. Regions that are shielded from oxygen are the anodes while the cathode reaction occurs where oxygen is plentiful.

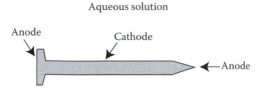

FIGURE 25.5   Regions that have been cold worked are anodic to regions that have not.

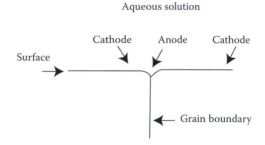

FIGURE 25.6   Because atoms at grain boundaries are in a higher energy state, the grain boundaries become anodic.

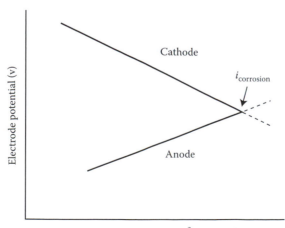

FIGURE 25.7   Difference between the anode and cathode potentials decreases with increasing current density corrosion current.

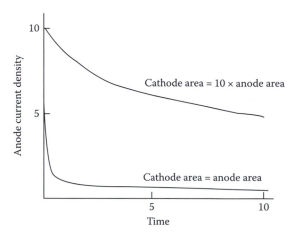

FIGURE 25.8 With a constant anode area, the corrosion current increases as the cathode becomes larger. The decrease of corrosion current with time is a result of polarization.

25.4   PASSIVATION

With some materials, when the anode potential reaches a critical value, the corrosion current drops abruptly to a degree that the corrosion rate is very small as shown in Figure 25.11. This condition is called passivation. A very thin oxygen layer on stainless steels is sufficient to cause passivation. Oxygen and a very small amount of corrosion are required to maintain the passive state. In Table 25.2, stainless steels occupy two places, depending on whether they are passive or not. Titanium alloys may be passive under special conditions.

25.5   CORROSION CONTROL

The means of controlling corrosion are means of disrupting or changing the corrosion cell. One is to remove the electrolyte (corrosion will not occur in the absence of

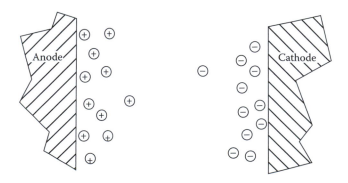

FIGURE 25.9 Polarization is caused by an accumulation of positive ions near the anode and negative ions near the cathode.

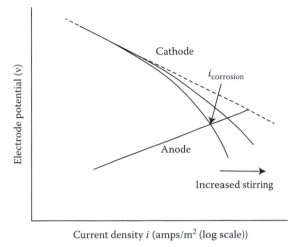

FIGURE 25.10 Polarization at the cathode decreases the cell potential. Increased convection decreases the polarization. The effects of anode polarization are similar.

water). Inert coatings such as paint serve this function. Another is to break the circuit. Cells caused by electrical contact of dissimilar metals can be interrupted by placing an insulator between the metals. Reversing the voltage by imposing opposite voltage with external circuit or sacrificial corrosion (Figure 25.12) are other means. With galvanized steel, the zinc is anodic to iron and hence protects the steel by sacrificial corrosion versus inert coatings (Figure 25.13).

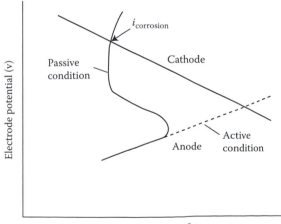

FIGURE 25.11 Above a critical anode potential, certain materials become passive. Their corrosion rate drops abruptly. Note that the current density is plotted on a logarithmic scale.

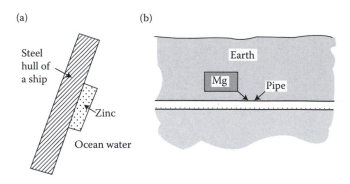

FIGURE 25.12   Corrosion protection by sacrificial corrosion of zinc (a) and magnesium (b).

25.6   RUST

Rust is ferric oxide, Fe_2O_3, ferric hydroxide, $Fe(OH)_3$. Ferrous ions Fe^{2+} are soluble, but further oxidation produces ferric ions, $3Fe^{2+} + 6OH^- \rightarrow 3Fe(OH)_3 + 3H_2O$. The ferric hydroxide that precipitates is insoluble. If dried, the ferric hydroxide turns to an oxide, $2Fe(OH)_3 \rightarrow Fe_2O_3 + 3H_2O$. Often the rust-producing reaction occurs at some distance from where the anode reaction occurs, so rust deposits may not be directly over the corroded region. This is illustrated in Figure 25.14.

Special steels containing small amounts of Cu and Ni form more protective rusts. These steels do rust, but more slowly than ordinary steels. This has led to their use in architecture and sculpture.

25.7   DIRECT OXIDATION

It might seem as though direct oxidation in air at high temperature would not involve an electrolytic cell. However, there is an anode and a cathode. The anode reaction

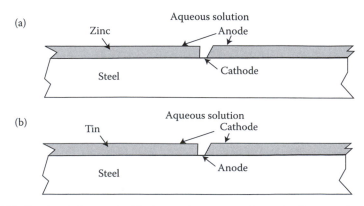

FIGURE 25.13   (a) Plating steel with zinc (galvanizing) offers cathodic protection to steel if the plating is scratched. (b) Tin plating offers no cathodic protection so the steel will occur if the plating is scratched.

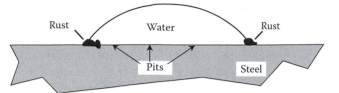

FIGURE 25.14   Rust formation away from corroded sites where the oxygen concentration is higher.

is $M \rightarrow M^{n+} + ne^-$ and the cathode reaction is $O_2 + 4e^- \rightarrow 2O^{2-}$. Either O^{2-} ions or M^{n+} ions and e^- must diffuse through the oxide. M^{n+} ions are smaller than O^{2-} ions and therefore diffuse faster. Hence their diffusion is rate controlling. Figure 25.15 illustrates the reactions and transport in direct oxidation.

Al_2O_3 and Cr_2O_3 are very defect-free oxides so diffusion and electron transport is very slow. Hence they are very protective. For an oxide to be protective, it must cover the surface. Hence volume of oxide $\geq$ volume of metal oxidized. If $nM + mO \rightarrow M_nO_m$, protection occurs if

$$\frac{(MW)_{oxide}}{r_{oxide}} \geq \frac{n(AW)_M}{\rho_M}. \tag{25.2}$$

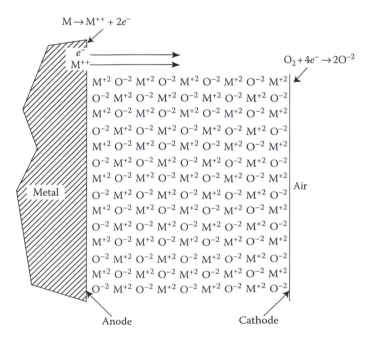

FIGURE 25.15   Direct oxidation. Oxide forms by diffusion of anions and electrons to the oxide-air surface.

Equation 25.2 is not satisfied for Na, K (alkali), and the alkaline earth metals, so they oxidize rapidly in hot air. If the ratio, $[(MW)_{oxide}/\rho_{oxide}]/[n(AW)_M/\rho_M]$, is too high, the compressive stresses in the oxide may cause it to spall off. This is the case for Fe_2O_3. Another necessary condition for a protective oxide is that it must be solid. Tungsten and molybdenum oxidize very rapidly at high temperatures because their oxides are volatile. Because vanadium pentoxide (V_2O_5) forms a low-melting eutectic with Fe_2O_3, which flows off the surface, fuels containing a small amount of vanadium have caused serious problems in power-generating turbines.

25.8 HYDROGEN EMBRITTLEMENT

High-strength steels can be embrittled by hydrogen. Two characteristics of hydrogen embrittlement are the following. (1) Fractures are not immediate. They occur sometime after application of the load. Hence, this is sometimes called *static fatigue*. (2) In a tension test, there is a lower strain to fracture. This loss of ductility increases with the amount of dissolved H and with the strength level of steel (Figure 25.16). Figure 25.17 shows a missile casing that failed by hydrogen embrittlement. At low strain rates, ductility is lower than that at high strain rates.

The sources of hydrogen in steel include pickling (H_2SO_4, HCl), electroplating (Cd, Cr, Zn), corrosion (especially NaOH), welding (especially with coated electrodes), and exposure to H_2 at high temperatures. Hydrogen dissolves monoatomically and diffuses very rapidly to regions of high hydrostatic tension (notches). Hydrogen can be baked out of steel as indicated in Figure 25.18.

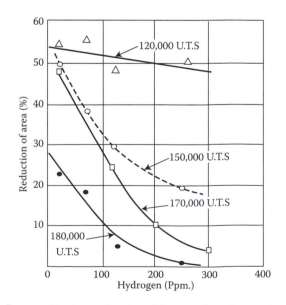

FIGURE 25.16 Increased hydrogen levels and increased steel strength result in a greater loss of ductility. (From H. M. Burke et al., *Metal Progress*, 67(5), 1955. With permission.)

FIGURE 25.17 Remains of a missile casing after failure by hydrogen embrittlement. (From Shank et al., *Metal Progress*. With permission.)

25.9 HYDRIDE FORMATION

Titanium, zirconium, niobium, and vanadium form hydrides. In zirconium, hydrides form as platelets. There is a tendency for the platelets to form parallel to (0001). The platelets are brittle. When stress is applied perpendicular to the platelet, there is a complete loss of ductility. Figure 25.19 shows the effect of hydrogen concentration on the ductility of Ti and Zr. Unlike the embrittlement of steels, this is not time dependent. Hydrides create a problem in zircaloy tubing used for cladding nuclear fuels.

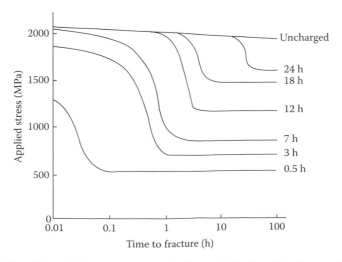

FIGURE 25.18 Effect of baking at 150°C on the fracture behavior of a hydrogen-charged 4340 steel. (Data from J. O. Morlettt, H. Johnson, and A. Troiano, *J. Iron. Steel Inst.*, 189, 1958.)

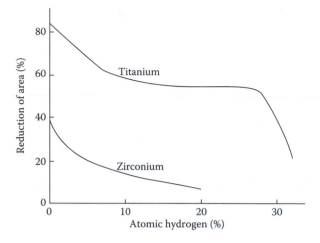

FIGURE 25.19 Effect of hydrogen on the ductility of titanium and zirconium. (Data from W. Betteridge and A. W. Franklin, *J. Inst. Metals*, 85 (1956–7), 475 (Zr) and R. I. Jaffe, *J. Metals*, 3 (1955), 247 (Ti).)

With internal pressure in a tube, the largest principal stress is in the hoop direction. The effect of hydrides is lessened if the texture can be controlled is such a way that (0001) lies in the tube wall. It is worst if (0001) is perpendicular to tube wall.

25.10 HYDROGEN IN COPPER

Hydrogen in tough-pitch copper, which contains oxygen, can lead to the reaction $2\underline{H} + Cu_2O \rightarrow 2Cu + H_2O$ when the copper is heated. The formation of water vapor can cause blistering and porosity, leading to a loss of ductility.

25.11 MISCELLANY

At the time of the First World War, the U.S. Army did not specify that brass cartridge cases be given stress-relief anneals. This caused a problem when cow barns in France were used to store ammunition. As anyone who has changed diapers knows, ammonia is one of the decomposition products of urine. Ammonia from cow urine caused stress-corrosion cracking of the brass cartridge cases that contained residual tensile stresses in the hoop direction. Since then the army has specified stress-relief annealing.

The role of oxygen in corrosion can at times seem anomalous. Oxygen is normally required for corrosion. However, the metal loss does not occur where the oxygen concentration is the greatest.

Oxygen is required for passivity of stainless steels. In one case, simply identifying stainless steel pipes by attaching Scotch tape to the outside of the pipes was responsible for corrosion when moisture condensed on them. The tape prevented the access of oxygen, so the regions under the pipe lost their passivity.

If conditions are oxidizing enough even carbon steel can become passive. Concentrated nitric acid can be shipped in mild steel containers. However, the containers would corrode extremely rapidly in dilute nitric acid.

PROBLEMS

1. True or false:
   a. Generally, in a corrosion cell it is better to have a large anode than a large cathode.
   b. To be passive, a stainless steel needs the presence of oxygen.
   c. The corrosion resistance of stainless steels generally increases with chromium content up to 12% C.
2. Steel is often plated with other metals. Which of the metals listed below would offer galvanic protection to steel?
   Ni, Zn, Cd, Sn, Cr
3. Copper is sometimes recovered from mine water by immersing steel scrap in the water and later recovering fine copper powder. Write the appropriate anode and cathode reactions.
4. Sometimes corrosion data is given in weight loss per area per time. For example, in the corrosion of steel, the loss might be 0.15 lb ft^{-2} year^{-1}. If uniform attack is assumed, what would be the corrosion in terms of mpy? The unit mpy means thousands of an inch (mil) per year.
5. Will the rate of corrosion of a piece of iron in tap water increase, decrease, or remain unaffected if
   a. NaCl is added to the water.
   b. Electrons are made to flow into the iron by means of a battery.
   c. Nickel is placed in contact with the iron.
   d. The water is frozen.
6. Which is worse from a corrosion standpoint, aluminum plates riveted together with steel rivets or steel plates riveted together with aluminum rivet?
7. If steel is galvanized with 0.50 oz of zinc ft^{-2}, what is the thickness of the zinc plate?
8. Stainless steel pipes were installed in a chemical plant. The pipes were identified with tapes. In humid weather, the pipes carrying cold water sweated. It was found that corrosion occurred under the tape. Explain.
9. Magnesium parts are heat treated in an atmosphere of 1% S_2, which forms a coat of $MgSO_4$ on the surface. Why is this atmosphere preferable to air? The densities of $MgSO_4$, MgO, and Mg are 2.66, 3.85, and 1.74, respectively.
10. True or false.
    a. Hydrogen embrittlement of steels is primarily caused by hydride formation.
    b. Hydrogen can be removed from steels by baking them in a low hydrogen atmosphere.
    c. The main source of hydrogen is steel from the reaction $H_2O \rightarrow O + 2H$.
    d. Plating with cadmium is an effective way to minimize hydrogen embrittlement.
    e. High-strength steels are more susceptible to hydrogen embrittlement than low-strength steels.
    f. The time delay between load application and fracture decreases with increasing levels of H.
    g. The time delay between load application and fracture decreases with increasing levels of stress.
    h. In zircaloy, the hydrogen embrittlement is related to the diffusion of H to dislocations.

    i. In aluminum alloys, dissolved hydrogen causes a negative rate sensitivity because it increases the rate at which dislocations move.

    j. The main problem with dissolved hydrogen in copper alloys is that it can react with oxygen to form water vapor.

REFERENCES

W. Betteridge and A. W. Franklin, *J. Inst. Metals*, 85, 475 (Zr), 1956–7.

H. M. Burke et al., *Metal Progress*, 67(5), 1955.

R. I. Jaffe, *J. Metals*, 3, 247 (Ti), 1955.

J. O. Morlettt, H. Johnson, and A. Troiano, *J. Iron. Steel Inst.*, 189, 1958.

Shank et al., *Metal Progress.*

Hydrogen in Metals, ASM, 1974.

L. H. Van Vlack, *Elements of Materials Science and Engineering*, 3rd ed., Addison–Wesley, 1974.

Appendix 1

Microstructural Analysis

A1.1 ASTM GRAIN SIZE NUMBER

One important characteristic of a microstructure is the size of its cells (grains). There are several ways to measure and characterize grain size. One is in terms of the number of grains per area. The ASTM grain size number, N, is defined by the relation

$$n = 2^{N-1}, \tag{A1.1}$$

where n is the number of grains per in^2 on a photomicrograph taken at 100× magnification. The simplest way to find n is to count the grains in a representative rectangular field as $n = (n_i + n_e/2 + n_c/4)/A_{100\times}$, where n_i is the number of grains entirely within the rectangular field, n_e is the number of grains that are cut by an edge of the field, and n_c (=4) is the number of grains on the corners. $A_{100\times}$ is the area (in.^{2}) of the field at 100× (or corrected to 100× for other magnifications).

A1.2 LINEAR INTERCEPT GRAIN SIZE

Another simple measure of grain size is the mean linear intercept, $\bar{\ell}$, of lines drawn randomly with respect to the microstructure. The quantity $\bar{\ell}$ is a measure of the grain diameter and is conventionally given in mm (in real space). The system is to lay a large length of line (or lines) randomly on the microstructure and count the number of intersections per length, N_L. The average linear intercept is then $\bar{\ell} = 1/N_L$. If the microstructure itself is random, the analysis lines may all be in the same direction, but if the grains are elongated, the analysis lines must be randomly oriented. For random microstructures, $\bar{\ell}$ and the ASTM grain size are related. An approximate relationship can be found by assuming that the grains are circles of radius, r. The area of a circular grain, πr^2, can also be expressed as the average height, $\bar{\ell}$, times its width, $2r$, as shown in Figure A1.1, so $\bar{\ell} \cdot 2r = \pi r^2$. Therefore

$$r = \left(\frac{2}{\pi}\right)\bar{\ell} \quad \text{or} \quad \bar{\ell} = \left(\frac{\pi}{2}\right)r. \tag{A1.2}$$

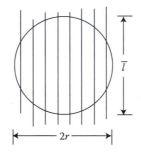

FIGURE A1.1 The area of a circle, πr^2, also equals the average intercept times twice the radius, $2\bar{\ell}$, so $\bar{\ell} = (\pi/2)r$.

Therefore the area per grain is $(4/\pi)\bar{\ell}^2$. The number of grains per in^2 at 100× is n, so the number of grains per mm^2 in real space (no magnification), n', is $n' = n(100/25.4)^2$ mm^{-2}, and the area per grain is $1/[n(100/25.4)^2]$. This must equal $(4/\pi)\bar{\ell}^2$, so

$$\bar{\ell} = \frac{0.225}{\sqrt{n}}. \tag{A1.3}$$

Substituting $n = 2^{N-1} = 2^{N/2}$,

$$\bar{\ell} = 0.254(2^{-N/2}). \tag{A1.4}$$

EXAMPLE PROBLEM A1.1

Consider the microstructure represented below at 200×. Determine independently the ASTM grain size number and the mean linear intercept, $\bar{\ell}$, and compare the two using Equation A1.4. The areas are 2.5 × 3 in.^{2} (Figure A1.2).

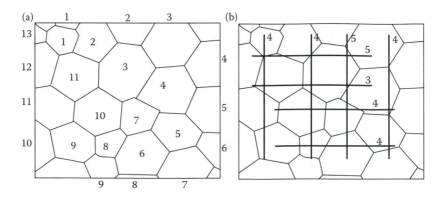

FIGURE A1.2 Analysis of a microstructure by counting grains (a) and by counting intercepts (b).

The area represented at 100× would be $2.5 \times 3 \times (100/200)^2 = 1.875$ in.^{2}. Counting grains, $n_i = 11$, $n_e = 13$; so $n = (11 + 13/2 + 1)/1.875 = 9.967$.

$$N = \ln(9.967)/\ln(2) + 1 = 4.3.$$

There are eight lines drawn, each of 50 mm length. In real dimensions, the total length of lines is $8 \times 50/200 = 2$ mm. There are a total of $4 + 4 + 5 + 4 + 5 + 3 + 4 + 4 = 33$ intersections, so $N_L = 33/2 = 16.5$ mm^{-1} or $\bar{\ell} = 2/33 = 0.0606$ mm.

With $N = 4.3$, Equation A1.4 gives $\bar{\ell} = 0.0713$, which is in reasonable agreement with 0.0606 considering the small length of count.

A1.3 RELATION OF GRAIN BOUNDARY AREA PER VOLUME TO LINEAR INTERCEPT

The grain boundary surface area, per spherical grain, is $2\pi R^2$, where R is the radius of the sphere. (The reason that it is not $4\pi R^2$ is that every grain boundary is shared by two neighboring grains.) The volume per spherical grain is $(4/3)\pi R^3$, so the grain boundary area per volume, S_V, is given by

$$S_V = \frac{2\pi R^2}{(4/3)\pi R^3} = \frac{3}{2R}. \tag{A1.5}$$

Now relate the spherical radius, R, to the linear intercept, $\bar{\ell}$ (Figure A1.3). The volume of a sphere can also be calculated by considering the circle through its center, which has an area πR^2. The volume equals this area times the average length of lines, $\bar{\ell}$, perpendicular to it, or πR^2. Therefore, the volume equals this length of lines, $\bar{\ell}$, perpendicular to it, or $V = \bar{\ell}pR^2$. Therefore, $(4/3)pR^3 = \bar{\ell}pR^2$. Substituting $R = (3/4)\bar{\ell}$ into $S_V = 3/(2R)$,

$$S_V = \frac{2}{\bar{\ell}} = 2N_L. \tag{A1.6}$$

A1.4 RELATION OF THE NUMBER OF INTERSECTIONS PER AREA AND LENGTH OF LINES PER VOLUME

Consider a box filled with a large number, n of randomly oriented lines, each of length, $\bar{\ell}$, as shown in Figure A1.4. Let the base of the box have an area, A, and let its height be h.

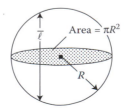

FIGURE A1.3 The volume of a sphere $= \bar{\ell}\pi R^3$.

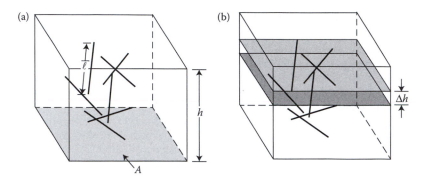

FIGURE A1.4   Box with n randomly oriented lines of length, $\bar{\ell}$ (a). Intersection of these lines with parallel planes separated by a distance, Δh (b).

The total length of line per volume is

$$L_V = \frac{n\ell}{hA}. \tag{A1.7}$$

Construct a large number of planes parallel to the base and separated from one another by a distance, Δh, so that the number of planes is $h/\Delta h$ and the total area is $A(h/\Delta h)$. The distance between intersections along lines inclined at an angle, θ, to the surface normal is $\Delta \ell = \Delta h/\cos \theta$, so the number of intersections per lines is $\ell/\Delta \ell = \ell \cos \theta/\Delta h$. For randomly oriented lines, the number oriented between θ and $\theta + d\theta$ is $dn = n\,df$, where $df = \sin\theta\,d\theta$ is the fraction so oriented. The number of intersections due to lines so oriented is $(\ell/\Delta h)\cos \theta\,dn = n(\ell/\Delta h)\cos\ell \sin \theta\,d\theta$. Integrating to include all orientations, (i.e., between 0 and $\pi/2$), the number of intersections $= \int n(\ell/\Delta h)\cos \theta \sin \theta\,d\theta = n\ell/(2\Delta h)$.

Since the total area is $A(h/\Delta h)$, the number of intersections per area is

$$N_A = \frac{n\ell}{2hA}. \tag{A1.8}$$

Comparing Equations A1.7 and A1.8,

$$N_A = \frac{L_V}{2}. \tag{A1.9}$$

Similarly, when a plane intersects a large number of randomly oriented surfaces, the length of intersection per area, L_A, is related to the surface area per volume, S_V, by

$$L_A = \frac{S_V}{2}. \tag{A1.10}$$

A1.5   DIHEDRAL ANGLES

A dihedral angle is the angle between two planes or equivalently the angle between the normals to two planes. The angle between two grain boundaries in a microstructure is in general not the true dihedral angle between the grain boundaries. The true angle is the angle that would be observed only if the intersection of the two grains is 90° to the surface of observation. Otherwise the observed angle may be either greater or less than the true dihedral angle. If a large number of observations were made the average observation should equal the true angle.

A1.6 MICROSTRUCTURAL RELATIONSHIPS

Microstructures consist of three-dimensional networks of cells or grains that fill space. Each cell is a polyhedron with faces, edges, and corners. Their shapes are strongly influenced by surface tension. However, before examining the nature of three-dimensional microstructures, the characteristics of two-dimensional networks will be treated.

A1.6.1 TWO-DIMENSIONAL RELATIONSHIPS

A two-dimensional network of cells consists of polygons, edges (sides), and corners. The number of each is governed by the simple relation

$$P - E + C = 1, \tag{A1.11}$$

where P is the number of polygons, E is the number of edges, and C is the number of corners. Figure A1.5 illustrates this relationship. If the microstructure is such that three and only three edges meet at every corner, $E = (3/2)C$, so

$$P - \frac{C}{2} = 1 \quad \text{and} \quad P - \frac{E}{3} = 1. \tag{A1.12}$$

For large numbers of cells, the 1 on the right-hand side of the equation becomes negligible, so $E = 3P$ and $C = 2P$. This restriction of three edges meeting at a corner also requires that the average angle at which the edges meet is 120°, and that the average number of sides per polygon is six.

If the edges were characterized by a line tension (in analogy to the surface tension of surfaces in three dimensions) and if the line tension for all edges were equal,

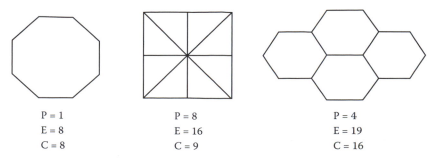

P = 1	P = 8	P = 4
E = 8	E = 16	E = 19
C = 8	C = 9	C = 16

FIGURE A1.5 Three networks of cells illustrating that $P - E + C = 1$.

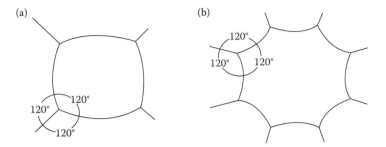

FIGURE A1.6 The sides of grains with fewer than six neighbors are inwardly concave (a). The sides of grains with more than six neighbors are outwardly concave (b).

equilibrium would require that all edges meet at 120°, so cells with more than six edges would have to be curved with the center of curvature away from the cell and those cells with fewer than six sides would be curved the opposite way as shown in Figure A1.6. Since boundaries tend to move toward their centers of curvature, the cells with large numbers of sides would tend to grow and those with few sides should shrink. Only a network in which all of the cells were regular hexagons would be stable.

A1.6.2 THREE-DIMENSIONAL FEATURES

Euler proposed that for a single body,

$$C - E + F - B = 1. \tag{A1.13}$$

And for any array of three-dimensional bodies,

$$C - E + F - B = 0. \tag{A1.14}$$

Table A1.1 illustrates this for several simple polyhedra for an infinite array of bodies. Here B is the number of bodies (grains), F is the number of faces, E is the number of edges, and C is the number of corners. Consider an isolated cube for

TABLE A1.1

Characteristics of Several Polyhedra

Polyhedron	Faces (F)	Edges (E)	Corners (C)	F – E + C
Tetrahedron	4	4	4	2
Cube	6	12	8	2
Octahedron	8	12	6	2
Dodecahedron (cubic)	12	24	14	2
Dodecahedron (pentagonal)	12	30	20	2
Tetrakaidecahedron	14	36	24	2

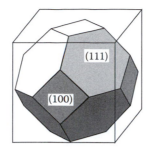

FIGURE A1.7   The Kelvin tetrakaidecahedron.

example. There is one body, there are six faces, 12 edges, and eight corners. B = 2, F = 6, E = 12, and C = 8. 8 − 12 + 6 − 1 = 1. For an infinite array of stacked cubes, each face is shared by two cubes, so F = B/2. Each edge is shared by four cubes, so E = B/4 and each corner is shared by eight cubes, so C = B/8. Substituting into Euler's equation, 8B/8 − 12B/4 + 6B/2 − B = 0.

Grains in a real material have certain restrictions: Each corner is shared by four grains, and each edge is shared by three grains. Furthermore, grains stack in such a way so as to fill space. Very few simple shapes fulfill these conditions.

One simple shape is the tetrakaidecahedron proposed by Lord Kelvin (1887). Figure A1.7 shows that it can be thought of as a cube with each corner truncated by an octahedron. Alternatively, it can be thought of as an octahedron with each corner truncated by a cube (Figure A1.8). There are 14 faces, 36 edges, and 24 corners. For an infinite array of these polyhedra, F = 14B/2 = 7B, C = 24B/4 = 6B, and E = 36B/3 = 12B, so C − E + F − B = 6B − 12B + 7B − B = 0. This shape is a useful approximation for analyzing grains in a polycrystal. For example, calculation of the surface area of the faces to the grain volume can be compared with other solid shapes and a sphere. The Kelvin tetrakaidecahedron has 14 faces. Six of these are squares

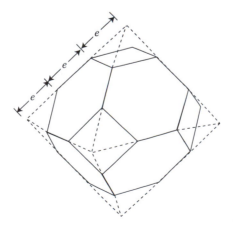

FIGURE A1.8   Construction of a tetrakaidecahedron by truncation of an octahedron by a cube. The edges of the tetrakaidecahedron are 1/3 as long as the edges of the octahedron.

parallel to {100} planes and eight are regular hexagons parallel to {111} planes. There are 24 corners and 36 edges.

Thus the total length of edges is $36e$, where e is the length of an edge, and the total surface area is the area of the six square faces plus the eight hexagonal faces, $6e^2 + 8(3\sqrt{3})e^2 = 47.569e^2$. The volume is the volume of the octahedron less the volume of the six truncated pyramids, $[9\sqrt{2} - 6(1/3\sqrt{2})]e^3 = 8\sqrt{2}e^3 = 11.314e^3$.

For the Kelvin tetrakaidecahedron, the ratio of surface area to that of a sphere is 1.099. Most other shapes have much higher ratios as shown in Figure A1.9.

Finding an exact expression for the mean linear intercept would be quite complicated. However, a good approximation is the mean linear intercept for a sphere of the same volume. Since the volume of a sphere is $(4/3)\pi r^3$, the radius of such a sphere can be found by equating the volumes,

$$V = \left(\frac{4}{3}\right)\pi r^3 = 11.314e^3 \quad \text{or} \quad r = 1.3927e. \tag{A1.15}$$

Since the mean linear intercept for a sphere is $\bar{\ell} = (4/3)r$,

$$\bar{\ell} = \left(\frac{4}{3}\right)(1.3927e) = 1.8569e \quad \text{or} \quad e = 0.5385\bar{\ell}. \tag{A1.16}$$

Now, using the expressions above, a number of quantities can be expressed in terms of $\bar{\ell}$: Volume of a grain

$$V = 11.314(0.5385\bar{\ell})^3 = 1.767\bar{\ell}^3. \tag{A1.17}$$

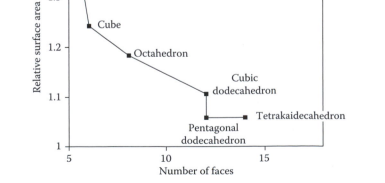

FIGURE A1.9 The ratio of surface area of several polyhedra to that of a sphere with the same volume.

Number of grains per volume $= 1/V = 1/[11.314(0.5385\bar{\ell})^3]$, so

$$\frac{\text{Grains}}{\text{Volume}} = \frac{0.5659}{\bar{\ell}^3}. \tag{A1.18}$$

Interfacial area per volume: Considering that every face is shared by two grains, $A_V = (1/2)(47.569e^2)/(11.314e^3) = 0.6438/e$ or

$$A_V = \frac{1.1956}{\bar{\ell}}. \tag{A1.19}$$

Number of grain corners per volume: Considering that each corner is shared by four grains, Number of corners per volume $= (1/4)(24/1.767)\bar{\ell}^3$, so

$$\frac{\text{Numbers of corners}}{\text{Volume}} = \frac{3.396}{\bar{\ell}^3}. \tag{A1.20}$$

Length of edge per volume: Consider that each edge is shared by three grains, the length of edge per volume $= (1/3)(36)(0.5385\bar{\ell})/(1.767\bar{\ell}^3)$, or

$$\frac{\text{Edge length}}{\text{Volume}} = \frac{3.657}{\bar{\ell}^2}. \tag{A1.21}$$

Substituting the simplifications that four edges meet at each corner, $E = 2C$, and three faces share a common edge, $E = \Sigma nF_n/3$, where n is the number of edges per face (i.e., $n = 5$ for pentagonal, etc.) into faces, $n = 4$ for triangle Equation A1.21,

$$F = -\frac{\Sigma nF_n}{6} = B + 1 \quad \text{or} \quad \Sigma(6 - n)P_n = 6(B + 1). \tag{A1.22}$$

For large numbers of cells $(B \to \infty)$, $C/B = \sqrt{n}/(6 - \sqrt{n})$,

$$\frac{C}{B} = \frac{\sqrt{n}}{6 - \sqrt{n}},$$

$$\frac{F}{B} = \frac{6}{6 - \sqrt{n}}, \tag{A1.23}$$

and

$$\frac{F}{B} - \frac{C}{B} = 1,$$

where $\sqrt{n}$ is the average number of edges per face. Smith (1952) states that C/B must be six or less and he points out that with C/B = 6, $\bar{\ell} = 5\frac{1}{7}$. Experimental studies of shapes of metal grains, soap bubbles, and vegetable cells (Figure A1.10) indicate average numbers of edges per face to be very nearly five, giving credence to C/B being very near to $5\frac{1}{7}$.

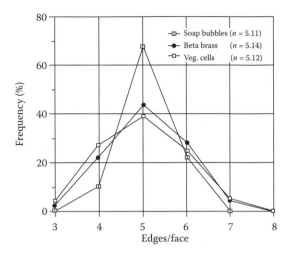

FIGURE A1.10 Experimental studies of the average number of edges per face in three-dimensional networks. (Data from C. S. Smith, in *Metal Interfaces*, ASM, Cleveland, OH, 1952.)

PROBLEMS

1. A soccer ball has 32 faces. They are all either pentagons or hexagons. How many are pentagons?
2. Figure A1.11 below is a microstructure at a magnification of 200×.
   a. Determine the ASTM grain size number.
   b. Determine the intercept grain size.
   c. Compare the answers of a and b using Equation A1.4.

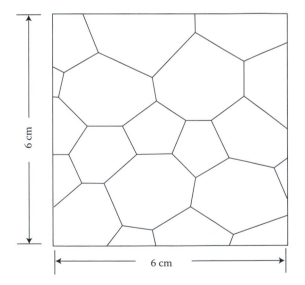
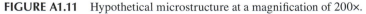

FIGURE A1.11 Hypothetical microstructure at a magnification of 200×.

3. What is the linear intercept grain size (in mm) corresponding to an ASTM grain size number of 8?

4. Dislocation density is often determined by counting the number of dislocations per area intersecting a polished surface. If the dislocation density in cold-worked copper is found to be $2 \times 10^{10} \text{cm}^{-2}$, what is the total length of dislocation line per volume?

REFERENCES

R. T. DeHoff and F. N. Rhine, Eds, *Quantitative Metallography*, McGraw-Hill, 1968.

Lord Kelvin, *Phil. Mag.*, 24, 503–514, 1887.

C. S. Smith, in *Metal Interfaces*, ASM, Cleveland, OH, 1952.

D. L. Waire and R. Phelan, *Science Notes*, *Phil. Mag, Letters,** 1994.

E. E. Underwood, *Quantitative Stereology*, Addison-Wesley, 1970.

*Addendum on space-filling w/arrays of six 14-face and two 12-face polyhedra. The 14-face has 12 pentagonal faces and two hexagonal faces, while the 12-face has distorted pentagons for faces. Waire and Phelan report that space filling is 0.3% more efficient than with the Kelvin tetrakaidecahedron. (This calculation allows faces in each to be curved.) An interesting question is of course, what are the dihedral angles between the faces? Another related question is whether the face curvatures are less than with Kelvin. The number of edges per face = $[2 \times 12 \times 5 + 6 \times (12 \times 5 + 2 \times 6)]/(6 \times 14 + 2 \times 12) = (120 + 432)/(84 + 24) = 5.111$ edges/face. The average number of faces per polyhedra = $6 \times 14 + 2 \times 12$.

Appendix 2

Miller–Bravais System of Indices for Hexagonal Crystals

A2.1   MILLER–BRAVAIS INDICES FOR HEXAGONAL CRYSTALS

The Miller–Bravais system uses four axes rather than three. The reason is that with four axes the symmetry is more apparent, as will be illustrated below. Three of the axes, a_1, a_2, and a_3 lie in the hexagonal (basal) plane at 120° to one another and the fourth or c-axis is perpendicular to it as shown in Figure A2.1.

A2.2   PLANAR INDICES

The rules are similar to those for Miller indices with three axes. To find the indices of a plane

1. Write its intercepts on the four axes in order (a_1, a_2, a_3, and c).
2. Take the reciprocals of these.
3. Reduce to the lowest set of integers with the same ratios.
4. Enclose in parentheses ($hki\ell$).

Commas are not used except in the rare case that one of the integers is larger than a one-digit number. (This is rare because we are normally interested only in directions with low indices.) If a plane is parallel to an axis, regard its intercept as ∞ and its reciprocal as 0. If the plane contains one of the axes or the origin, either draw a parallel plane or translate the axes before finding indices. This is permissible since all parallel planes have the same indices. Figure A2.2 shows several examples.

In the four-digit system, the third digit, i, can always be deduced from the first two, $i = -h - k$, and is therefore redundant. With the three-digit systems, it may either

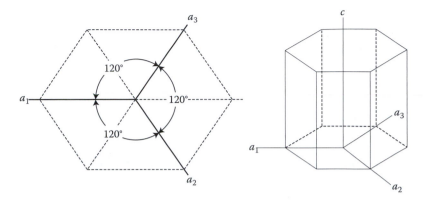

FIGURE A2.1 Axis system for hexagonal crystals.

be replaced by a dot, (hk_1), or omitted entirely, (hk_1). However, the disadvantage of omitting the third index is that the hexagonal symmetry is not apparent. In the four-digit (Miller–Bravais) system, a family of planes is apparent from the indices. For example,

$\{01\bar{1}0\} = (01\bar{1}0)$, $(\bar{1}010)$, and $(1\bar{1}00)$. The equivalence of the same family is not so apparent in the three-digit system, $\{010\} = (010)$, $(\bar{1}00)$, and $(1\bar{1}0)$.

Also compare $\{\bar{2}110\} = (\bar{2}110)$, $(1\bar{2}10)$, and $(11\bar{2}0)$ with $\{\bar{2}10\} = (\bar{2}10)$, $(1\bar{2}0)$, and (110).

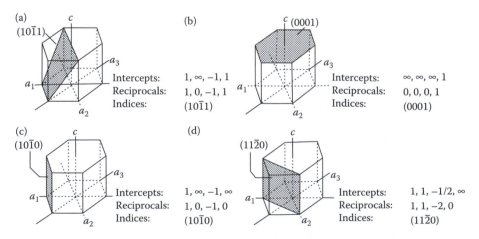

FIGURE A2.2 Examples of planar indices for hexagonal crystals. Note that the sum of the first three indices is always zero, $h + k + i = 0$.

A2.3 DIRECTION INDICES

The direction indices are the translations parallel to the four axes that produce the direction under consideration. The first three indices must be chosen so that they sum to zero and are the smallest set of integers that will express the direction. They are enclosed without commas in brackets $[hki_l]$. Examples are shown in Figure A2.3.

There is also a three-digit system for directions in hexagonal crystals. It uses the translations along the a_1, a_2, and a_3 axes (U, V, and W, respectively). The four-digit $[uvtw]$ and three-digit $[UVW]$ systems are related by

$$U = u - tu = \frac{2U - V}{3},$$

$$V = v - tv = \frac{2V - U}{3},$$

$$W = w,$$

$$(A2.1)$$

and

$$t = -(u + v) = \frac{-(U + V)}{3}.$$

Comparison of the four- and three-digit systems is illustrated in Figure A2.4. The four-digit Miller–Bravais system is used in this text.

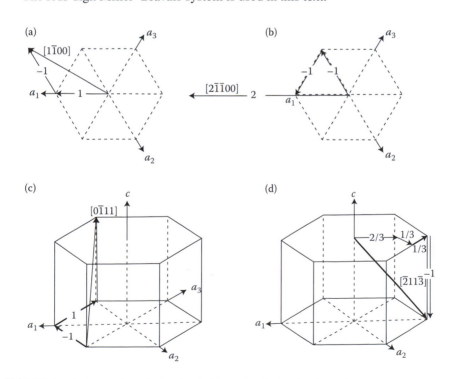

FIGURE A2.3 Examples of direction indices with the Miller–Bravais system.

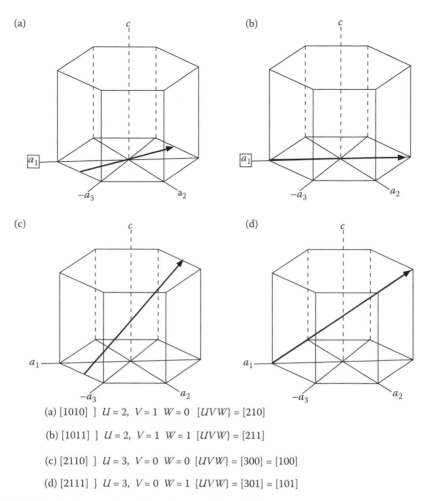

(a) [1010]] $U = 2,\ V = 1\ \ W = 0\ \ [UVW\} = [210]$

(b) [1011]] $U = 2,\ V = 1\ \ W = 1\ \ [UVW\} = [211]$

(c) [2110]] $U = 3,\ V = 0\ \ W = 0\ \ [UVW\} = [300] = [100]$

(d) [2111]] $U = 3,\ V = 0\ \ W = 1\ \ [UVW\} = [301] = [101]$

FIGURE A2.4   Comparison of the four- and three-digit systems.

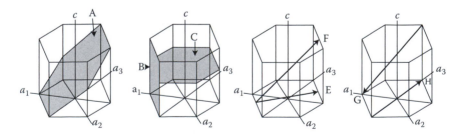

FIGURE A2.5   Several planes and directions for Problem 1.

PROBLEM

1. Write the correct direction indices, [ ], and planar indices, ( ), for the directions and planes sketched in Figure A2.5.

Index